Stephanie Catanzaro

Basic Mathematical Skills

A Guided Approach

Basic Mathematical Skills

A Guided Approach

Robert A. Carman
Santa Barbara City College

Marilyn J. Carman
Santa Barbara High School District

John Wiley & Sons; Inc.
New York London Sidney Toronto

Library of Congress Cataloging in Publication Data:

Carman, Robert A

Basic mathematical skills.

Includes index.

1. Arithmetic—1961- 2. Arithmetic—Programmed instruction. I. Carman, Marilyn J., joint author. II. Title.

QA107.C57 513'.07'7 74-20779

ISBN 0-471-13495-3

Printed in the United States of America

10-9 8 7

Design by Karin Batten and Elliot Kreloff

To
Patty, Laurie, Mary, and Eric
. . . our other works
in collaboration

Preface

Many students enrolled in our schools find themselves frustrated. They are eager, ambitious, and often quite capable of succeeding in their planned careers. They want to learn, but they find themselves handicapped because they do not have the basic mathematics skills needed to continue. They are not interested in "new math"; They need help with basic arithmetic computation skills – very old math indeed, but very used and very vital. This book is intended for such students.

Those who have difficulty with mathematics will find that this book has a number of special features designed to make it most effective for them. These include the following.

- Careful attention has been given to readability and reading specialists have helped plan both the written text and the visual organization.
- A diagnostic pretest and performance objectives keyed to the text are given at the start of each unit. These show the student what is expected of him and provide a sense of direction.
- Each unit ends with a self-test covering the work of that unit.
- The format is clear and easy to follow. It respects the individual needs of each reader, providing immediate feedback at each step to assure their understanding and continued attention.
- The *why* as well as the *how* of every mathematical operation is carefully explained. (Sometimes we even explain our explanations.)
- Numerous problem sets are included so that the student has abundant opportunity for practicing the mathematical operations being learned. Both routine drill and more imaginative and challenging word problems are included. Answers to all of these problems are given in the back of the text.
- Supplementary problem sets are included at the end of each unit and additional problem sets are available in the accompanying teacher's manual. Answers to all supplementary problem sets are given in the teacher's manual.
- A light, lively conversational style of writing and a pleasant, easy to understand visual approach are used.

Students who have worked with the material in field test studies tell us it is helpful, interesting, even fun, to work through. And it works—they learn. In addition to the usual topics taught in a course in developmental arithmetic, we have included optional sections on measurement and beginning algebra. The unit on measurement includes a timely introduction to the Metric System and other topics of value to students enrolled in vocational or occupational programs. The algebra unit provides a head start for students who plan to enroll in a beginning algebra course or who need some interpretive and manipulative skill in algebra for science or technical courses.

A preliminary version of this text was tested with students in three community colleges for more than a year. Prior to testing, the materials were developed, revised, and rerevised on the basis of responses of more than 5000 community college students of all ability levels.

Flexibility of use was a major design criterion, and field testing indicates that the book can be used successfully in a variety of course formats. It can be used as a textbook in traditional lecture-oriented courses. It is very effective in situations where an instructor wishes to modify a traditional course by devoting a portion of class time to independent study. The text is especially useful in programs of individualized study, whether in a learning lab situation, with tutors, or in totally independent study.

An accompanying Teacher's Resource Book provides

- information on how to set up and operate several varieties of individualized or partly individualized courses, including learning labs and tutorial programs;
- multiple forms of all unit tests, brief quizzes, and final examinations;
- additional problem sets;
- answers to all problems in the unit tests, quizzes, exams and problem sets.

It is a pleasure for us to acknowledge our debts to the many people who have contributed to the development of this book. We are greatly indebted to the following teachers who read the preliminary version and offered many helpful criticisms and suggestions: W. Royce Adams, Santa Barbara City College; Chris Avery, De Anza College; Vernon C. Barker, Palomar College; Ben Bockstege, Broward Community College; John Carr, Merritt College; Robert A. Davies, John Jay College of Criminal Justice; Harold Englesohn, Kingsborough Community College; Don Farris, Williamsport Area Community College; Loren Gaither, Fresno City College; John Lavelle, Millersville State College; Tom Munro, San Francisco City College; John O'Connor, Ohlone College; Bill H. Ross, Texas State Technological Institute; James W. Snow, Lane Community College; Dick Spangler, Tacoma Community College; Sister Clarice Sparkman, San Jose City College; Evelyn Stephens, Fashion Institute of Technology; Larry Weiner, Ohlone College; and Sue Verity, Los Angeles Southwest College. The book has benefited from their excellence as teachers.

The talented and imaginative staff of John Wiley & Sons has added a very great deal to the effectiveness of this book and made its publication a pleasure for us. Particular thanks are due Wiley Mathematics Editor Gary Ostedt for his unfailing support and enthusiasm.

Over the last few years, students and teachers at Santa Barbara City College have had the good fortune to work with a group of skilled and dedicated tutors. Their insight into the complex human problems of teaching developmental mathematics concepts is reflected everywhere in this book. Especially helpful were Andy Aull, Tim Hall, Lynne Brown, Irma Herrera, Craig Turek, and David Castro. They taught us to listen and to care, and we are grateful.

Robert A. Carman
Santa Barbara, California — Marilyn J. Carman

Contents

ABOUT THIS BOOK xi

UNIT 1 ARITHMETIC OF WHOLE NUMBERS

Preview **1**
Addition 7
Subtraction 25
Multiplication 36
Division 51
Factors 65
Exponents and Square Roots 81
Self-Test on Unit 1 **91**
Supplementary Problem Sets **93**

UNIT 2 FRACTIONS

Preview **105**
Renaming Fractions 107
Multiplication and Division of Fractions 121
Addition and Subtraction of Fractions 135
Word Problems with Fractions 157
Self-Test on Unit 2 **169**
Supplementary Problem Sets **171**

UNIT 3 DECIMALS

Preview **181**
Addition and Subtraction of Decimals 183
Multiplication and Division of Decimal Numbers 195
Decimal Fractions 213
Signed Numbers 225
Square Roots 235
Self-Test on Unit 3 **247**
Supplementary Problem Sets **249**

UNIT 4 PERCENT

Preview **259**
Numbers and Percent 261
Percent Problems 271
Applications of Percent Calculations 289
Self-Test on Unit 4 **305**
Supplementary Problem Sets **307**

UNIT 5 MEASUREMENT

Preview **313**
Unit Conversion 315

Arithmetic with Measurement Numbers 339
Metric Units 347
Area 385
Volume 391
Self-Test on Unit 5 **399**
Supplementary Problem Sets **401**

UNIT 6 **INTRODUCTION TO ALGEBRA**

Preview **413**
Algebra Words and Operations 429
Translating English to Algebra 431
Evaluating Literal Expressions 441
Solving Simple Equations 451
Ratio and Proportion 471
Self-Test on Unit 6 **483**
Supplementary Problem Sets **485**

Answers 495

Table of Square Roots 517

Study Cards 518

Index 527

About This Book

Sally is one of the many people who go through life afraid of mathematics and upset by numbers. She will bumble along miscounting her marbles, bouncing checks, and eventually trying to avoid college courses that require even simple math. Most such people need to return and make a fresh start. Few get the chance. This book presents fresh start math. It is designed so that you can:

★ start at the beginning or where you need to start,
★ work on only what you need to know,
★ move as fast or as slow as you wish,
★ skip material you already understand,
★ do as many practice problems as you need,
★ take self-tests to measure your progress.

In other words, if you find mathematics difficult and want a fresh start, this book is designed for you.

This is no ordinary book. You cannot browse in it; you don't read it. You *work* your way through it. The ideas are arranged step-by-step in short portions or *frames.* Each frame contains information, careful explanations, examples, and questions to test your understanding. Read the material in each frame carefully, follow the examples, and answer the questions that lead to the next frame. Correct answers move you quickly through the book. Incorrect answers lead you to frames that provide further explanation. You move through this book frame by frame, sometimes forward, sometimes backward. Because we know that every person is different and has different needs, each major section of the book starts with a preview that will help you determine those parts on which you need to work.

As you move through the book you will notice that material not directly connected to the frames appears in boxes and margins. Read these at your leisure. They contain information that you may find useful, interesting, and even fun.

Most students hesitate to ask questions. They would rather risk failure than look foolish by asking "dumb questions." To relieve you of worry over dumb questions (or DAQs), we'll ask and answer them for you. Thousands of students have taught us that "dumb questions" can produce smart students. Watch for DAQ.

In 1846, the Reverend H. W. Adams described what happened when the 10 year old math whiz Truman Safford was asked to multiply, in his head, the number 365,365,365,365,365,365 by itself: "He flew around the room like a top, pulled his pantaloons over the tops of his boots, bit his hands, rolled his eyes in their sockets, sometimes smiling and talking, and then, seeming to be in agony, in not more than one minute, he said 133,491,850,208,566,925,016,658, 299,941,583,225." In this book we will show you a way to do arithmetic that is not so strenuous, quite a bit slower, and not nearly so much fun to watch.

Now, turn to page 1 and let's begin.

Arithmetic of Whole Numbers

PREVIEW 1

Objective	Sample Problems		Where To Go for Help Page	Frame
Upon successful completion of this program you will be able to:				
1. Add, subtract, multiply, and divide whole numbers.	(a) $6341 + 14{,}207 + 635 =$	________	3	1
	(b) $64{,}508 - 37{,}629$	= ________	25	16
	(c) 4328×407	= ________	36	24
	(d) 672×2009	= ________	36	24
	(e) $46{,}986 \div 745$	= ________	51	35
	(f) $37\overline{)3003}$	= ________	51	35
	(g) $\frac{1541}{23}$	= ________	51	35
	(h) 12×0	= ________	36	24
	(i) $16 \div 1$	= ________	51	35
2. Write a whole number as a product of its prime factors.	(a) 3780	= ________	65	44
	(b) 1848	= ________	65	44
3. Calculate integer powers of a whole number.	(a) 2^3	= ________	81	59
	(b) 42^2	= ________	81	59
4. Find the square root of a perfect square.	(a) $\sqrt{169}$	= ________	81	59

(Answers to these sample problems are on the back of this sheet. Don't peek.)

If you are certain you can work all of these problems correctly, turn to page 91 for a self-test. If you want help with any of these objectives or if you cannot work one of the sample problems, turn to the page indicated. Super-students—those who want to be certain they learn all of this—will turn to frame **1** and begin work there.

____
Date

____
Name

____
Course/Section

PREVIEW 1

Answers to Sample Problems

1. (a) 21,183
   (b) 26,879
   (c) 1,761,496
   (d) 1,350,048
   (e) 63 with a remainder of 51
   (f) 81 with a remainder of 6
   (g) 67
   (h) 0
   (i) 16
2. (a) $2^2 \cdot 3^3 \cdot 5 \cdot 7$
   (b) $2^3 \cdot 3 \cdot 7 \cdot 11$
3. (a) 8
   (b) 1764
4. (a) 13

Arithmetic of Whole Numbers

1 Charlie Brown and his friends use numbers to tell time, count marbles, and keep track of their lunch money. Even for these simple operations "tooty-two" won't do the job. We may be living in an age of electronic computers and calculating machines, but most of the arithmetic used in industry, business, and school is still done by hand. For most educated adults, working with numbers is as important a part of their job as being able to read or write. In this unit we will take a how-to-do-it look at the basic operations of arithmetic: addition, subtraction, multiplication, and division.

What is a number? It is a way of thinking, an idea, that enables us to compare very different sets of objects. It is the idea behind the act of counting. The number three is the idea that describes any collection of three objects: 3 people, 3 trees, 3 colors, 3 dreams. We recognize that these collections all have the quality of "threeness" even though they may differ in every other way.

We use *numerals* to name numbers. For example, the number of corners on a square is four, or 4, or IV in Roman numerals, or 四 in Chinese numerals, or "tooty-two" for Sally in the cartoon.

In our modern number system we use the ten *digits* 0, 1, 2, 3, 4, 5, 6, 7, 8, and 9 to build numerals, just as we use the twenty-six letters of the alphabet to build words.

Is 10 a digit?

Think about it. Then turn to **3** to continue.

---

2 Hi.

What are you doing in here? Lost? Window shopping? Just passing through? Nowhere in this book are you directed to frame **2**. (Notice that little **2** to the left above? That's a frame number.) Remember, in this book you move from frame to frame as directed, but not necessarily in 1-2-3 order. Follow directions and you'll never get lost.

Now, return to **1** and keep working.

---

3 No, 10 is a numeral formed from the two digits 1 and 0.

Remember: a number is an idea related to counting,
a numeral is a symbol used to represent a number,
a digit is one of the ten symbols 0, 1, 2, 3, 4, 5, 6, 7, 8, and 9 that we use to form numerals.

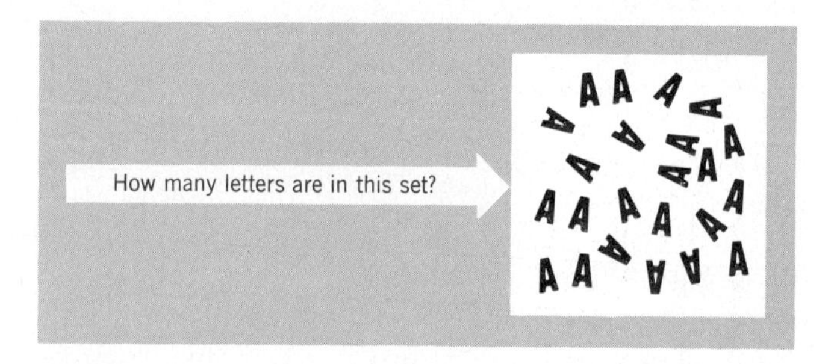

Count them. Write your answer, then turn to frame **4**.

---

4 We counted 23, and of course we write it in the ordinary, everyday manner. Leave the Roman, Chinese, and other numeral systems to Romans, Chinese, and people who enjoy the history of mathematics.

The basis of our system of numeration is grouping into sets of ten or multiples of ten.

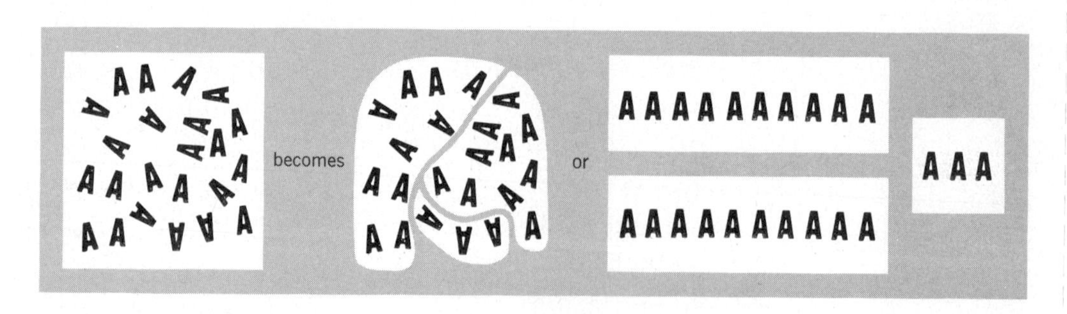

2 tens + 3 ones
20 + 3 or 23.

Any number may be written in this *expanded form.* For example,

$46 = 40 + 6 = 4$ tens $+ 6$ ones
$274 = 200 + 70 + 4 = 2$ hundreds $+ 7$ tens $+ 4$ ones
$305 = 300 + 0 + 5 = 3$ hundreds $+ 0$ tens $+ 5$ ones

Write out the following in expanded form:

362 = ______ + ______ + ______ = ___ hundreds + ___ tens + ___ ones

425 = ______ + ______ + ______ = ___ hundreds + ___ tens + ___ ones

208 = ______ + ______ + ______ = ___ hundreds + ___ tens + ___ ones

Check your work in **5**.

5

NAMING LARGE NUMBERS

Any large number given in numerical form may be translated to words by using the following diagram.

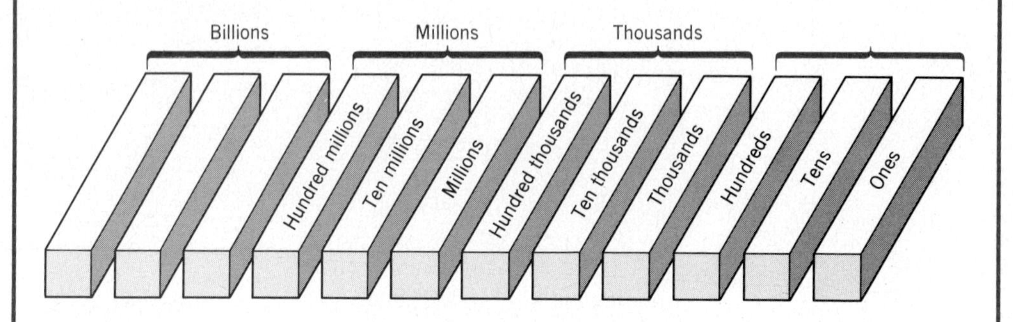

The number 14,237 can be placed in this diagram like this

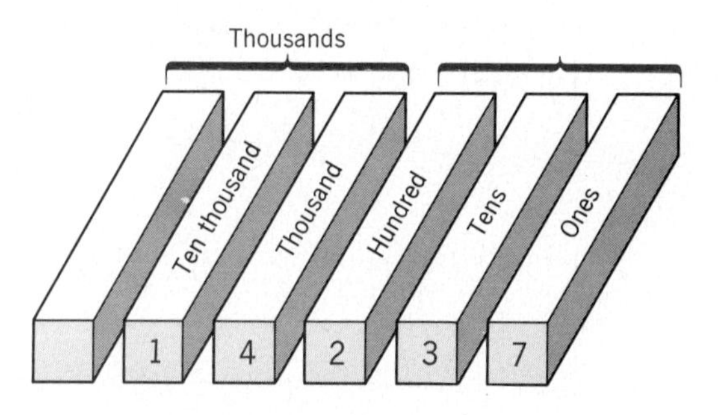

and is read "fourteen thousand, two hundred thirty-seven."

The number 47,653,290,866 becomes

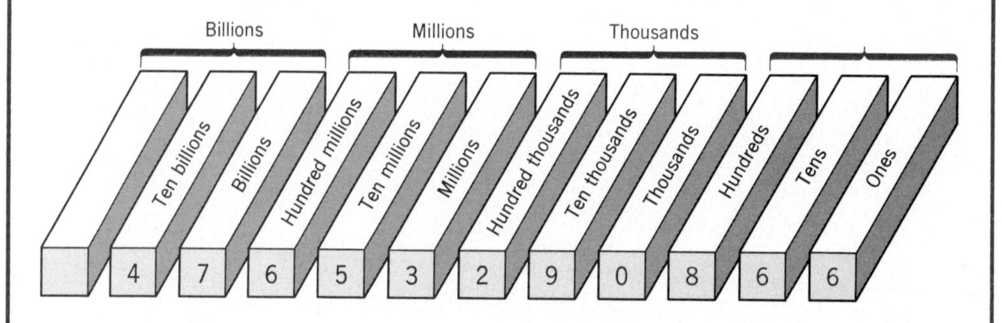

and is read "forty-seven billion, six hundred fifty-three million, two hundred ninety thousand, eight hundred sixty-six."

In each block of three digits read the digits in the normal way ("forty-seven," "six hundred fifty-three") and add the name of the block ("billion," "million"). Notice that the word "and" is not used in naming these numbers.

5 $362 = 300 + 60 + 2 = 3$ hundreds $+ 6$ tens $+ 2$ ones
$425 = 400 + 20 + 5 = 4$ hundreds $+ 2$ tens $+ 5$ ones
$208 = 200 + 0 + 8 = 2$ hundreds $+ 0$ tens $+ 8$ ones

Notice that the 2 in 362 means something very different from the 2 in 425 or 208. In 362 the 2 signifies two ones. In 425 the 2 signifies two tens. In 208 the 2 signifies two hundreds. Ours is a *place value* system of naming numbers: the value of any digit depends on the place where it is located.

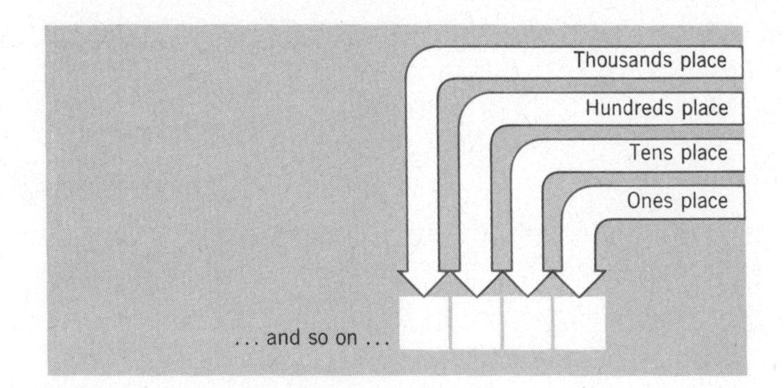

Writing numbers in expanded form will be helpful later when you want to understand how arithmetic operations work.

ADDITION

Addition is the simplest arithmetic operation.

■■ + ■■ = ■■■■
■■   ■   ■■■

$4 + 3 =$ ______

Complete the calculation and go to 7.

6 Here is the completed addition square:

Add	4	2	8	7	5	6	1	3	9
2	6	4	10	9	7	8	3	5	11
4	8	6	12	11	9	10	5	7	13
7	11	9	15	14	12	13	8	10	16
5	9	7	13	12	10	11	6	8	14
1	5	3	9	8	6	7	2	4	10
9	13	11	17	16	14	15	10	12	18
6	10	8	14	13	11	12	7	9	15
8	12	10	16	15	13	14	9	11	17
3	7	5	11	10	8	9	4	6	12

Did you notice that changing the order in which you add numbers does not change their sum?

$4 + 3 = 7$ and $3 + 4 = 7$
$2 + 4 = 6$ and $4 + 2 = 6$ and so on.

This is true for any addition problem involving whole numbers. It is known as the *commutative* property of addition. A commuter is a person who changes location daily, moving back and forth between suburbs and city. The commutative property says that changing the location or order of the numbers being added does not change their sum.

If you have not already memorized the addition of one-digit numbers, it is time you did so. To help you, a study card for addition facts is provided in the back of this book. Use it if you need it.

If you want more practice on adding one-digit numbers, go to **9**. Otherwise continue in **8**.

---

7 ■■ ■■ + ■■■ = ■■■■ ■■■

$4 + 3 = 7$ of course

We add collections of objects by combining them into a single set and then counting and naming that new set. The numbers being added, 4 and 3 in this case, are called *addends* and 7 is the *sum* of the addition.

There are a few simple addition facts you should have stored in your memory and be ready to use. Complete the following table by adding the number at the top to the number at the side and placing their sum in the proper square. We have added $1 + 2 = 3$ and $4 + 3 = 7$ for you.

Add	4	2	8	7	5	6	1	3	9
2									
4								7	
7									
5									
1		3							
9									
6									
8									
3									

Check your answer in **6**.

---

8 Now, let's try a more difficult addition problem:

$35 + 42 =$ ______

The first step in any arithmetic problem is to *guess* at the answer. Never work a problem until you know roughly what the answer is going to be. Always know where you are going.

Don't expect your guess to be exactly correct. Make it quick and reasonably close.

Make a guess at the answer to the problem above. Write your guess here ______ then turn to **10** and continue.

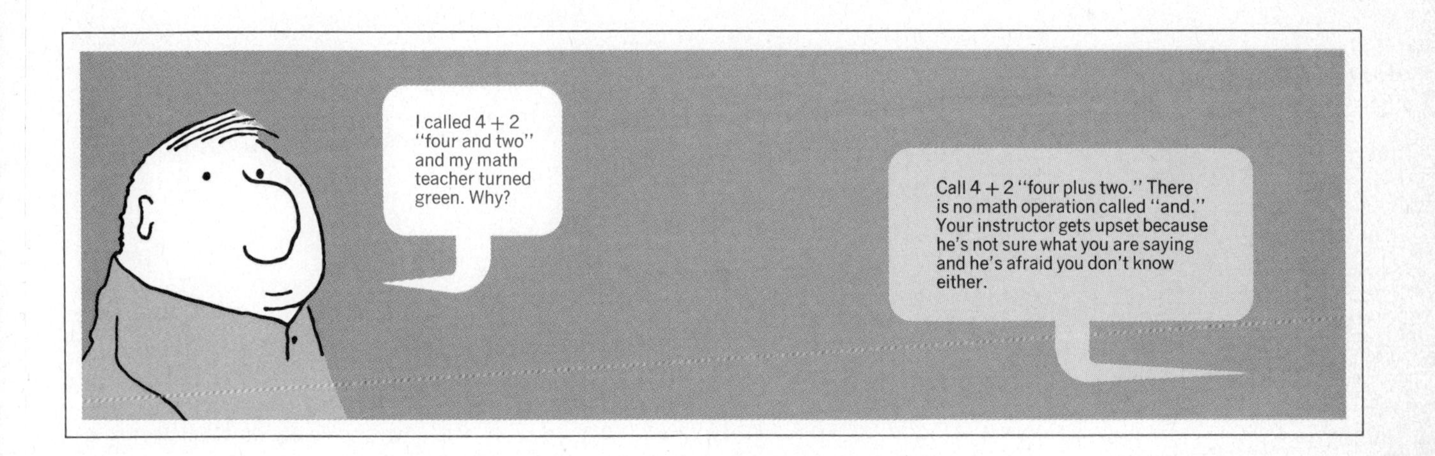

---

9 Practice problems for one-digit addition

Add the following. Work quickly, you should be able to answer all problems in a set correctly in the time indicated. (These times are for community college students enrolled in a developmental math course.)

A. Add:

3	7	3	8	5	3	9	2	6	8
5	9	3	5	6	8	4	7	7	4

7	9	4	7	8	6	7	9	8	5
7	8	2	5	7	3	4	3	8	4

9	2	6	2	8	4	2	5	9	8
6	8	4	9	6	3	6	9	9	7

8	5	7	8	9	6	9	7	8	6
6	9	5	4	7	6	4	6	5	9

7	8	6	3	8	7	6	4	9	5
4	9	5	9	3	8	7	8	9	7

Average time = 90 sec Record: 32 sec

B. Add:

7	5	2	5	8	4	3	8	9	7
3	6	9	7	8	5	6	4	3	6

6	8	9	3	7	2	9	9	7	4
4	5	6	5	7	7	4	9	8	7

8	9	8	5	8	4	9	6	4	8
3	7	6	5	9	5	5	6	3	2

5	6	7	7	5	6	2	3	6	9
8	7	5	9	4	5	8	7	8	8

7	5	9	4	3	8	7	8	5	7
4	9	2	6	8	6	9	4	8	8

Average time = 90 sec   Record: 35 sec

C. Add. Try to do all addition mentally:

2	7	3	4	2	6	3	5	9	5
5	3	6	5	7	7	4	7	6	2
4	2	5	8	9	8	4	8	3	8

6	5	4	8	6	9	7	4	8	1
2	4	2	1	8	3	1	9	4	8
7	5	9	9	8	5	6	1	6	7

1	9	3	1	7	2	9	9	8	5
9	9	1	6	9	9	8	5	3	4
2	1	4	3	6	1	2	1	3	7

Average time = 90 sec   Record: 41 sec

The answers are on page 495.

When you have had the practice you need, turn to **8** and continue.

ROMAN NUMERALS

A number is an idea. A numeral is a symbol that enables us to express that idea in writing and use it in counting and calculating. Roman numerals were used by the ancient Romans almost 2000 years ago and are still seen on clock faces, building inscriptions, and textbooks. Seven symbols are used:

I	V	X	L	C	D	M
1	5	10	50	100	500	1000

Notice that the numbers represented are 1, 5, and multiples of 5 and 10, the number of fingers on one hand and on two hands. There is no zero. We write numerals with these symbols by placing them in a row and adding or subtracting. For example,

1 = I
2 = II
3 = III
4 = IV (I subtracted from V)
5 = V
6 = VI (V + I)
7 = VII (V + I + I)
8 = VIII
9 = IX (I subtracted from X)
10 = X
17 = XVII   152 = CLII   and so on

The Romans used only addition and they wrote 4 as IIII, but in order to keep numerals smaller, later mathematicians used subtraction to form

IV	IX	XL	XC	CD	CM
4	9	40	90	400	900

Only these six subtractions are allowed. From them, other combinations can be made. For example XIX = X + IX or 10 + 9 or 19

LXIV = LX + IV or 60 + 4 or 64

Roman numerals are a bit more difficult to write than the ones we use, and they are a headache to multiply or divide, but they are very easy to add or subtract. For example, CXI + XVI = CXXVII. The numerals we use now (0, 1, 2, 3, . . .) were first seen in Europe in about the thirteenth century, but Roman numerals were used by bankers and bookkeepers until the eighteenth century. They did not trust symbols like 0 that could easily be changed to 6, 8, or 9 by a dishonest clerk.

10 35 + 42 is approximately 30 + 40 or 70. The correct answer will be about 70, not 7 or 700 or 7000. Once you have a reasonable estimate of the answer, you are ready to do the arithmetic work.

Calculate 35 + 42 = ______

Work it out and check your answer in **11**.

11 You should have set it up like this:

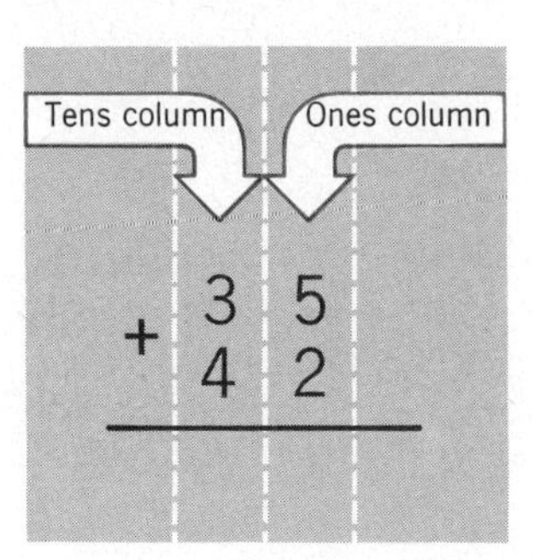

1. The numbers to be added are arranged vertically in columns.
2. The right end or ones digits are placed in the ones column, the tens digits are placed in the tens column, and so on.

Avoid the confusion of

```
 35            35
+ 42    or   + 42
-----       ------
```

The most frequent cause of errors in arithmetic is carelessness, especially in simple procedures such as lining up the digits correctly.

Once the digits are lined up the problem is easy.

```
 35
+42
---
 77
```

Does your answer agree with your original guess? Yes. The guess, 70, is roughly equal to the actual sum, 77.

What we have just shown you is the *guess n' check* method of doing mathematics calculations.

★ Step 1. *Guess* at the answer.
★ Step 2. Work the problem carefully.
★ Step 3. *Check* your answer against your guess. If they are very different, repeat both steps 1 and 2.

Most students hesitate at guessing, afraid they might guess incorrectly. Relax. You are the only one who will know your guess. Do it in your head, do it quickly, and make it reasonably accurate. Step 3 helps you detect incorrect answers before you finish the problem. The guess n' check method means you never work in the dark, you always know where you are going. Use this approach on every math calculation and you need never have an incorrect answer again.

Here is a slightly more difficult problem:

$27 + 48 =$ ______

Try it, then go to 12.

12 First, guess. 27 + 48 is roughly 30 + 50 or about 80. The answer is closer to 80 than to 8 or 800.

Second, line up the addends vertically.

$$\begin{array}{r} 27 \\ +48 \\ \hline \end{array}$$

Third, work it out carefully.

$$\begin{array}{r} {}^{1} \\ 27 \\ +48 \\ \hline 75 \end{array}$$

Finally, check your answer against your guess. The guess, 80, is roughly equal to the actual answer, 75.

What does that little 1 above the tens digit column mean? What really happens when you "carry" a digit? Let's look at it in detail.

In expanded notation,

$$\begin{array}{rl} 27 = & 2 \text{ tens} + \ \ 7 \text{ ones} \\ +48 = & \underline{4 \text{ tens} + \ \ 8 \text{ ones}} \\ & 6 \text{ tens} + 15 \text{ ones} = \underbrace{6 \text{ tens} + 1 \text{ ten}} + 5 \text{ ones} \\ & \qquad\qquad\qquad\ \ = 7 \text{ tens} + 5 \text{ ones} \\ & \qquad\qquad\qquad\ \ = 75 \end{array}$$

The 1 that is "carried" over to the tens column is really a 10!

Here is another way to see this:

Step 1

$$\begin{array}{r} 27 \\ +48 \\ \hline 15 \end{array}$$

Add ones $7 + 8 = 15$

15 is the first partial sum.

Step 2

$$\begin{array}{r} 27 \\ +48 \\ \hline 15 \\ 60 \end{array}$$

Add tens $20 + 40 = 60$

60 is the second partial sum.

Step 3

$$\begin{array}{r} 27 \\ +48 \\ \hline 15 \\ 60 \\ \hline 75 \end{array}$$

Add partial sums $15 + 60 = 75$

Again, you should see that the "carry 1" is the 10 in 15.

Using partial sums is the long way to do it, so usually we take a short-cut and write

$$\begin{array}{r} \overset{1}{2}7 \\ +48 \\ \hline 5 \end{array} \quad \begin{array}{l} 7+8=15 \\ \text{Write 5,} \\ \text{carry 1 ten} \end{array} \qquad \begin{array}{r} \overset{1}{2}7 \\ +48 \\ \hline 75 \end{array} \quad 1+2+4=7$$

You will learn the short-cut method here, but it is important that you know why it works.

Use the partial sum method to add

429 + 758 = ______

Set it up step by step as we did above, then turn to **13**.

★★★ (Don't forget to guess n' check.)

---

13 429 + 758 = ______

Guess: 400 + 700 = 1100

Line up the addends:

```
  429
+ 758
```

Use partial sums:

429	
+758	
17	Add ones: 9 + 8 = 17
70	Add tens: 20 + 50 = 70
1100	Add hundreds: 400 + 700 = 1100
1187	Add the partial sums

Check: 1100 (the guess) roughly equals 1187.

And of course you would use the short-cut method once you understand this process:

Step 1

```
   1
  429
+ 758    9 + 8 = 17
    7    Write 7,
         carry 1 ten
```

Step 2

```
   1
  429
+ 758    1 + 2 + 5 = 8
   87
```

Step 3

```
   1
  429
+ 758
 1187
```

Now, try these problems. Work them first using partial sums, then using the short-cut. Be sure to guess n' check.

(a) 256 + 867 = ______

(b) 2368 + 754 = ______

(c) 980 + 456 = ______

Check your answers in **14** when you are finished.

HOW TO ADD LONG LISTS OF NUMBERS

Very often, especially in business and industry, it is necessary to add long lists of numbers. The best procedure is to break the problem down into a series of simpler additions. First add sets of two or three numbers, then add these sums to obtain the total.

For example,

9		
3	12	
7		
6	13	25
12		
4	16	
17		
+ 5	22	38
		63

You do a little more writing but carry fewer numbers in your head; the result is fewer mistakes.

Better yet, keep your eye peeled for combinations that add to 10 or 15, and work with mental addition of three addends.

9		
3	10	
7		19
6		
12	10	
4		22
17		
5		22
		63

Try these for practice:

8	7	3	11	3	13
17	6	5	7	5	17
3	8	7	2	12	11
4	5	6	5	7	14
11	9	5	6	6	15
9	3	1	7	4	8
16	7	3	13	1	9
7	12	4	6	2	16
11	8	2	5	18	12
5	16	7	14	9	7
		3	16	7	8

The answers are on page 495.

14 (a) **Guess:** 200 + 900 = 1100

```
  256
 +867
 ----
   13   Add ones:              6 +   7 =   13
  110   Add tens:             50 +  60 =  110
 1000   Add hundreds:        200 + 800 = 1000
 ----
 1123   Add partial sums:
```

Check: 1100 is roughly equal to 1123.

Short-cut method:

Step 1

```
   1
  256
 +867   6 + 7 = 13
 ----
    3   Write 3,
        carry 1 ten
```

Step 2

```
  11
  256
 +867   1 + 5 + 6 = 12
 ----
   23   Write 2,
        carry 1 hundred
```

Step 3

```
  11
  256
 +867   1 + 2 + 8 = 11
 ----
 1123
```

(b) **Guess:** 2300 + 700 = 3000

```
  2368
 + 754
 -----
    12   Add ones:              8 +   4 =   12
   110   Add tens:             60 +  50 =  110
  1000   Add hundreds:        300 + 700 = 1000
  2000   Add thousands:      2000
 -----
  3122   Add partial sums:
```

Check: the guess, 3000, is roughly equal to the answer 3122.

Short-cut method:

Step 1

```
    1
  2368
 + 754   8 + 4 = 12
 -----
     2   Write 2,
         carry 1 ten
```

Step 2

```
   11
  2368
 + 754   1 + 6 + 5 = 12
 -----
    22   Write 2,
         carry 1 hundred
```

Step 3

```
  111
  2368
 + 754   1 + 3 + 7 = 11
 -----
   122   Write 1,
         carry 1 thousand
```

Step 4

```
  111
  2368
 + 754   1 + 2 = 3
 -----
  3122
```

(c) **Guess:** 1000 + 400 = 1400

```
 980
+456
----
   6   Add ones:        0 +   6 =    6
 130   Add tens:       80 +  50 =  130
1300   Add hundreds:  900 + 400 = 1300
----
1436
```

Check: 1400 is approximately equal to 1436.

Short-cut method:

Step 1		Step 2		Step 3	
980 +456 6	0 + 6 = 6	$\overset{1}{9}80$ +456 36	8 + 5 = 13 Write 3, carry 1 hundred	$\overset{1}{9}80$ +456 1436	1 + 9 + 4 = 13

If you had difficulty with any of these you should return to **5** and review. Otherwise go to **15** for a set of practice addition problems.

16 Subtraction is the reverse of the process of addition.

Addition: $3 + 4 = \square$
Subtraction: $3 + \square = 7$

Written this way, a subtraction problem asks the question, "How much must be added to a given number to produce a required amount?"

Most often, however, the numbers in a subtraction problem are written using a minus sign ($-$):

$17 - 8 = \square$ means that there is a number $\square$ such that $8 + \square = 17$.

Written this way, a subtraction problem asks you to find the difference between two numbers.

Write in the answer to this subtraction problem, then turn to **17**.

17

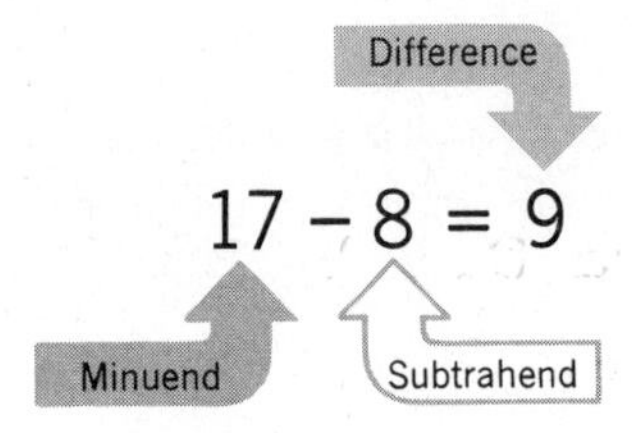

Special names are given to the numbers in a subtraction problem, and it will be helpful if you know them.

★ The *minuend* is the larger of the two numbers in the problem. It is the number that is being decreased.

★ The *subtrahend* is the number that is being subtracted from the minuend.

★ The *difference* is the amount that must be added to the subtrahend to produce the minuend. It is the answer to the subtraction problem.

The ability to solve simple subtraction problems depends on your knowledge of the addition of one digit numbers. For example, find the difference

$9 - 4 =$ ______

Do it and then turn to **18**.

---

18 $9 - 4 = \underline{\quad 5 \quad}$

Solving this problem probably involved a chain of thoughts something like this:

> "Nine minus four. Four added to what number gives nine? Five? Try it: four plus five equals nine. Right."

If you have memorized the addition of one-digit numbers, as shown in frame **6** or on the study card on page 518, subtraction problems involving small whole numbers will be easy for you. If you haven't memorized these, do it now.

Now try a more difficult subtraction problem:

$47 - 23 = \underline{\qquad}$

What is the first step?

Work the problem and continue in **19**.

19 The *first* step is to guess at the answer—remember?

$47 - 23$ is roughly $40 - 20$ or 20.

The difference, your answer, will be about 20—not 2 or 10 or 200.

The *second* step is to write the numbers in a vertical format as you did with addition. Be careful to keep the ones digits in line in one column, the tens digits in a second column, and so on.

$$\begin{array}{r} 47 \\ -\,23 \\ \hline \end{array}$$

Notice that the minuend is written above the subtrahend—larger number on top.

Once the numbers have been arranged in this way, the difference may be written immediately.

Step 1

$$\begin{array}{r} 47 \\ -23 \\ \hline 4 \end{array}$$ Ones digits: $7 - 3 = 4$

Step 2

$$\begin{array}{r} 47 \\ -23 \\ \hline 24 \end{array}$$ Tens digits: $40 - 20 = 20$

The difference is 24, and this agrees with our first guess.

With some problems it is necessary to rewrite the minuend (larger) number before the problem can be solved. For example, the number 24 means $20 + 4$ or 2 tens + 4 ones. It can be rewritten as $10 + 14$. The difference $24 - 8$ is not easy to find because 8 is larger than 4. To perform the subtraction we write 24 as $10 + 14$ and then subtract 8.

$24 - 8 = (10 + 14) - 8 = 10 + (14 - 8) = 10 + 6 = 16$

Try this one:

$64 - 37 =$ ______

Check your work in **20**.

Can you do a problem like 37 −64?
In this section we are considering only subtraction of a smaller number from a larger number. See Unit 3 for information on that problem.

20 First, guess. 64 − 37 is roughly 60 − 40 or 20.

Second, arrange the numbers vertically in columns.

$$\begin{array}{r} 64 \\ -37 \\ \hline \end{array}$$

Third, write them in expanded form to understand the process.

$$\begin{array}{rcrcr} 64 & = & 6 \text{ tens} + 4 \text{ ones} & = & 5 \text{ tens} + 14 \text{ ones} \\ -37 & = & -(3 \text{ tens} + 7 \text{ ones}) & = & -(3 \text{ tens} + \ \ 7 \text{ ones}) \\ \hline & & & = & 2 \text{ tens} + \ \ 7 \text{ ones} \\ & & & = & 20 \quad + \ \ 7 \\ & & & = & 27 \text{ which agrees with our guess.} \end{array}$$

Because 7 is larger than 4 we must "borrow" 1 ten from the 6 tens in the minuend. We are actually regrouping or rewriting the minuend.

In actual practice we do not write it out in expanded form. Our work might look like this:

Step 1

$$\begin{array}{r} 64 \\ -37 \\ \hline \end{array}$$

Step 2

$$\begin{array}{r} {}^{5\,14} \\ \not{6}\not{4} \\ -37 \\ \hline 7 \end{array}$$

Borrow one ten, change the 6 in the tens place to 5, change 4 to 14, and subtract 14 − 7 = 7

Step 3

$$\begin{array}{r} {}^{5\,14} \\ \not{6}\not{4} \\ -37 \\ \hline 27 \end{array}$$

50 − 30 = 20, write 2

★ Double check subtraction problems by adding the answer and the subtrahend; their sum should equal the minuend.

$$\begin{array}{r} {}^{1} \\ 37 \\ +27 \\ \hline 64 \end{array}$$

Try these problems for practice:

(a) $\begin{array}{r} 71 \\ -39 \\ \hline \end{array}$ (b) $\begin{array}{r} 263 \\ -127 \\ \hline \end{array}$ (c) $\begin{array}{r} 426 \\ -128 \\ \hline \end{array}$ (d) $\begin{array}{r} 902 \\ -465 \\ \hline \end{array}$

Solutions are in **21**.

---

21 **Guess:** $70 - 40 = 30$

(a) **Step 1** **Step 2**

$$\begin{array}{r} 71 \\ -39 \\ \hline \end{array} \qquad \begin{array}{r} \scriptstyle 6\,11 \\ \not{7}\not{1} \\ -39 \\ \hline 32 \end{array}$$

Borrow one ten from 70,
change the 7 in the tens place to 6,
change the 1 in the ones place to 11,
$11 - 9 = 2$, write 2,
$60 - 30 = 30$, write 3.

Check: The answer 32 is approximately equal to the guess, 30.

Guess: $200 - 100 = 100$

(b) **Step 1** **Step 2**

$$\begin{array}{r} 263 \\ -127 \\ \hline \end{array} \qquad \begin{array}{r} \scriptstyle 5\,13 \\ 2\not{6}\not{3} \\ -127 \\ \hline 136 \end{array}$$

Borrow one ten from 60,
change the 6 in the tens place to 5,
change the 3 in the ones place to 13,
$13 - 7 = 6$, write 6,
$50 - 20 = 30$, write 3,
$200 - 100 = 100$, write 1.

Check: The answer is approximately equal to the guess.

Guess: $400 - 100 = 300$

(c) **Step 1** **Step 2** **Step 3**

$$\begin{array}{r} 426 \\ -128 \\ \hline \end{array} \qquad \begin{array}{r} \scriptstyle 1\,16 \\ 4\not{2}\not{6} \\ -128 \\ \hline 8 \end{array} \qquad \begin{array}{r} \scriptstyle 3\,11\,16 \\ \not{4}\not{2}\not{6} \\ -128 \\ \hline 298 \end{array}$$

$16 - 8 = 8$
$110 - 20 = 90$, write 9
$300 - 100 = 200$, write 2

Notice that in this case we must borrow twice. Borrow 1 ten from the 20 in 426 to make 16, then borrow 1 hundred from the 400 in 426 to make 110.

Guess: $900 - 500 = 400$

(d) **Step 1** **Step 2** **Step 3**

$$\begin{array}{r} 902 \\ -465 \\ \hline \end{array} \qquad \begin{array}{r} \scriptstyle 8\,10 \\ \not{9}\not{0}2 \\ -465 \\ \hline \end{array} \qquad \begin{array}{r} \scriptstyle 8\,9\,12 \\ \not{9}\not{0}\not{2} \\ -465 \\ \hline 437 \end{array}$$

$12 - 5 = 7$
$90 - 60 = 30$, write 3
$800 - 400 = 400$, write 4

In problem (d) we first borrow 1 hundred from 900 to get a 10 in the tens place, then we borrow one 10 from the tens place to get a 12 in the ones place.

In expanded form this last problem is

$$\begin{array}{r} 902 \\ -465 \\ \hline \end{array} \quad \longrightarrow \quad \begin{array}{r} 9 \text{ hundreds} + 0 \text{ tens} + 2 \text{ ones} \\ -(4 \text{ hundreds} + 6 \text{ tens} + 5 \text{ ones}) \\ \hline \end{array}$$

$$\begin{array}{r} 8 \text{ hundreds} + 10 \text{ tens} + 2 \text{ ones} \\ -(4 \text{ hundreds} + 6 \text{ tens} + 5 \text{ ones}) \\ \hline \end{array}$$

$$\begin{array}{r} 8 \text{ hundreds} + 9 \text{ tens} + 12 \text{ ones} \\ -(4 \text{ hundreds} + 6 \text{ tens} + 5 \text{ ones}) \\ \hline 4 \text{ hundreds} + 3 \text{ tens} + 7 \text{ ones} \end{array}$$

$$= 400 + 30 + 7$$
$$= 437$$

Now, do you want more worked examp[illegible] of problems containing zeros, similar to this last one? If so, go to 23[illegible]

Otherwise, go to 22 for a set of prac[illegible] problems.

---

23 Let's work through a few examples.

(a) **Step 1** **Step 2** **Step 3**

$$\begin{array}{r} \\ 400 \\ -167 \\ \hline \\ \end{array} \qquad \begin{array}{r} {}^{3\;10} \\ \not{4}\not{0}0 \\ -167 \\ \hline \\ \end{array} \qquad \begin{array}{r} {}^{3\;9\;10} \\ \not{4}\not{0}\not{0} \\ -167 \\ \hline 233 \end{array}$$

Check: $\begin{array}{r} 167 \\ +233 \\ \hline 400 \end{array}$ Do you see that we have rewritten 400 as 300 + 90 + 10?

(b) **Step 1** **Step 2** **Step 3** **Step 4**

$$\begin{array}{r} \\ 5006 \\ -2487 \\ \hline \\ \end{array} \qquad \begin{array}{r} {}^{4\;10} \\ \not{5}\not{0}06 \\ -2487 \\ \hline \\ \end{array} \qquad \begin{array}{r} {}^{4\;9\;10} \\ \not{5}\not{0}\not{0}6 \\ -2487 \\ \hline \\ \end{array} \qquad \begin{array}{r} {}^{4\;9\;9\;16} \\ \not{5}\not{0}\not{0}\not{6} \\ -2487 \\ \hline 2519 \end{array}$$

Check: $\begin{array}{r} 2487 \\ +2519 \\ \hline 5006 \end{array}$

(c) **Step 1** **Step 2** **Step 3** **Step 4**

$$\begin{array}{r} \\ 24{,}632 \\ -\;\;5{,}718 \\ \hline \\ \end{array} \qquad \begin{array}{r} {}^{2\;12} \\ 246\not{3}\not{2} \\ -\;\;5718 \\ \hline 14 \end{array} \qquad \begin{array}{r} {}^{3\;16\;2\;12} \\ 2\not{4}\not{6}\not{3}\not{2} \\ -\;\;5718 \\ \hline 914 \end{array} \qquad \begin{array}{r} {}^{1\;13\;16\;2\;12} \\ \not{2}\not{4}\not{6}\not{3}\not{2} \\ -\;\;5718 \\ \hline 18914 \end{array}$$

Check: $\begin{array}{r} 5718 \\ +18914 \\ \hline 24632 \end{array}$

Any subtraction problem that involves borrowing should *always* be checked in this way. It is very easy to make a mistake in this process.

Go to **22** for a set of practice problems on subtraction.

---

24 In a certain football game, the West Newton Waterbugs scored five touchdowns at six points each. How many total points did they score through touchdowns? We can answer the question several ways:

1. Count points:

   ....

2. Add touchdowns:

   $6 + 6 + 6 + 6 + 6 = ?$

or

3. Multiply:

   $5 \times 6 = ?$

We're not sure about the mathematical ability of the West Newton scorekeeper, but most people would multiply. Multiplication is a short-cut method of counting or repeated addition.

How many points did they score?

Work it out one way or another, then go to **25**.

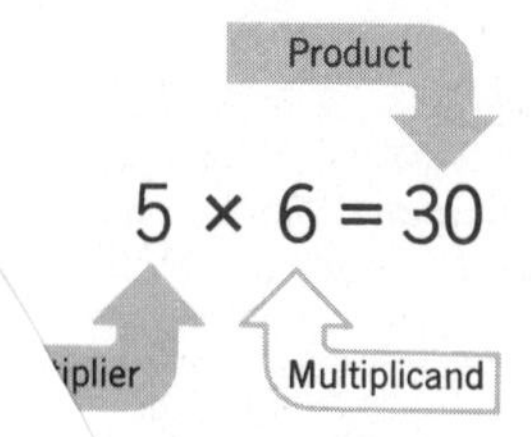

In ltiplication statement the *multiplicand* is the number to be multiplied, the *iplier* is the number multiplying the multiplicand, and the *product* is the lt of the multiplication. The multiplier and multiplicand are called the *fac* of the product.

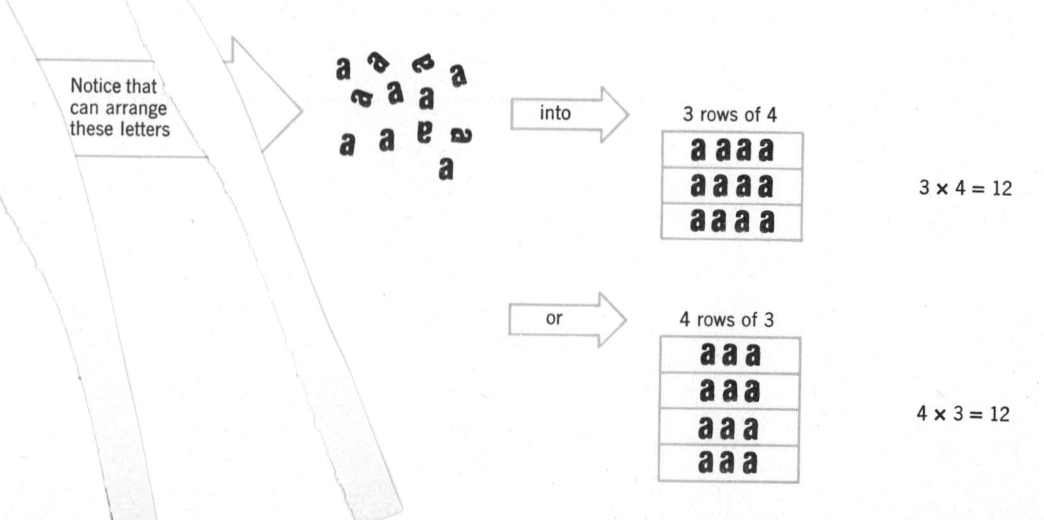

Changing the order of the factors does not change their product. This is the *commutative* property of multiplication.

In order to become skillful at multiplication, you must know the one-digit multiplication table from memory.

Complete the table below by multiplying the number at the top by the number at the side and placing their product in the proper square. We have multiplied $3 \times 4 = 12$ and $2 \times 5 = 10$ for you.

Multiply	2	5	8	1	3	6	9	7	4
1									
7									
5	10								
4					12				
9									
2									
6									
3									
8									

Check your answers in **27**.

26 Practice problems for one-digit multiplication

Multiply as shown. Work quickly. You should be able to answer all problems in a set correctly in the time indicated. (These times are for community college students enrolled in a developmental math course.)

A. Multiply:

6	4	9	6	3	9	7	8	2	8
2	8	7	6	4	2	0	3	7	1
12	32	63	36	12	18	0	24	14	8

6	8	5	5	2	3	9	7	3	0
8	2	9	6	5	3	8	5	6	4
48	16	45	30	10	9	72	35	18	0

7	5	4	7	4	8	6	9	8	6
4	3	9	7	2	5	7	6	8	4
28	15	36	49	8	40	42	54	64	24

5	3	5	9	9	6	1	8	4	7
4	0	5	3	9	1	1	6	4	9
20	0	25	27	81	6	1	48	16	63

Average time: 100 sec Record: 37 sec

B. Multiply:

2	6	3	5	6	4	4	8	2	7
8	5	3	7	3	5	7	6	6	9
16	30	9	35	18	20	28	48	12	63

8	0	2	3	1	5	6	9	5	8
4	6	9	8	9	5	4	5	2	9
32	0	18	24	9	25	24	45	10	72

3	7	5	6	9	2	7	8	9	2
5	7	8	9	4	4	6	8	0	2
15	49	40	54	36	8	42	64	0	4

5	9	1	8	6	4	9	0	2	7
5	3	7	7	6	3	9	4	1	8
25	27	7	56	36	12	81	0	2	56

Average time: 100 sec Record: 36 sec

Check your answers on page 497.

When you have had the practice you need, turn to **28** and continue.

THE SEXY SIX

Here are the six most often missed one-digit multiplications:

"inside"

$9 \times 8 = 72$
$9 \times 7 = 63$
$9 \times 6 = 54$
$8 \times 7 = 56$
$8 \times 6 = 48$
$7 \times 6 = 42$

It may help you to notice that in these multiplications the "inside" digits, such as 8 and 7, are consecutive and the digits of the answer add to nine: $7 + 2 = 9$. This is true for *all* one-digit numbers multiplied by 9.

Be certain you have these memorized.

(There is nothing very sexy about them, but we did get your attention, didn't we?)

27 Here is the completed multiplication table.

Multiply	2	5	8	1	3	6	9	7	4
1	2	5	8	1	3	6	9	7	4
7	14	35	56	7	21	42	63	49	28
5	10	25	40	5	15	30	45	35	20
4	8	20	32	4	12	24	36	28	16
9	18	45	72	9	27	54	81	63	36
2	4	10	16	2	6	12	18	14	8
6	12	30	48	6	18	36	54	42	24
3	6	15	24	3	9	18	27	21	12
8	16	40	64	8	24	48	72	56	32

If you are not able to perform these one-digit multiplications quickly from memory, you should practice until you can do so. A study card multiplication table is provided in the back of this book (page 518). Cut it out and use it if you need it.

Notice that the product of any number and 1 is that same number. For example,

$1 \times 2 = 2$
$1 \times 6 = 6$

or even

$1 \times 753 = 753$

Zero has been omitted from the table because the product of any number and zero is zero. For example,

$0 \times 2 = 0$
$0 \times 7 = 0$
$0 \times 395 = 0$

This is reasonable:

3×7 (three 7s) $= 21$
2×7 (two 7s) $= 14$
1×7 (one 7) $= 7$
0×7 (no 7s) $= 0$

If you want more practice in one-digit multiplication, go to **26**. Otherwise, go to **28**.

---

28 The multiplication of larger numbers is based on the one-digit number multiplication table. Find the product of

$34 \times 2 =$ ______

Remember the procedure you followed for addition. What are the first few steps in this multiplication?

Try it, then go to **29**.

---

29 First, **guess.** $30 \times 2 = 60$. The actual product of the multiplication will be about 60.

Second, arrange the factors to be multiplied vertically, with ones digits in a single column, tens digits in a second column, and so on.

$$\begin{array}{r} 34 \\ \times\ \ 2 \\ \hline \end{array}$$

Finally, to make the process clear, let's write it in expanded form.

$$\begin{array}{r} 34 \\ \times\ 2 \\ \hline \end{array} \longrightarrow \begin{array}{l} 3 \text{ tens} + 4 \text{ ones} \\ \qquad\quad \times 2 \\ \hline 6 \text{ tens} + 8 \text{ ones} = 60 + 8 = 68 \end{array}$$

Check: The guess 60 is roughly equal to the answer 68.

Notice that when a single number multiplies a sum, it forms a product with each addend in the sum. For example,

$2 \times (3 + 4) = (2 \times 3) + (2 \times 4)$

In the expanded multiplication above, the multiplier 2 forms a product with each addend in the sum (3 tens + 4 ones).

Write the following multiplication in expanded form.

$$\begin{array}{r} 28 \\ \times\ 3 \\ \hline \end{array}$$

Check your work in **30**.

30 **Guess:** $30 \times 3 = 90$

$$\begin{array}{r} 28 \\ \times\ 3 \\ \hline \end{array} \rightarrow \begin{array}{l} 2 \text{ tens} + 8 \text{ ones} \\ \qquad \times 3 \\ \hline 6 \text{ tens} + 24 \text{ ones} \end{array}$$

$$\begin{aligned} &= 6 \text{ tens} + 2 \text{ tens} + 4 \text{ ones} \\ &= 8 \text{ tens} + 4 \text{ ones} \\ &= 80 + 4 \\ &= 84 \end{aligned}$$

Check: 90 is roughly equal to 84.

Of course we do not normally use the expanded form; instead we simplify the work like this:

★ **Step 1**

$$\begin{array}{r} 28 \\ \times\ 3 \\ \hline 24 \end{array}$$

Multiply ones $8 \times 3 = 24$

★ **Step 2**

$$\begin{array}{r} 28 \\ \times\ 3 \\ \hline 24 \\ 60 \\ \hline \end{array}$$

Multiply $3 \times 20 = 60$

★ **Step 3**

$$\begin{array}{r} 28 \\ \times\ 3 \\ \hline 24 \\ 60 \\ \hline 84 \end{array}$$

Add partial products $24 + 60 = 84$

When you are certain about how to do this, you can take a short cut and write

$$\begin{array}{r} {\scriptstyle 2} \\ 28 \\ \times\ 3 \\ \hline 84 \end{array}$$

$3 \times 8 = 24$ write 4 and carry 20

3×2 tens $= 6$ tens $\quad$ 6 tens + 2 tens = 8 tens write 8

Try these to be certain you understand the process. Work all four problems using both the step-by-step and short-cut methods.

(a) $\begin{array}{r} 43 \\ \times\ 5 \\ \hline \end{array}$ (b) $\begin{array}{r} 73 \\ \times\ 4 \\ \hline \end{array}$ (c) $\begin{array}{r} 29 \\ \times\ 6 \\ \hline \end{array}$ (d) $\begin{array}{r} 258 \\ \times\ 7 \\ \hline \end{array}$

Check your work in **31**.

Where do the + and − signs come from?
They were used in the fifteenth century to show that boxes of merchandise were overweight (+) or underweight (−). Within 40 years or so bookkeepers and mathematicians started using them. The + symbol came from the Latin word *et* meaning *and*.

31 (a) **Guess:** $5 \times 40 = 200$

★ **Step 1**

$$\begin{array}{r} 43 \\ \times\ 5 \\ \hline 15 \end{array} \quad 5 \times 3 = 15$$

★ **Step 2**

$$\begin{array}{r} 43 \\ \times\ 5 \\ \hline 15 \\ 200 \\ \hline \end{array} \quad 5 \times 40 = 200$$

★ **Step 3**

$$\begin{array}{r} 43 \\ \times\ 5 \\ \hline 15 \\ 200 \\ \hline 215 \end{array}$$

Short-Cut Method

$$\begin{array}{r} \scriptstyle 1 \\ 43 \\ \times\ 5 \\ \hline 215 \end{array}$$

$5 \times 3 = 15$
5×4 tens $= 20$ tens
20 tens $+ 1$ ten $= 21$ tens

(b) **Guess:** $70 \times 4 = 280$

★ **Step 1**

$$\begin{array}{r} 73 \\ \times\ 4 \\ \hline 12 \end{array} \quad 3 \times 4 = 12$$

★ **Step 2**

$$\begin{array}{r} 73 \\ \times\ 4 \\ \hline 12 \\ 280 \\ \hline \end{array} \quad 4 \times 70 = 280$$

★ **Step 3**

$$\begin{array}{r} 73 \\ \times\ 4 \\ \hline 12 \\ 280 \\ \hline 292 \end{array}$$

Short-Cut Method

$$\begin{array}{r} \scriptstyle 1 \\ 73 \\ \times\ 4 \\ \hline 292 \end{array}$$

$4 \times 3 = 12$
4×7 tens $= 28$ tens
28 tens $+ 1$ ten $= 29$ tens

(c) **Guess:** $30 \times 6 = 180$

★ **Step 1**

$$\begin{array}{r} 29 \\ \times\ 6 \\ \hline 54 \end{array}$$

★ **Step 2**

$$\begin{array}{r} 29 \\ \times\ 6 \\ \hline 54 \\ 120 \\ \hline \end{array}$$

★ **Step 3**

$$\begin{array}{r} 29 \\ \times\ 6 \\ \hline 54 \\ 120 \\ \hline 174 \end{array}$$

Short-Cut Method

$$\begin{array}{r} \scriptstyle 5 \\ 29 \\ \times\ 6 \\ \hline 174 \end{array}$$

$6 \times 9 = 54$
6×2 tens $= 12$ tens
12 tens $+ 5$ tens $= 17$ tens

(d) **Guess:** $300 \times 7 = 2100$

★ **Step 1**

$$\begin{array}{r} 258 \\ \times 7 \\ \hline 56 \end{array} \quad 7 \times 8 = 56$$

★ **Step 2**

$$\begin{array}{r} 258 \\ \times 7 \\ \hline 56 \\ 350 \end{array} \quad 7 \times 50 = 350$$

★ **Step 3**

$$\begin{array}{r} 258 \\ \times 7 \\ \hline 56 \\ 350 \\ 1400 \\ \hline \end{array} \quad 7 \times 200 = 1400$$

★ **Step 4**

$$\begin{array}{r} 258 \\ \times 7 \\ \hline 56 \\ 350 \\ 1400 \\ \hline 1806 \end{array} \quad \text{Add}$$

Short-Cut

$$\begin{array}{r} \scriptstyle 4\,5\ \ \\ 258 \\ \times 7 \\ \hline 1806 \end{array}$$

Calculations involving two-digit multipliers are done in the same way. Apply this method to

$$\begin{array}{r} 57 \\ \times 24 \\ \hline \end{array}$$

The worked example is in **32**.

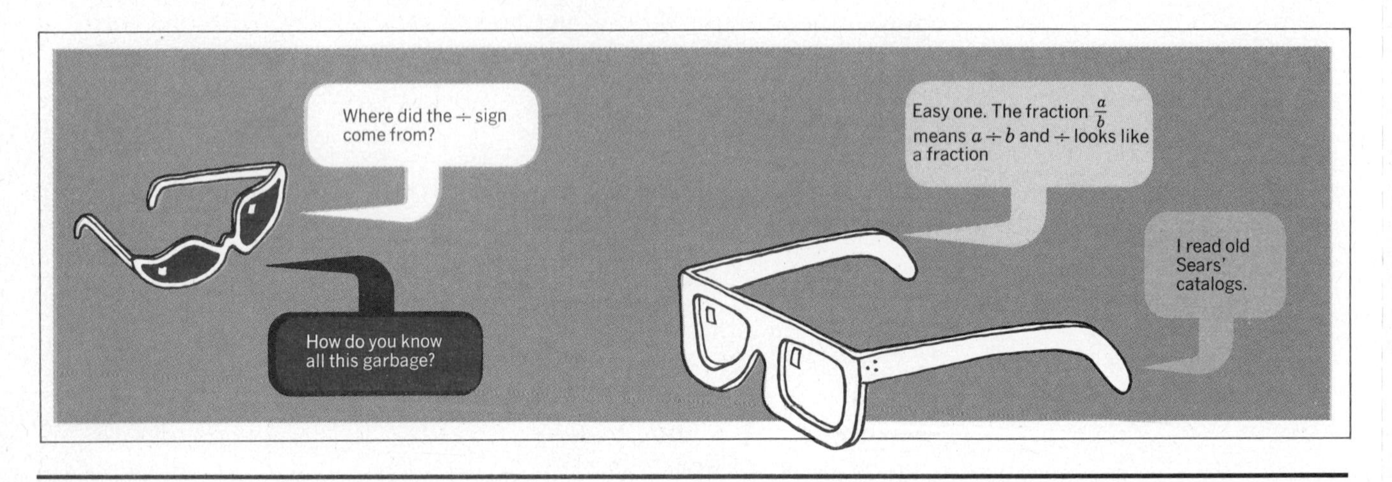

---

32 **Guess:** $60 \times 20 = 1200$

★ **Step 1**

57	Multiply by the ones
$\times$24	digit (4)
28	$4 \times 7 = 28$
200	$4 \times 50 = 200$

★ **Step 2**

57	Multiply by the tens
$\times$24	digit (2)
28	
200	
140	$20 \times 7 = 140$
1000	$20 \times 50 = 1000$
1368	Add

Use the short-cut method to reduce the written work.

$$\begin{array}{r} {\scriptstyle 1} \\ {\scriptstyle 2} \\ 57 \\ \times 24 \\ \hline 228 \\ 1140 \\ \hline 1368 \end{array}$$

- 228 ← $4 \times 7 = 28$, write 8, carry 2 tens; $4 \times 50 = 200$, add carry 20 to get 220, write 22
- 1140 ← $20 \times 7 = 140$, write 40, carry 1 hundred; $20 \times 50 = 1000$, add carry 100 to get 1100, write 11
- 1368 ← Add

The zero in 1140 is usually omitted to save time.

Try these:

(a) $\begin{array}{r} 64 \\ \times 37 \\ \hline \end{array}$ (b) $\begin{array}{r} 327 \\ \times 145 \\ \hline \end{array}$ (c) $\begin{array}{r} 342 \\ \times 102 \\ \hline \end{array}$

Work each problem as shown above. Use the short-cut method if possible. Check your answers in **33**.

MULTIPLICATION SHORT-CUTS

There are hundreds of quick ways to multiply various numbers. Most of them are only quick if you are already a math whiz and will confuse you more than help you. Here are a few that are easy to do and easy to remember.

1. To multiply by 10, annex a zero on the right end of the multiplicand. For example,

   $34 \times 10 = 340$
   $256 \times 10 = 2560$

   Multiplying by 100 or 1000 is similar.

   $34 \times 100 = 3400$
   $256 \times 1000 = 256000$

2. To multiply by a number ending in zeros, carry the zeros forward to the answer. For example,

   $$\begin{array}{r} 26 \\ \times 20 \\ \hline \end{array} \rightarrow \begin{array}{r} 26 \\ \times 20 \\ \hline 520 \end{array}$$

   Multiply 26×2 and attach the zero on the right. The product is 520.

   $$\begin{array}{r} 34 \\ \times 2100 \\ \hline \end{array} \rightarrow \begin{array}{r} 34 \\ \times 2100 \\ \hline 34 \\ 68 \\ \hline 71400 \end{array}$$

3. If both multiplier and multiplicand end in zeros, bring all zeros forward to the answer.

   $$\begin{array}{r} 230 \\ \times 200 \\ \hline \end{array} \rightarrow \begin{array}{r} 230 \\ \times 200 \\ \hline 46000 \end{array}$$

   Attach three zeros to the product of 23×2.

   "Another example:"

   $$\begin{array}{r} 1000 \\ \times 100 \\ \hline 100{,}000 \end{array}$$

   This sort of multiplication is mostly a matter of counting zeros.

33 (a) **Guess:** $60 \times 40 = 2400$

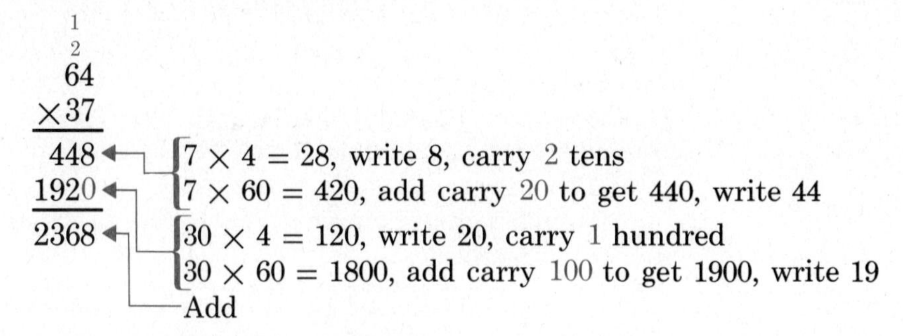

1
2
64
×37
448
1920
2368
7 × 4 = 28, write 8, carry 2 tens
7 × 60 = 420, add carry 20 to get 440, write 44
30 × 4 = 120, write 20, carry 1 hundred
30 × 60 = 1800, add carry 100 to get 1900, write 19
Add

(b) **Guess:** $300 \times 100 = 30{,}000$

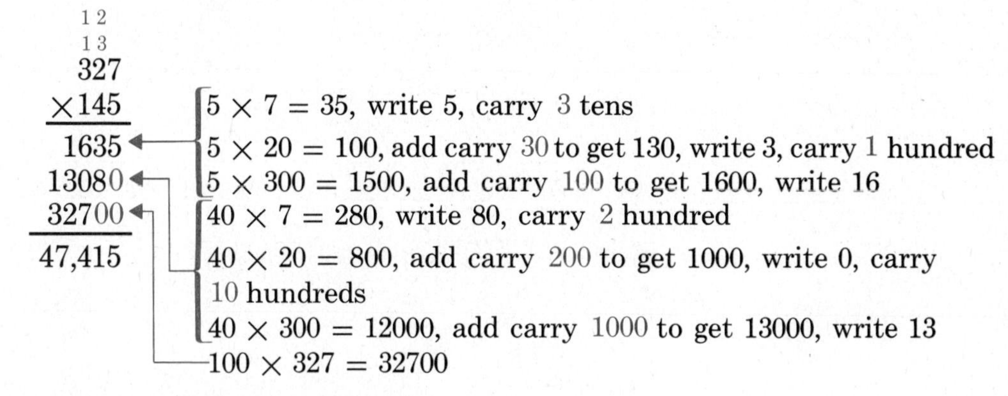

1 2
1 3
327
×145
1635
13080
32700
47,415
5 × 7 = 35, write 5, carry 3 tens
5 × 20 = 100, add carry 30 to get 130, write 3, carry 1 hundred
5 × 300 = 1500, add carry 100 to get 1600, write 16
40 × 7 = 280, write 80, carry 2 hundred
40 × 20 = 800, add carry 200 to get 1000, write 0, carry 10 hundreds
40 × 300 = 12000, add carry 1000 to get 13000, write 13
100 × 327 = 32700

(c) **Guess:** $300 \times 100 = 30{,}000$

```
   342
  ×102
   684     2 × 342 = 684
   000     0 × 342 = 000
 34200   100 × 342 = 34200
 34884
```

Be very careful when there are zeros in the multiplier; it is very easy to misplace one of those zeros. Do not skip any steps and be sure to guess n' check.

Go to **34** for a set of practice problems on multiplication of whole numbers.

---

35 DIVISION

Division is the reverse process for multiplication. It enables us to separate a given quantity into equal parts. The mathematical phrase $12 \div 3$ is read "twelve divided by three," and it asks us to separate a collection of 12 objects into 3 equal parts. The mathematical phrases

$12 \div 3 \qquad 3\overline{)12} \qquad \frac{12}{3} \qquad \text{and} \qquad 12/3$

all represent division and are all read "twelve divided by three."

Perform this division: $12 \div 3 =$ ______

Turn to **36** to continue.

36

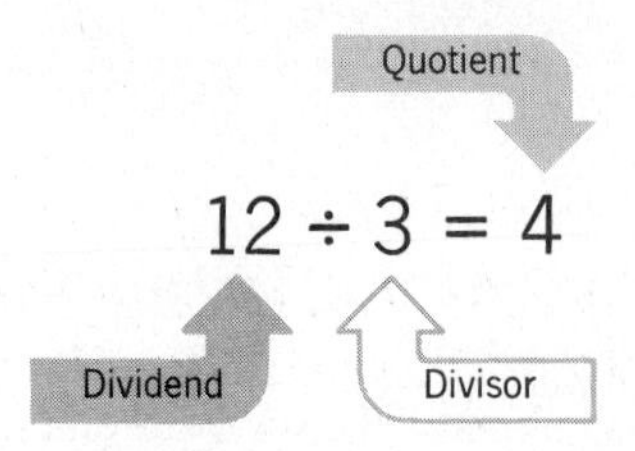

In this division problem 12, the number being divided, is called the *dividend;* 3, the number used to divide, is called the *divisor;* and 4, the result of the division is called the *quotient,* from a Latin word meaning "how many times."

One way to perform divison is to reverse the multiplication process.

$24 \div 4 = \square$ means that $4 \times \square = 24$

If the one-digit multiplication tables are firmly in your memory, you will recognize immediately that $\square = 6$. By placing a 6 in $\square$ you make the statement $4 \times \square = 24$ true.

Try these:

$35 \div 7 =$ ______ $42 \div 6 =$ ______

$28 \div 4 =$ ______ $56 \div 7 =$ ______

$45 \div 5 =$ ______ $18 \div 3 =$ ______

$70 \div 10 =$ ______ $63 \div 9 =$ ______

$30 \div 5 =$ ______ $72 \div 8 =$ ______

Check your answers in **37**.

37 $35 \div 7 = \underline{5}$ $\quad$ $42 \div 6 = \underline{7}$

$28 \div 4 = \underline{7}$ $\quad$ $56 \div 7 = \underline{8}$

$45 \div 5 = \underline{9}$ $\quad$ $18 \div 3 = \underline{6}$

$70 \div 10 = \underline{7}$ $\quad$ $63 \div 9 = \underline{7}$

$30 \div 5 = \underline{6}$ $\quad$ $72 \div 8 = \underline{9}$

You should be able to do all of these quickly by working backward from the one-digit multiplication tables.

How do we divide dividends that are larger than 9×9 and therefore not in the multiplication table? Obviously, we need a better procedure.

One way to learn how many times the divisor divides the dividend is to subtract it repeatedly. For example, in $12 \div 3$

$$\begin{array}{r}12\\-\ 3\\\hline 9\end{array}\quad\begin{array}{r}9\\-3\\\hline 6\end{array}\quad\begin{array}{r}6\\-3\\\hline 3\end{array}\quad\begin{array}{r}3\\-3\\\hline 0\end{array}$$

3 is subtracted from 12 four times, so that $12 \div 3 = 4$.

Try it. Perform the division $138 \div 23$ using repeated subtraction.

Check your answer in **38**.

AVERAGES

The *average* of a set of numbers is the single number that best represents the whole set. One simple kind of average is the arithmetic average or arithmetic mean defined as

$$\text{Arithmetic average} = \frac{\text{sum of measurements}}{\text{number of measurements}}$$

For example, the average weight of the five middle linemen on our college football team is

$$\text{Average weight} = \frac{215 \text{ lb} + 235 \text{ lb} + 224 \text{ lb} + 212 \text{ lb} + 239 \text{ lb}}{5}$$

$$= \frac{1125 \text{ lb}}{5} = 225 \text{ lb}$$

Try these problems for practice.

1. In a given week a student worked in the college library for the following hours each day: Monday 2 hours, Tuesday 3 hours, Wednesday $2\frac{1}{2}$ hours, Thursday $3\frac{1}{4}$ hours, Friday $2\frac{1}{4}$ hours, and Saturday 4 hours. What average amount of time does the student work per day?
2. On four weekly quizzes in her history class, a student scores 84, 74, 92, and 88 points. What is her average score?
3. A salesman sells the following amounts in successive weeks: $647.20, $705.17, $1205.65, $349.34, and $409.89. What is his average weekly sales?

The answers are on page 497.

38 $138 \div 23$

138	115	92	69	46	23
$-\ 23$	$-\ 23$	-23	-23	-23	-23
115	92	69	46	23	0

23 may be subtracted from 138 six times; therefore $138 \div 23 = 6$.

We could also divide by simply guessing. For example, to find $245 \div 7$ by guessing, we might go through a mental conversation with ourselves something like this:

"7 into 245 goes how many times?" **Lots.**

"How many? Pick a number." **Maybe 10.**

"Let's try it: $7 \times 10 = 70$, so 10 is much too small. Try a larger number." **How about 50? Does 7 go into 245 50 times?**

"Well, $7 \times 50 = 350$, which is larger than 245. Try again." **I'm getting tired. Will 30 do it?**

"$7 \times 30 = 210$. That is quite close to 245. Try something a little larger than 30." **31? 32? . . .**

Sooner or later the tired little guesser in your head will arrive at 35 and find that $7 \times 35 = 245$, so

$245 \div 7 = \underline{35}$

With pure guessing even a simple problem could take all afternoon. We need a short-cut. The best division process combines one-digit multiplication, repeated subtraction, and educated guessing. For example, in the problem

$96 \div 8 =$

start with a guess: the answer is about 10, since $8 \times 10 = 80$.

★ **Step 1**

Tens column ↓ ↓ Ones column

$8\overline{)\ 96}$

Arrange the divisor and dividend horizontally.

$$\begin{array}{r} 10 \\ 8\overline{)96} \end{array}$$

Can 8 be subtracted from 9? Yes, once.
Write 1 in the tens column and place a zero in the ones column.
10 is your first guess at the quotient.

★ **Step 2**

$$\begin{array}{r} 10 \\ 8\overline{)\ 96} \\ -80 \\ \hline 16 \end{array}$$

Multiply $8 \times 10 = 80$.
Subtract $96 - 80 = 16$.

★ Step 3

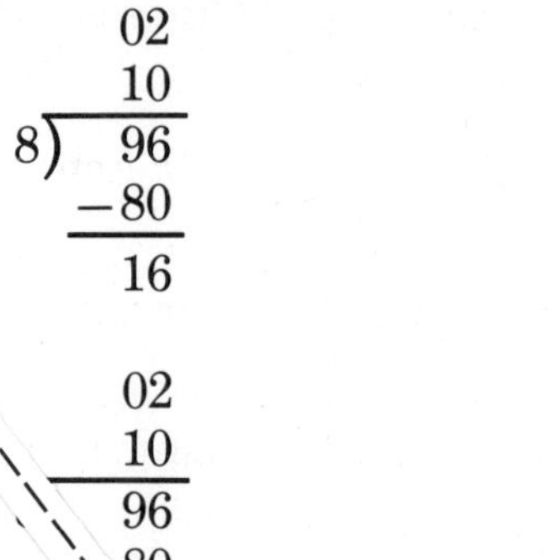

Use 16 as the new dividend.
Can 8 be subtracted from 1? No.
Write a zero in the tens column above.
Can 8 be subtracted from 16? Yes, twice.
Write a 2 in the ones column above.

★ Step 4

```
   02
   10
8) 96
  -80
   16   <———— Multiply 8 × 2 = 16
  -16   <———— Subtract 16 − 16 = 0
    0
```

The quotient is the sum of the numbers in the answer space: 10 + 2 = **12,** so that 96 ÷ 8 = 12.

Always check your answer: 8 × 12 = 96.

Now you try one: [illegible]2 ÷ 7 = ______

Work this problem using the method shown above, then go to **39**.

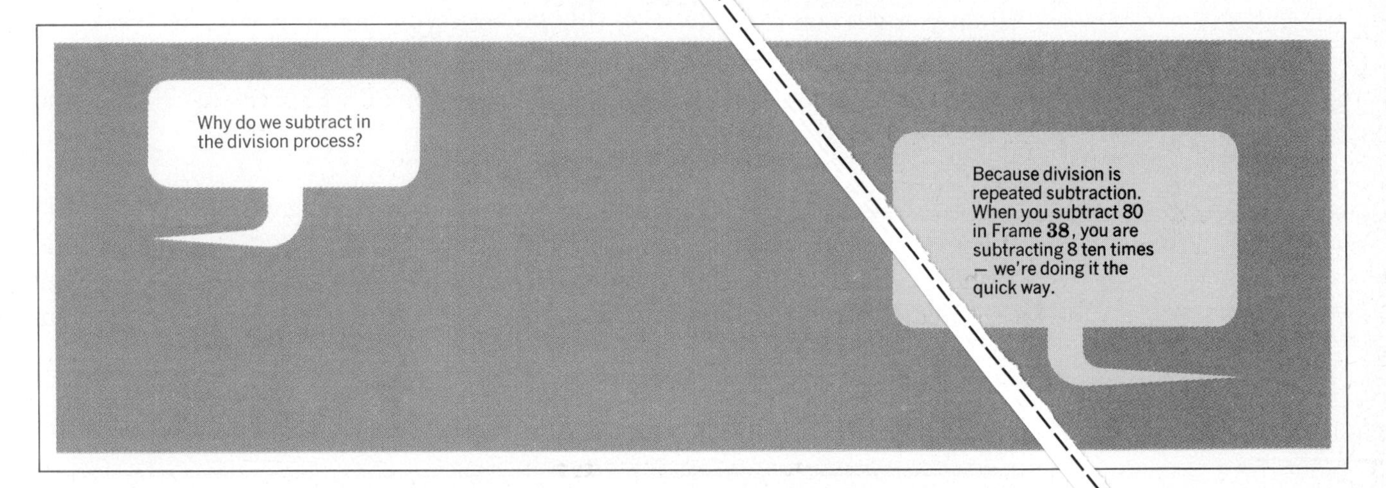

Why do we subtract in the division process?
Because division is repeated subtraction. When you subtract 80 in Frame 38, you are subtracting 8 ten times — we're doing it the quick way.

39 $112 \div 7$ **Guess:** $7 \times 10 = 70$

$7 \times 20 = 140$, so the answer is between 10 and 20.

★ **Step 1**

```
   010
7)112
```

Can 7 be subtracted from 1? No.
Write a 0 above the 1 in the hundreds column.
Can 7 be subtracted from 11? Yes, once.
Write a 1 in the tens column.
Place a 0 in the ones column.

★ **Step 2**

```
      6
    010
7)  112
   - 70
     42
   - 42
```

Multiply $7 \times 10 = 70$
and subtract $112 - 70 = 42$.

★ **Step 3**

Use 42 as the new dividend and repeat the process.
$7 \times 6 = 42$, $42 - 42 = 0$

Quotient $= 10 + 6 = 16$
Remainder $= 0$

Check: $7 \times 16 = 112$.

When you get comfortable with this process, you can omit writing zeros and your work will look like this:

```
  16
7)112   7 into 11 once, write 1
  7     11 - 7 = 4, bring down the 2
  42    7 into 42, 6 times, write 6
  42    7 × 6 = 42, 42 - 42 = 0.
   0
```

Here are a few problems for practice in division.

(a) $976 \div 8$ (b) $3174 \div 6$ (c) $204 \div 6$

Check your work in **40**.

SPECIAL DIVISORS

A few divisors require special attention. Remember that, for any two numbers a and b, $a \div b = \square$ means that $b \times \square = a$. That is, $21 \div 3 = 7$ means $3 \times 7 = 21$.

★ 1. If any number is divided by one, the quotient is the original number.

$6 \div 1 = 6$ and $\frac{6}{1} = 6$ because $6 \times 1 = 6$

★ 2. If any number is divided by itself, the quotient is one.

$6 \div 6 = 1$ and $\frac{6}{6} = 1$ because $6 \times 1 = 6$

★ 3. If zero is divided by any non-zero number, the quotient is zero.

$0 \div 6 = 0$ and $\frac{0}{6} = 0$ because $6 \times 0 = 0$

★ 4. If any number is divided by zero, the quotient is not defined in mathematics.

If $6 \div 0 = \square$ then $0 \times \square = 6$, but 0 times any number equals zero. There is no number $\square$ that will make this equation true.

The division $0 \div 0$ never appears in mathematics because it can have any value whatever. If $0 \div 0 = \square$, then $0 = 0 \times \square$, and this equation is true no matter what number we put in $\square$.

40 (a) **Guess:** $8 \times 100 = 8$

The answer is something close to 800.

$$
\begin{array}{r}
122 \\
8\overline{)\ 976} \\
-8 \\
\hline
17 \\
-16 \\
\hline
16 \\
-16 \\
\hline
0
\end{array}
$$

★ **Step 1**

8 into 9? Once, write 1.

★ **Step 2**

$8 \times 1 = 8$, subtract $9 - 8 = 1$.

★ **Step 3**

Bring down 7. 8 into 17 twice, write 2.

★ **Step 4**

$8 \times 2 = 16$, subtract $17 - 16 = 1$.

★ **Step 5**

Bring down 6. 8 into 16, twice, write 2.

★ **Step 6**

$8 \times 2 = 16$, subtract $16 - 16 = 0$.

Quotient = 122

Check: $8 \times 122 = 976$

(b) **Guess:** $6 \times 500 = 3000$

The answer is roughly 500.

$$
\begin{array}{r}
529 \\
6\overline{)\ 3174} \\
-30 \\
\hline
17 \\
-12 \\
\hline
54 \\
-54 \\
\hline
0
\end{array}
$$

★ **Step 1**

6 into 3? No.
6 into 31? 5 times. Write 5.

★ **Step 2**

$6 \times 5 = 30$. Subtract $31 - 30 = 1$.

★ **Step 3**

Bring down 7. 6 into 17? Twice.
Write 2.

★ **Step 4**

$6 \times 2 = 12$. Subtract $17 - 12 = 5$.

★ **Step 5**

Bring down 4. 6 into 54? 9 times.
Write 9.

★ **Step 6**

$6 \times 9 = 54$. Subtract $54 - 54 = 0$.

Quotient = 529

Check: $6 \times 529 = 3174$

(c) **Guess:** $6 \times 30 = 180$ The quotient is about 30.

```
    34
6)  204
   -18
    24
   -24
     0
```

★ **Step 1**

6 into 2? No.
6 into 20? Three times. Write 3.

★ **Step 2**

$6 \times 3 = 18$. Subtract $20 - 18 = 2$.

★ **Step 3**

Bring down 4. 6 into 24? 4 times. Write 4.

Quotient = 34

Check: $6 \times 34 = 204$

★ **Step 4**

$6 \times 4 = 24$. Subtract $24 - 24 = 0$.

Now try a problem using a two-digit divisor.

$5084 \div 31 =$ ______

The procedure is the same as above. Check your answer in **41**.

---

41 $5084 \div 31$

Guess: This is roughly the same as $500 \div 3$ or about 200. The quotient will be about 200.

```
      164
31)  5084
    -31
     198
    -186
     124
    -124
       0
```

★ **Step 1**

31 into 5? No.
31 into 50? Yes, once. Write 1.

★ **Step 2**

$31 \times 1 = 31$. Subtract $50 - 31 = 19$.

★ **Step 3**

Bring down 8. 31 into 198? (That is about the same as 3 into 19.) Yes, 6 times. Write 6.

★ **Step 4**

$31 \times 6 = 186$. Subtract $198 - 186 = 12$.

★ **Step 5**

Bring down 4. 31 into 124? (That is about the same as 3 into 12.) Yes, 4 times. Write 4.

Quotient: 164

Check: $31 \times 164 = 5084$

★ **Step 6**

$31 \times 4 = 124$. Subtract $124 - 124 = 0$.

Notice that in Step 3 it is not at all obvious how many times 31 will go into 198. Again, you must make an educated guess and check your guess as you go along. If you guess n' check on every problem you will always get the correct answer.

So far, we have looked only at division problems that "come out even"—division problems where the remainder is zero. Obviously not all division problems are of this kind. What would you do with these problems?

(a) $59 \div 8 =$ ______ (b) $341 \div 43 =$ ______ (c) $7528 \div 37 =$ ______

Look in **42** for our answers.

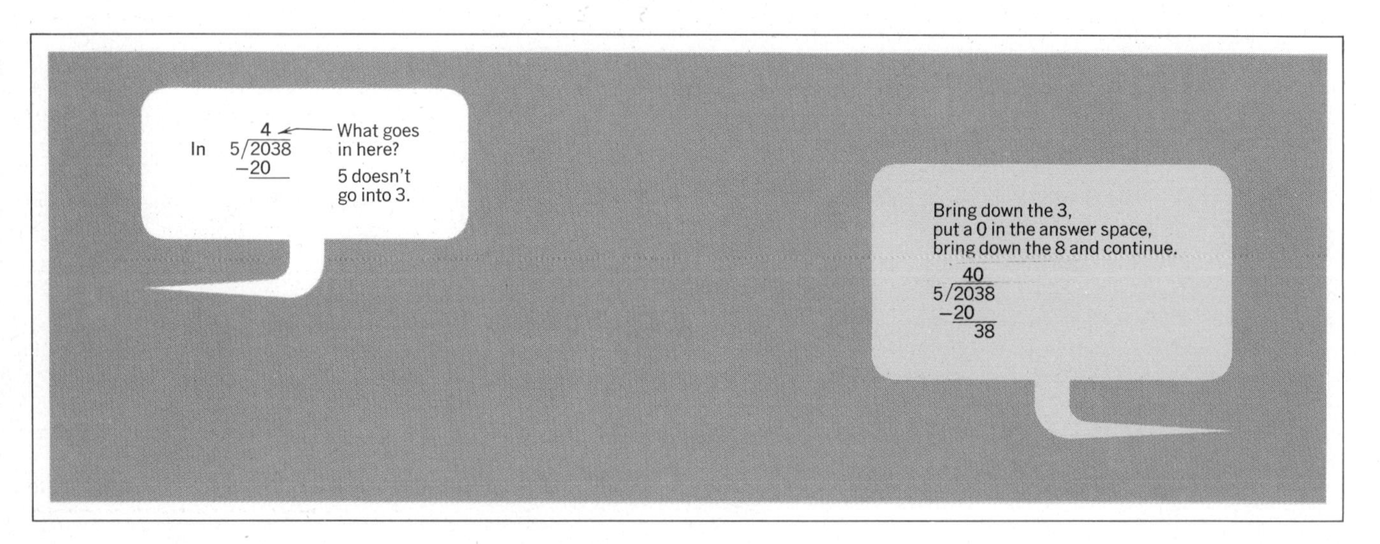

42

(a) $\begin{array}{r} 7 \\ 8\overline{)\ 59} \\ -56 \\ \hline 3 \end{array}$ The quotient is 7 with a remainder of 3.

(b) $\begin{array}{r} 7 \\ 43\overline{)\ 341} \\ -301 \\ \hline 40 \end{array}$ Your first guess would probably be that 43 goes into 341 8 times (try 4 into 34) but $43 \times 8 = 344$, which is larger than 341. The quotient is therefore 7 with a remainder of 40.

(c) $\begin{array}{r} 203 \\ 37\overline{)\ 7528} \\ -74 \\ \hline 128 \\ -111 \\ \hline 17 \end{array}$ On this step notice that 37 cannot be subtracted from 12. Write a zero in the answer space and bring down the 8. The quotient is 203 and the remainder is 17.

Now turn to **43** for a set of practice problems on division of whole numbers.

---

7. The planet Pluto travels once around sun in approximately 90,464 days. If one earth year is equal to 365 days, how many earth years is a Pluto year?

____

8. The noise of an explosion travels 6200 meters through the air in 18 seconds. What is the speed of this sound? (Your answer should be in meters per second.)

____

9. A number is said to be evenly divisible by a second number if the second number divides it with no remainder.

(a) Show that 2520 is evenly divisible by the numbers 2, 3, 4, 5, 6, 7, 8, 9, 10, 12, 15, and 18. (It has 36 more divisors also!)
(b) Show that the product of any three consecutive whole numbers is evenly divisible by 6.

10. Divide (a) $123321\overline{)111{,}111{,}111{,}111}$ = _______________

(b) $111{,}111{,}111\overline{)12{,}345{,}678{,}987{,}654{,}321}$ = _______________

(c) $8\overline{)64}$ = ____ $68\overline{)4624}$ = ____ $668\overline{)446224}$ = _______________

11. If your yearly income is $13,780, what is your average weekly income?

____

12. What is the average of the numbers 1256, 1742, 1654, and 1500?

____

When you have had the practice you need, continue by going to frame 44 where we talk about the factoring of whole numbers or by returning to the preview for this unit.

Date

Name

Course/Section

44 The symbols 6, VI, and 𝍸 | are all names for the number six. We can also write any number in terms of arithmetic operations involving other numbers. For example, $(4 + 2)$, $(7 - 1)$, (2×3), and $(18 \div 3)$ are also ways of writing the number six. It is particularly useful to be able to write any whole number as a product of other numbers. If we write

$6 = 2 \times 3$

2 and 3 are called the *factors* of 6. Of course we could write

$6 = 1 \times 6$

and see that 1 and 6 are also factors of 6, but this does not tell us anything new about the number 6. The factors of 6 are 1, 2, 3, and 6.

What are the factors of 12? (Choose an answer.)

(a) 2, 3, 4, and 6 — Go to **45**
(b) 1, 2, 3, 4, 6, and 12 — Go to **46**
(c) 0, 1, 2, 3, 4, 6, and 12 — Go to **47**

---

45 Not quite.

Any two whole numbers whose product is 12 are factors of 12. It is easy to see that

$1 \times 12 = 12$

Therefore 1 and 12 are factors of 12.

Return to **44** and choose a better answer.

---

46 Right you are.

$1 \times 12 = 12$,
$2 \times 6 = 12$,
and $3 \times 4 = 12$ are all ways of writing 12 as the product of two numbers; therefore 1, 2, 3, 4, 6, and 12 are all factors of 12.

Any number is *evenly divisible* by its factors, that is, every factor divides the number with zero remainder. For example,

$12 \div 1 = 12$	$12 \div 4 = 3$
$12 \div 2 = 6$	$12 \div 6 = 2$
$12 \div 3 = 4$	$12 \div 12 = 1$

List the factors of

(a) 18 (b) 20 (c) 24 (d) 48

Check your answers in **48**.

---

47 Not correct.

Zero is never a factor of any number. There is no number $\square$ such that

$0 \times \square = 12$

The product of 0 and any number is always 0.

Return to **44** and choose a better answer.

---

48 The factors of 18 are 1, 2, 3, 6, 9, and 18.
The factors of 20 are 1, 2, 4, 5, 10, and 20.
The factors of 24 are 1, 2, 3, 4, 6, 8, 12, and 24.
The factors of 48 are 1, 2, 3, 4, 6, 8, 12, 16, 24, and 48.

For some numbers the only factors are 1 and the number itself. For example, the factors of 7 are 1 and 7 because

$1 \times 7 = 7$

There are no other numbers that divide 7 with remainder zero. Such numbers are known as *prime numbers.* A prime number is one for which there are no factors other than 1 and the prime number itself.

Here is a list of the first few prime numbers.

2	3	5	7	11
13	17	19	23	29
31	37	41	43	47

Notice that 1 is not listed. All prime numbers have two distinct, unequal factors: 1 and the number itself. The number 1 has only one factor—itself. The number 1 is not a prime.

There are 25 prime numbers less than 100, 168 less than 1000, and no limit to the total number. Mathematicians have tried for centuries to find a simple pattern that would enable them to write down the primes in order and predict if any given number is a prime. As yet no one has succeeded.

How then does one determine if some given number is a prime? There is no magic way to decide. You must divide the number in question by each whole number in order, starting with 2. If the division has no remainder, the original number is not a prime. Continue dividing until the quotient obtained is less than the divisor.

For example, is 53 a prime number? Try to work it out, then turn to **49** and continue.

---

49 To decide if 53 is a prime, we must perform a series of divisions:

$2\overline{)53}$ = 26 Remainder = 1 $5\overline{)53}$ = 10 Remainder = 3

$3\overline{)53}$ = 17 Remainder = 2 $6\overline{)53}$ = 8 Remainder = 5

$4\overline{)53}$ = 13 Remainder = 1 $7\overline{)53}$ = 7 Remainder = 4

We can stop the search for a divisor here because dividing by 8 gives a quotient (6) less than the divisor (8). All divisors produce a non-zero remainder; therefore 53 is prime.

Notice that it is really not necessary to test divide by 4 or 6. If the number is not evenly divisible by 2, it cannot be evenly divided by 4 or 6 because these are multiples of 2. In testing for primeness we need test divide only by primes. This will save time, but you must have the first few primes memorized. For most of your work it will be sufficient if you remember the first 8 or 10 primes listed in **48**. A study card listing primes to be memorized is included in the back of this book. Use it.

Which of the following numbers are prime?

(a) 103 (b) 114 (c) 143 (d) 223
(e) 289 (f) 449 (g) 527 (h) 667

Test divide by primes as shown above, then check your answers in **50**.

THE SIEVE OF ERATOSTHENES

More than 2000 years ago, Eratosthenes a Greek geographer-astronomer, devised a way of locating pri s that is still the most effective known. His procedure separate he primes out of the set of all whole numbers. Here is one ve on of what he did. First, arrange the whole numbers in six colu ns starting with 2 as shown. Second, circle the primes and cross t all multiples of 2; circle the next number 3 and cross out all mul ples of 3; circle the next remaining number 5 and cross out all mu iples of 5; and so on. The circled numbers remaining are the prim .

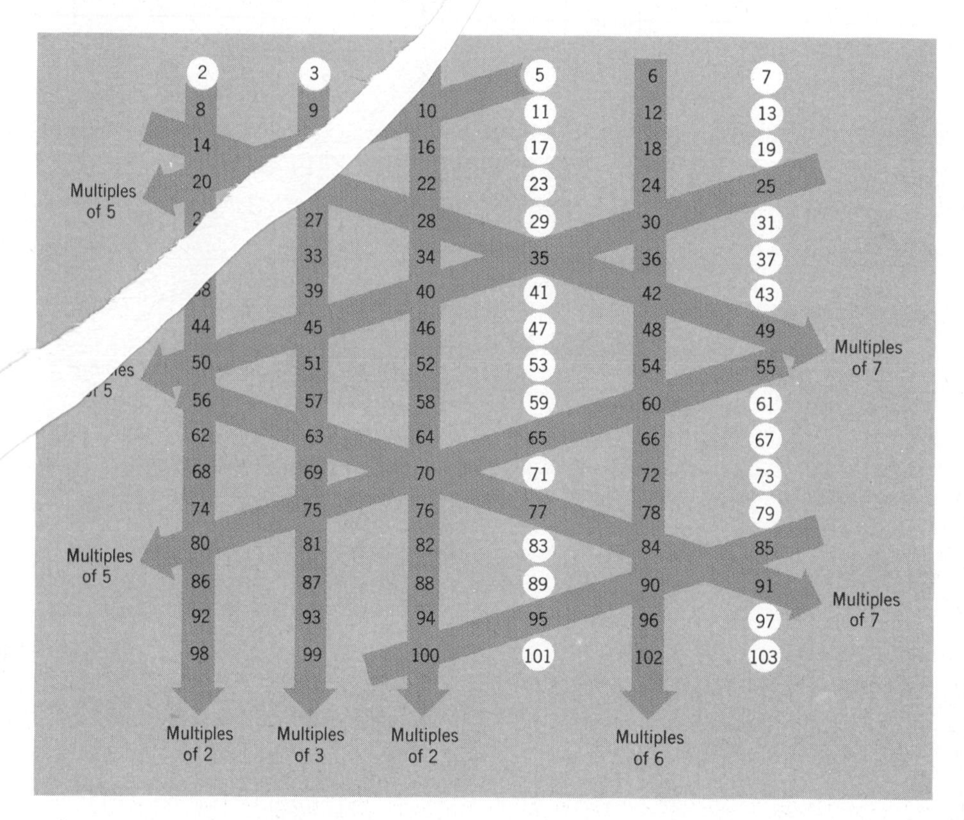

Mathematicians call this procedure a *sieve* because it is a way of sorting out or separating the primes from the other whole numbers.

Notice that all primes greater than 3 are either in the 5 or 7 column; they are either one less or one more than a multiple of 6. Pairs of primes, such as 5 and 7 or 11 and 13, separated by one integer, are called "twin primes." Can you find any other interesting patterns?

50 (a) $103 \div 2 = 51$ Remainder 1
$103 \div 3 = 34$ Remainder 1
$103 \div 5 = 20$ Remainder 3
$103 \div 7 = 14$ Remainder 5
$103 \div 11 = 9$ Remainder 4

103 is a prime.

No need to continue, because the quotient (9) is less than the divisor (11).

(b) $114 \div 2 = 57$ with no remainder. Because 114 is evenly divisible by 2, it is *not* a prime.
(c) $143 \div 11 = 13$ with no remainder. Because 114 is evenly divisible by 11 and 13, it is *not* a prime.
(d) 223 is a prime. Test divide it by 2, 3, 5, 7, and 11.
(e) $289 = 17 \times 17$. It is *not* a prime.
(f) 449 is a prime. Test divide it by 2, 3, 5, 7, 11, 13, 17, and 19.
(g) $527 = 17 \times 31$. It is *not* a prime.
(h) $667 = 23 \times 29$. It is *not* a prime.

★ The *prime factors* of any numbers are those factors that are prime numbers. The prime factors of 6 are 2 and 3. The prime factors of 21 are 3 and 7.

Two factors of 42 are 6 and 7, but the *prime* factors of 42 are 2, 3, and 7. The number 6 is not a prime factor; it is not a prime.

To factor a number means to find its prime factors. Finding the prime factors of a number is a necessary skill if you want to learn how to add and subtract fractions.

What are the prime factors of 30?

(a) $5 \times 6 = 30$ Go to **51**
(b) 2, 3, 5, 6, 10, 15 Go to **52**
(c) 2, 3, 5 Go to **53**

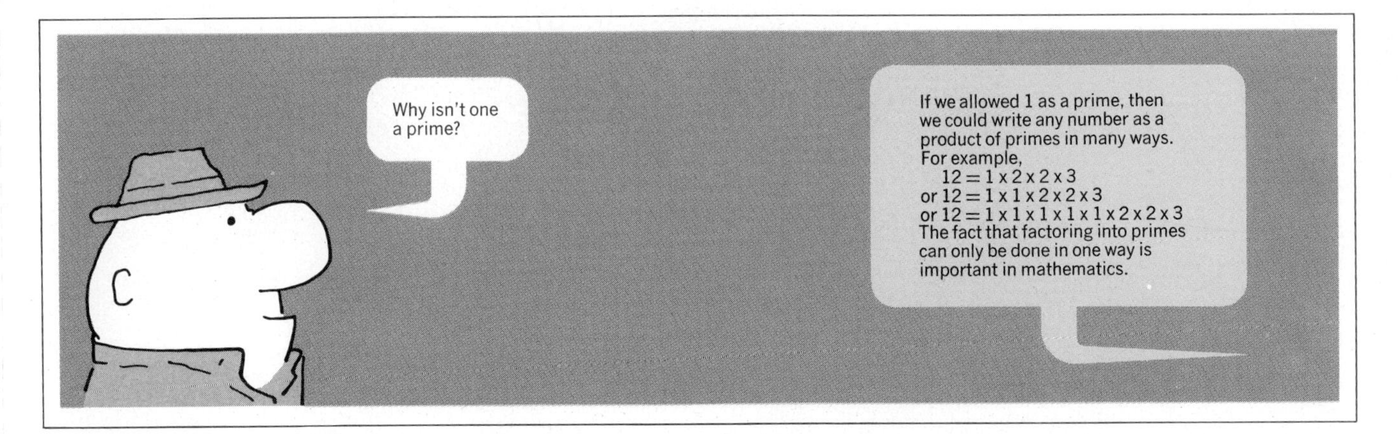

Why isn't one a prime?
If we allowed 1 as a prime, then we could write any number as a product of primes in many ways. For example,
12 = 1 x 2 x 2 x 3
or 12 = 1 x 1 x 2 x 2 x 3
or 12 = 1 x 1 x 1 x 1 x 1 x 2 x 2 x 3
The fact that factoring into primes can only be done in one way is important in mathematics.

51 You goofed on this one.

The numbers 5 and 6 are factors of 30. Their product equals 30. But they are not both prime numbers. Prime factors must be prime numbers. Return to **50** and choose a different answer.

---

52 These are some of the factors of 30, but not all of them are prime. Return to **50** and choose the set of *prime* factors.

---

53 Right!

The prime factors of 30 are 2, 3, and 5.

Prime numbers are especially interesting because any number can be written as the product of primes in only one way.

$6 = 2 \times 3$ and $6 = 3 \times 2$ count as only one way of writing 6 as a product of primes. The order is not important. The key idea is that 2 and 3 are the *only* primes whose product is 6.

$9 = 3 \times 3$	as a product of primes.
$10 = 2 \times 5$	
$11 = 11$	An easy one; 11 is itself a prime. We would not write 1×11 because 1 is not a prime . . . remember?
$12 = 2 \times 2 \times 3$	There are several ways to write 12 in terms of its factors, $12 = 2 \times 6$ or $12 = 3 \times 4$, but only one way with primes. Notice that the prime factor 2 must appear *twice*.

Write 60 as the product of its prime factors.

$60 =$ ______

Check your work in **54**.

Why are prime numbers called "prime"?
The Latin word primus means first in importance and the primes are the important main ingredient of numbers. Every number is either a prime or a product of primes.

---

54 $60 = 2 \times 2 \times 3 \times 5$ Write them in any order you like, $2 \times 5 \times 3 \times 2$ or $5 \times 2 \times 3 \times 2$ or whatever. This is the *only* set of primes whose product is 60.

It is this property of prime numbers that prompted the Greek mathematicians 30 centuries ago to call them "primes"—the "first" numbers from which the rest could be built.

Being able to write any number as a product of primes is a valuable skill. You will need this skill when you work with fractions in Unit 2.

How can we rewrite a number in terms of its prime factors? Let's work through an example.

Find the prime factors of 315.

First, divide by the primes in order starting with 2.

2 | 315 → 2 is not a factor because it does not divide the number evenly. Bring the number down and try the next prime.

3 | 315 → 3 is a factor. Write it to the side → [3]
3 | 105 and divide by 3 again.

3 | 35 → 3 is a factor again. Write it to the side → [3] and divide by 3 again.

→ 3 does not divide 35 evenly. Bring down 35 and try the next prime.

5 | 35 → 5 is a factor. Write it to the side. → [5]
7 → 7 is a factor. Write it to the side. → [7]

Then $315 = 3 \times 3 \times 5 \times 7$ written as a product of prime factors.

Finally, check this multiplication to be certain that you have missed no factors.

$3 \times 3 \times 5 \times 7 = 9 \times 5 \times 7 = 45 \times 7 = 315$

Find the prime factors of 693.

Check your work in **55**.

THE PRIMES LESS THAN 100

2	3	5	7	11
13	17	19	23	29
31	37	41	43	47
53	59	61	67	71
73	79	83	89	97

55 Your work should look something like this:

$2 \,|\, \underline{693}$ → 2 does not divide 693 evenly.

$3 \,|\, \underline{693}$ → $693 \div 3 = 231$ → $\boxed{3}$

$3 \,|\, \underline{231}$ → $231 \div 3 = 77$ → $\boxed{3}$

$3 \,|\, \underline{77}$ → 3 does not divide 77 evenly.

$5 \,|\, \underline{77}$ → 5 does not divide 77 evenly.

$7 \,|\, \underline{77}$ → $77 \div 7 = 11$ → $\boxed{7}$

11 → $\boxed{11}$

$693 = 3 \times 3 \times 7 \times 11$ **Check:** $3 \times 3 \times 7 \times 11$

$= 9 \times 7 \times 11$

$= 63 \times 11 = 693$

Use this method to find the prime factors of

(a) 570 (b) 792 (c) 945

Our work is in **56**.

THE FACTOR TREE

A very helpful way to think about fractions is by using a *factor tree*. For example, factor 1764.

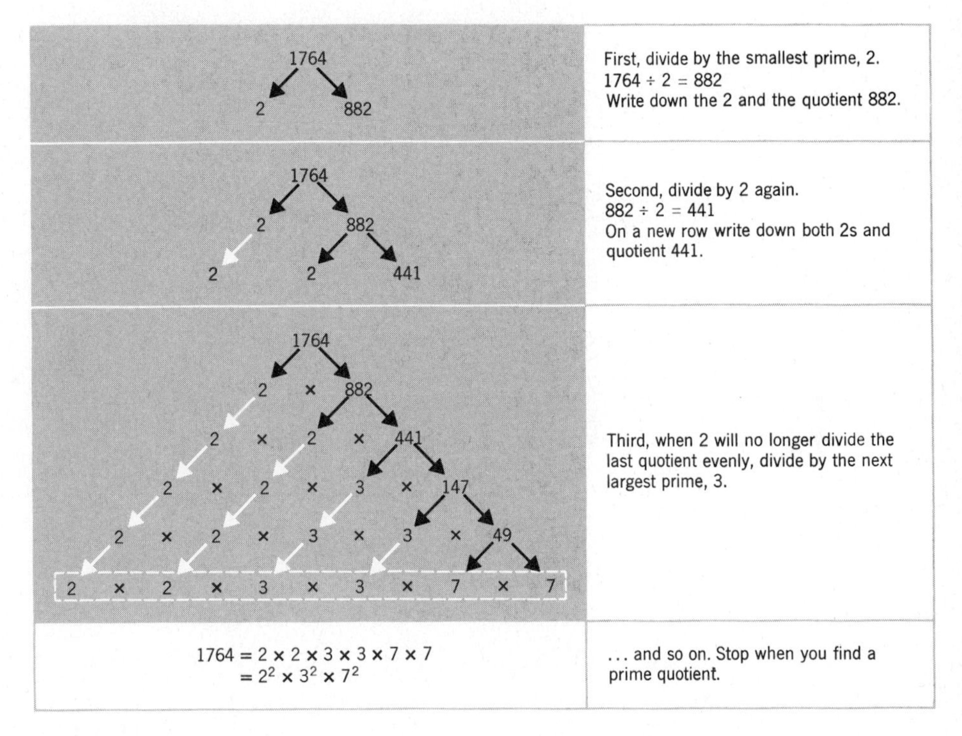

1764
2
882
First, divide by the smallest prime, 2.
1764 ÷ 2 = 882
Write down the 2 and the quotient 882.
1764
2
882
2
2
441
Second, divide by 2 again.
882 ÷ 2 = 441
On a new row write down both 2s and quotient 441.
1764
2 × 882
2 × 2 × 441
2 × 2 × 3 × 147
2 × 2 × 3 × 3 × 49
2 × 2 × 3 × 3 × 7 × 7
Third, when 2 will no longer divide the last quotient evenly, divide by the next largest prime, 3.
1764 = 2 × 2 × 3 × 3 × 7 × 7
= 2² × 3² × 7²
... and so on. Stop when you find a prime quotient.

At each level of the tree the product of the horizontal numbers is equal to the original number to be factored. The last row gives the prime factors.

56 (a) 2 | 570 → 570 ÷ 2 = 285 → [2]
2 | 285 → Not a divisor

3 | 285 → 285 ÷ 3 = 195 → [3]
3 | 95 → Not a divisor

5 | 95 → 95 ÷ 5 = 19 → [5]
19 → 19 is a prime → [19]

570 = 2 × 3 × 5 × 19

(b) 2 | 792 → 792 ÷ 2 = 396 → [2]
2 | 396 → 396 ÷ 2 = 198 → [2]
2 | 198 → 198 ÷ 2 = 99 → [2]
2 | 99 → Not a divisor

3 | 99 → 99 ÷ 3 = 33 → [3]
3 | 33 → 33 ÷ 3 = 11 → [3]
11 → 11 is a prime → [11]

792 = 2 × 2 × 2 × 3 × 3 × 11

(c) 2 | 945 → Not a divisor

3 | 945 → 945 ÷ 3 = 315 → [3]
3 | 315 → 315 ÷ 3 = 105 → [3]
3 | 105 → 105 ÷ 3 = 35 → [3]
3 | 35 → Not a divisor

5 | 35 → 35 ÷ 5 = 7 → [5]
7 → 7 is a prime → [7]

945 = 3 × 3 × 3 × 5 × 7

It is possible to tell at a glance, without actually dividing, if any number is evenly divisible by 2, 3, or 5. Knowing how to do so can save you a bit of work. Use these divisibility rules.

★ 1. Any number is evenly divisible by 2 if its ones digit is 2, 4, 6, 8, or 0.

Example: 12 is divisible by 2; it ends in 2.
46 is divisible by 2; it ends in 6.
7498 is divisible by 2; it ends in 8.

★ 2. Any number is evenly divisible by 3 if the sum of its digits is divisible by 3.

Example: 18 is divisible by 3 since $1 + 8 = 9$, and 9 is divisible by 3.
471 is divisible by 3 since $4 + 7 + 1 = 12$, and 12 is divisible by 3.
72,954 is divisible by 3 since $7 + 2 + 9 + 5 + 4 = 27$, and 27 is divisible by 3.
215 is *not* divisible by 3 since $2 + 1 + 5 = 8$, and 8 is not divisible by 3.

★ 3. Any number is evenly divisible by 5 if its ones digit is 5 or 0.

Example: 25 is divisible by 5; it ends in 5.
370 is divisible by 5; it ends in 0.
73,495 is divisible by 5; it ends in 5.

There are divisibility rules for other numbers (shown on the next page, if you are interested) but the few above are all you really need to remember.

Use these rules to decide which of the following numbers are divisible by 2, 3, or 5. Do all work mentally. Check ✓ those evenly divisible by 2. Mark with an x those evenly divisible by 3. Circle those evenly divisible by 5.

16 23 27 39 45 111 132 210 223 231
330 335 372 453 498 785 921 1017 2111
73,908 123,456 4,271,305

Turn to **57** to check your answers.

DIVISIBILITY RULES

Consider the following telephone number: 2260830. In a few seconds, and without using pencil and paper, can you show that it is evenly divisible by 2, 3, 5, 6, 10, and 11 but not by 4, 7, 8 or 9? The trick is to use the following divisibility rules.

★ 2 Any number is divisible by 2 if its ones digit is 2, 4, 6, 8, or 0.

Example: 14, 96, and 378 are all divisible by 2.

★ 3 Any number is evenly divisible by 3 if the sum of its digits is divisible by 3.

Example: 672 is divisible by 3, since $6 + 7 + 2 = 15$ and 15 is divisible by 3.

★ 4 Any number is evenly divisible by 4 if the number formed by its two rightmost digits is divisible by 4.

Example: 716 is divisible by 4, since 16 is divisible by 4.

★ 5 Any number is evenly divisible by 5 if its ones digit is 0 or 5.

Example: 35, 90, and 1365 are all divisible by 5.

★ 6 Any number is divisible by 6 if it is divisible by both 2 and 3.

Example: 822 is divisible by 6, since its ones digit is 2 and $8 + 2 + 2 = 12$, which is divisible by 3.

★ 8 Any number is divisible by 8 if its last three digits are divisible by 8.

Example: 1160 is divisible by 8 since 160 is divisible by 8.

★ 9 Any number is divisible by 9 if the sum of its digits is divisible by 9.

Example: 9243 is divisible by 9 since $9 + 2 + 4 + 3 = 18$, which is divisible by 9.

★ 10 Any number whose ones digit is 0 is divisible by 10.

Example: 60, 210, and 19,830 are all divisible by 10.

continued on Page 77

continued on Page 76

There are no really simple rules for 7, 11, or 13. Here are the least complicated rules known.

★ 7 Divide the number in question by 50. Add the quotient and remainder. The original number is divisible by 7 if the sum of quotient and remainder is divisible by 7.

Example: $476 \div 50 = 9$; remainder $= 26$.
Add $9 + 26 = 35$.
35 is divisible by 7; therefore 476 is also divisible by 7.

★ 11 Divide the number in question by 100. Add the quotient and remainder. The original number is divisible by 11 if the sum of quotient and remainder is divisible by 11.

Example: $1562 \div 100 = 15$, remainder $= 62$.
Add $15 + 62 = 77$.
77 is divisible by 11; therefore 1562 is also divisible by 11.

★ 13 Proceed as with 11, but divide by 40.

Example: $1170 \div 40 = 29$, remainder 10.
Add $29 + 10 = 39$.
39 is divisible by 13; therefore 1170 is also divisible by 13.

57 √ means divisible by 2. x means divisible by 3. Circle means divisible by 5.

√		x	x	x	x	√x	√x		x
16	23	27	39	(45)	111	132	(210)	223	231

√x		√x	x	√x		x	x	
(330)	(335)	372	453	498	(785)	921	1017	2111

√x	√x	
73,908	123,456	(4,271,305)

Now you should be ready for some practice problems on factoring. Turn to **58** and continue.

---

65 In the pattern

$10^5 = 100{,}000$
$10^4 = 10{,}000$
$10^3 = 1{,}000$
$10^2 = 100$

each product on the right decreases by a factor of 10. Therefore the next two lines must be

$10^1 = 10$
$10^0 = 1$

Of course this is true for any base.

$2^1 = 2 \qquad 2^0 = 1$
$3^1 = 3 \qquad 3^0 = 1$
$4^1 = 4 \qquad 4^0 = 1$ and so on.

If we factor the number 2592 into its prime factors, we find that

$2592 = 2 \times 2 \times 2 \times 2 \times 2 \times 3 \times 3 \times 3 \times 3$

Write this using exponents.

$2592 =$ $2^5 \times 3^4$

Check your work in **66**.

Why isn't 2^0 equal to 0?
Remember, we're not multiplyin 2 x 0 in 2^0. We define $2^0 = 1$ so that each power of 2 is one factor of 2 larger than the last: 1, 2, 4, 8, 16, . . .

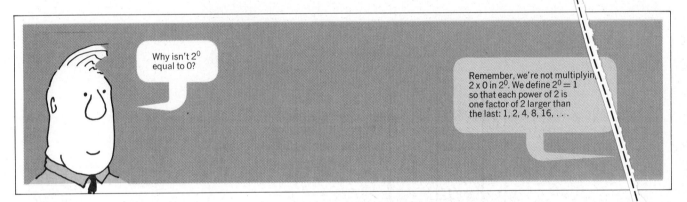

66 $2592 = \underline{2^5 \times 3^4}$

Using exponents provides a simple and compact way to write any number as a product of its prime factors.

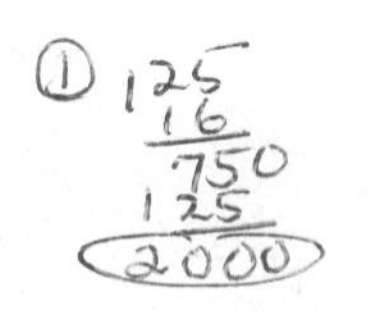

Find the following products by multiplying:

$2^4 \times 5^3 =$ ______	$3^4 \times 2^7 \times 1^6 =$ ______
$6^2 \times 7^3 =$ ______	$5^3 \times 8^2 \times 2^0 =$ ______
$2^3 \times 3^2 \times 5^4 =$ ______	$7^2 \times 9^1 \times 3^5 =$ ______
$2^2 \times 3^3 \times 4^4 =$ ______	$3^4 \times 5^2 \times 7^1 =$ ______

Go to **67** to check your answers.

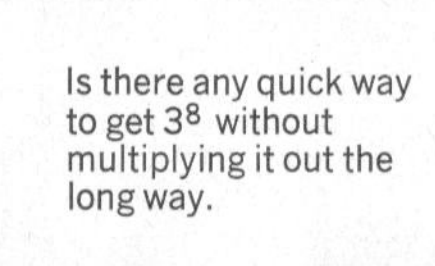

Is there any quick way to get 3^8 without multiplying it out the long way.

No. But powers of 10 can be written down with no work — see frame **65**.

67

$2^4 \times 5^3 = \underline{2000}$	$3^4 \times 2^7 \times 1^6 = \underline{10{,}368}$
$6^2 \times 7^3 = \underline{12{,}348}$	$5^3 \times 8^2 \times 2^0 = \underline{8{,}000}$
$2^3 \times 3^2 \times 5^4 = \underline{45{,}000}$	$7^2 \times 9^1 \times 3^5 = \underline{107{,}163}$
$2^2 \times 3^3 \times 4^4 = \underline{27{,}648}$	$3^4 \times 5^2 \times 7^1 = \underline{14{,}175}$

What is interesting about the numbers

1, 4, 9, 16, 25, 36, 49, 64, 81, 100, . . . ?

Do you recognize them? See **68**.

---

68 These numbers are the squares or second powers of the counting numbers,

$1^2 = 1$
$2^2 = 4$
$3^2 = 9$
$4^2 = 16$, and so on.

1, 4, 9, 16, 25, . . . are called *perfect squares.* If you have memorized the multiplication table for one-digit numbers, you will recognize them immediately. If you do not remember them, it will be helpful to you to memorize them. A study card for them is provided in the back of this book. Here are the first 20 perfect squares:

PERFECT SQUARES

$1^2 = 1$	$6^2 = 36$	$11^2 = 121$	$16^2 = 256$
$2^2 = 4$	$7^2 = 49$	$12^2 = 144$	$17^2 = 289$
$3^2 = 9$	$8^2 = 64$	$13^2 = 169$	$18^2 = 324$
$4^2 = 16$	$9^2 = 81$	$14^2 = 196$	$19^2 = 361$
$5^2 = 25$	$10^2 = 100$	$15^2 = 225$	$20^2 = 400$

The number 3^2 is read "three squared." What is "square" about $3^2 = 9$? The name comes from an old Greek idea about the nature of numbers. Ancient Greek mathematicians called certain numbers "square numbers" or "perfect squares" because they could be represented by a square array of dots

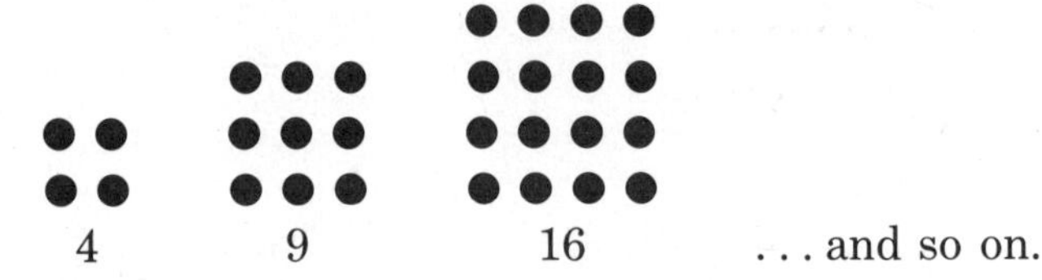

. . . and so on.

The number of dots along the side of the square was called the "root" or origin of the square number. We call it the *square root.* For example, the square root of 16 is 4, since $4 \times 4 = 16$.

What is the square root of 64?

(a) 32 Go to **69**
(b) 8 Go to **70**

---

69 Sorry, you are incorrect. We cannot simply divide 64 in half to find its square root!

The square root of 64 is some number $\square$ such that $\square \times \square = 64$. For example, the square root of 25 is equal to 5 because $5 \times 5 = 25$. To "square" means to multiply by itself, and to find a square root means to find a number that when multiplied by itself gives the original number.

Now, return to **68** and choose a better answer.

---

70 Right! The square root of 64 is equal to 8 because $8 \times 8 = 64$.

The sign $\sqrt{}$ is used to indicate the square root.

$\sqrt{16} = 4$ Read it "square root of 16"
$\sqrt{9} = 3$ Read it "square root of 9"
$\sqrt{169} = 13$ Read it "square root of 169"

Find

$\sqrt{81} =$ 9 $\sqrt{361} =$ 19 $\sqrt{289} =$ 17

Try using the table in **68** if you do not recognize these.

Continue in **71**.

---

71 $\sqrt{81} = 9$ Check: $9 \times 9 = 81$
$\sqrt{361} = 19$ Check: $19 \times 19 = 361$
$\sqrt{289} = 17$ Check: $17 \times 17 = 289$

Always check your answer as shown.

How do you find the square root of any whole number? The surest and simplest way is to consult a table of square roots. There is no easy way to recognize or identify perfect squares. You will find such a table of square roots on page 517 of this book. However, most square roots listed there are not whole numbers. Modern mathematicians have extended the Greek idea of square roots from perfect squares to all numbers. Methods for finding square roots of numbers that are not perfect squares will be explained in Unit 3.

Now, for a set of practice problems on exponents and square roots, turn to **72**.

Unit I
Arithmetic of Whole Numbers

73

Self-Test

1. $37 + 46$ = 83
2. $4372 + 849$ = 5221
3. $81 + 47 + 36$ = 164
4. $43 - 17$ = 26
5. $702 - 416$ = 286
6. $6001 - 3973$ = 2028
7. $8300 - 4605$ = 3695
8. 43×36 = 1548
9. 237×204 = 48348
10. 406×137 = 55622
11. 2153×304 = 654512
12. $1081 \div 47$ = ________
13. $27\overline{)21687}$ = ________
14. $2008 \div 4$ = ________
15. $\dfrac{26 \times 48}{39}$ = ________
16. Factor: 408 = ________
17. Factor: 9438 = ________
18. Factor: 1000 = ________
19. Factor: 3264 = ________
20. 3^4 = ________
21. $3^0 \times 4^2 \times 5^3$ = ________
22. $2^1 \times 3^4 \times 7^2$ = ________
23. 123^2 = ________
24. $\sqrt{225}$ = ________
25. $\sqrt{324}$ = ________

Answers are on page 513.

Date 2/23/78.

Name Stephanie Catanzaro

Course/Section ________

SUPPLEMENTARY PROBLEM SET 1A

Addition

UNIT 1 ARITHMETIC OF WHOLE [illegible]

A. Add:

1.	2.	3.	4.	5.	6.	7.	8.
38	73	43	59	38	[illegible]	[illegible]	43
14	12	25	47	16	[illegible]	[illegible]	92

9.	10.	11.	12.	13.	14.	15.	16.
78	33	59	52	37	[illegible]	[illegible]	28
26	67	56	79	64	[illegible]	[illegible]	94

B. Add:

1.	2.	3.	4.	5.	6.	7.
107	633	346	451	934	[illegible]	[illegible]
482	481	729	809	518	[illegible]	[illegible]

8.	9.	10.	11.	12.	13.	14.
802	335	458	1064	3409	6728	8491
419	894	623	5346	5763	3404	6487

C. Arrange vertically and add:

1. $932 + 73 + 85 =$ ______
2. $319 + 487 + 36 + 43 =$ ______
3. $645 + 386 + 904 =$ ______
4. $78 + 564 + 346 + 804 =$ ______
5. $4875 + 9067 + 6074 =$ ______
6. $12 + 304 + 1768 + 406 + 57 =$ ______
7. $4258 + 6304 + 1002 + 90104 =$ ______
8. $237 + 10007 + 6999 =$ ______

D. Add:

1.	2.	3.	4.	5.
6	34	305	4097	7364
9	27	76	1908	21
3	72	42	600	582
8	8	815	53	906
2	45	521	726	5731
9	66	90	231	1085
7	8	314	5043	379

Date ______

Name ______

Course/Section ______

SUPPLEMENTARY PROBLEM SET 2A

Subtraction

UNIT 1 ARITHMETIC OF WHOLE NUMBERS

A. Subtract:

1. 14 − 6
2. 12 − 7
3. 19 − 8
4. 8 − 6
5. 11 − 3
6. 14 − 9
7. 12 − 8
8. 15 − 7
9. 18 − 9
10. 9 − 6
11. 14 − 3
12. 17 − 9
13. 12 − 6
14. 8 − 5
15. 7 − 0
16. 16 − 7

B. Subtract:

1. 32 − 18
2. 54 − 23
3. 48 − 19
4. 35 − 28
5. 92 − 47
6. 70 − 53
7. 81 − 39
8. 75 − 58
9. 748 − 435
10. 380 − 56
11. 458 − 149
12. 621 − 337
13. 317 − 88
14. 900 − 348

C. Subtract:

1. 3017 − 1643
2. 9230 − 185
3. 12048 − 695
4. 8632 − 4088
5. 70000 − 358
6. 3704 − 978

D. Word Problems:

1. In a June graduating class of 523 there were 287 girls. How many boy graduates were there?

   ____

2. A plane flying at 13,000 feet would clear a 9846 mountain peak by how much?

   ____

3. If Dave's yearly income was $6041 and $1847 of it went for taxes, how much was left?

   ____

4. The car odometer read 43,198 at the beginning of a trip and 45,105 as it started across the Golden Gate Bridge. How many miles were traveled?

   ____

5. The population of a certain city was 52,874 in 1970, 60,740 in 1970. What was the population increase in this five year period?

   ____

Date ______________________

Name ______________________

Course/Section ______________________

SUPPLEMENTARY PROBLEM SET 3A

Multiplication

UNIT 1 ARITHMETIC OF WHOLE NUMBERS

A. Multiply:

1.	2.	3.	4.	5.	6.	7.	8.
37	65	74	31	46	53	86	46
4	7	8	6	8	7	6	12

9.	10.	11.	12.	13.	14.	15.	16
38	63	76	54	75	84	93	87
14	26	63	98	27	35	54	86

B. Multiply:

1.	2.	3.	4.	5.	6.	7.
215	307	614	447	907	47	407
364	412	507	123	406	368	368

8.	9.	10.	11.	12.	13.	14.
4007	1204	4672	4008	1142	4582	7546
368	543	3208	3009	8007	6040	5121

C. Multiply:

1. 1,437,651 × 235 = ______
2. 8652 × 3,275,803 = ______
3. 8888 × 4444 = ______
4. 1,234,567 × 7,654,321 = ______
5. 14,215,469 × 5 = ______
6. 15,873 × 63 = ______

D. Word Problems:

1. If bus fare is computed at 6¢ per mile, what is the fare for a trip of 265 miles?

   ____

2. Student workers earn $1.85 per hour. What are they paid for a 15 hour week?

   ____

3. An assembly line at the Ace Widgit Co. turns out 2,457 widgets per hour. How many widgets are produced in a 40 hour week?

   ____

Date ______________

Name ______________

Course/Section ______________

4. In a certain school district each of the 23 school buses can carry a maximum of 47 passengers per trip. How many passengers can the school move each day if each bus makes one trip?

____

5. A gross of pencils contains 144 pencils. How many pencils are in 17 gross?

____

6. How much Vitamin C is contained in 650 tablets each containing 250 mg?

____

7. A garden is laid out into 54 rows with 37 tomato plants in a row. How many tomato plants are in the garden?

____

SUPPLEMENTARY PROBLEM SET 4A

Division

UNIT 1 ARITHMETIC OF WHOLE NUMBERS

A. Divide:

1. $63 \div 9 =$ ______
2. $76 \div 4 =$ ______
3. $96 \div 8 =$ ______
4. $54 \div 6 =$ ______
5. $56 \div 7 =$ ______
6. $111 \div 3 =$ ______
7. $\frac{92}{4} =$ ______
8. $476 \div 7 =$ ______
9. $441 \div 9 =$ ______
10. $568 \div 8 =$ ______
11. $576 \div 6 =$ ______
12. $2106 \div 9 =$ ______
13. $9216 \div 4 =$ ______
14. $\frac{2247}{7} =$ ______
15. $9\overline{)6003} =$ ______

B. Divide:

1. $468 \div 16 =$ ______
2. $877 \div 27 =$ ______
3. $893 \div 19 =$ ______
4. $861 \div 21 =$ ______
5. $469 \div 67 =$ ______
6. $990 \div 45 =$ ______
7. $989 \div 23 =$ ______
8. $931 \div 49 =$ ______
9. $992 \div 31 =$ ______
10. $\frac{969}{57} =$ ______
11. $\frac{972}{36} =$ ______
12. $13\overline{)936} =$ ______
13. $43\overline{)946} =$ ______
14. $73\overline{)803} =$ ______
15. $82\overline{)738} =$ ______

C. Word Problems:

1. The total day's sales for Jocko's Sporting Goods Store was \$1320. What was the average amount spent by each of the 165 customers who visited the store that day?

______

2. The total weight of the 18 linemen on the Tru Blu U. football team is 3852 pounds. What is the average weight of a lineman on the team?

______

3. The enrollment at the five schools in the local school district is 1686, 1207, 1485, 2006, and 1701. What is the average enrollment per school in the school district?

______

4. If you drive 462 miles in 7 hours, how many miles per hour did you average?

______

Date ______

Name ______

Course/Section ______

5. A bicycle touring group plans to cycle from Chicago to Montreal, a distance of 745 miles, in 5 days. How far must they travel in an average day?

____

6. Belgium has a population of 9,659,600 people and an area of 11780 square miles. What is the population density of Belgium in people per square miles?

____

7. George Perdon of Australia ran the 2303 miles from Perth, Australia to Melbourne, Australia in 47 days. How many miles per day did he run on the average?

____

SUPPLEMENTARY PROBLEM SET 5A

Factoring

UNIT 1 ARITHMETIC OF WHOLE NUMBERS

A. Write the following as products of primes:

1. 22 = ____ 2. 35 = ____ 3. 48 = ____ 4. 110 = ____

5. 113 = ____ 6. 152 = ____ 7. 203 = ____ 8. 225 = ____

9. 231 = ____ 10. 441 = ____ 11. 315 = ____ 12. 337 = ____

13. 378 = ____ 14. 540 = ____ 15. 648 = ____ 16. 782 = ____

17. 784 = ____ 18. 897 = ____ 19. 1110 = ____ 20. 1365 = ____

B. Circle the primes among the following numbers:

4, 5, 6, 7, 9, 13, 21, 36, 37, 38, 39, 41, 43, 49, 51, 53, 57, 59, 61, 63, 73, 77, 79, 81, 83, 87, 89, 91, 93, 97, 101, 107

C. Circle the numbers divisible by 3:

6, 12, 24, 26, 31, 39, 46, 57, 69, 94, 92, 114, 204, 307, 14,462, 1,991,132, 10,002, 124,345

Date

Name

Course/Section

Fractions

PREVIEW 2

Objective	Sample Problems		Where To Go for Help: Page	Frame
Upon successful completion of this program you will be able to:			Page	Frame
1. **Rename Fractions:** Write fractions equivalent to any given fraction.	(a) $\frac{3}{4} = \frac{?}{12}$	______	107	**1**
Reduce a fraction to lowest terms.	(b) $\frac{18}{39} =$	______	107	**1**
Write a mixed number as a fraction.	(c) $4\frac{2}{3} =$	______	107	**1**
Compare fractions.	(d) Which is larger $\frac{3}{13}$ or $\frac{4}{17}$?	______	107	**1**
2. **Multiply and Divide Fractions**	(a) $1\frac{1}{3} \times 2\frac{3}{5} =$	______	121	**19**
	(b) $\left(2\frac{2}{3}\right)^2 =$	______	121	**19**
	(c) $3\frac{1}{2} \div 1\frac{3}{4} =$	______	129	**26**
	(d) $6 \div 2\frac{1}{3} =$	______	129	**26**
	(e) Divide 6 by $2\frac{2}{5}$	______	129	**26**
3. **Add and Subtract Fractions**	(a) $4\frac{2}{3} +$ [illegible] $\frac{3}{[illegible]} =$	______	141	**34**
	(b) $5\frac{1}{8} - 3\frac{1}{3}$ [illegible]	______	141	**34**
	(c) $9 - 1\frac{3}{8} =$	______	141	**34**
4. **Word Problems**	(a) What fraction of $\frac{[illegible]}{3}$ is $1\frac{3}{4}$?	______	157	**51**
	(b) Find $2\frac{1}{3}$ of $1\frac{7}{8}$.	______	157	**51**
	(c) If 7 apples cost 91¢ what will 4 cost?	______	157	**51**
	(d) Find a number such that $\frac{7}{[illegible]}$ of it is $2\frac{1}{2}$.	______	157	**51**

______
Date

Stephanie Catanzaro
Name

Math 50
Course/Section

(Answers to these sample problems are given below.)

If you are certain you can work all of these problems correctly turn to page 169 for a self-test. If you want help with any of these objectives or if you cannot work one of the sample problems, turn to the page indicated. Superstudents—those who want to be certain they learn all of this—will join us in frame 1.

PREVIEW 2

Answers to Sample Problems

1. (a) 9

   (b) $\frac{6}{13}$

   (c) $\frac{14}{3}$

   (d) $\frac{4}{17}$

2. (a) $\frac{52}{15}$ or $3\frac{7}{15}$

   (b) $\frac{64}{9}$ or $7\frac{1}{9}$

   (c) 2

   (d) $\frac{18}{7}$ or $2\frac{4}{7}$

   (e) $\frac{5}{2}$ or $2\frac{1}{2}$

3. (a) $\frac{77}{12}$ or $6\frac{5}{12}$

   (b) $\frac{43}{24}$ or $1\frac{19}{24}$

   (c) $\frac{61}{8}$ or $7\frac{5}{8}$

4. (a) $\frac{21}{8}$ or $2\frac{5}{8}$

   (b) $\frac{35}{8}$ or $4\frac{3}{8}$

   (c) 52¢

   (d) $\frac{20}{7} = 2\frac{6}{7}$

Fractions

By permission of John Hart and Field Enterprises, Inc.

1 Every caveman knows that measuring sticks must have numbers to label their marks, but only very smart ones like our friend in the cartoon learn to talk about the space between markers. We use counting numbers to count measurement units: 1 inch, 2 inches, 3 inches, or even 4 centimeters if you are a very modern caveman. Fractions enable us to label some of the points between counting numbers. The word *fraction* comes from the Latin *fractus* meaning *to break,* and we use these numbers to describe subdivisions of standard measurement units, for length, time, money, or whatever we choose to measure.

RENAMING FRACTIONS

Consider the rectangular area below. What happens when we break it into equal parts?

2 parts, each one-half of the whole

3 parts, each one-third of the whole

4 parts, each one-fourth of the whole

Divide this area into fifths by drawing vertical lines.

Try it, then go to 2.

2 Notice that the five parts or "fifths" are equal in area.

A fraction is normally written as the division of two whole numbers: $\frac{2}{3}$, $\frac{3}{4}$, or $\frac{26}{12}$. One of the five equal areas above would be "one-fifth" or $\frac{1}{5}$ of the entire area.

$\frac{1}{5} = \frac{\text{1 shaded part}}{\text{5 parts total}}$

How would you label this portion of the area? = ? $\frac{3}{5}$

Continue in 3.

---

3 $\frac{3}{5} = \frac{\text{3 shaded total}}{\text{5 parts total}}$

The fraction $\frac{3}{5}$ implies an area equal to three of the original portions.

$$\frac{3}{5} = 3 \times \left(\frac{1}{5}\right)$$

There are three equal parts and the name of each part is $\frac{1}{5}$ or one-fifth.

In this collection of letters **HHHHSST**, what fraction are **H**s?

Count the total number of letters and decide what portion are **H**s.

Check your answer in 4.

---

4 Fraction of **H**s $= \frac{\text{number of Hs}}{\text{total number of letters}} = \frac{4}{7}$ (read it "four-sevenths")

The fraction of **S**s is $\frac{2}{7}$ and the fraction of **T**s is $\frac{1}{7}$.

The two numbers that form a fraction are given special names to simplify talking about them. In the fraction $\frac{3}{5}$ the upper number 3 is called the *numerator* from the Latin *numero* meaning *number*. It is a count of the number of parts. The lower number 5 is called the *denominator* from the Latin *nomen* or *name*. It tells us the name of the part being counted.

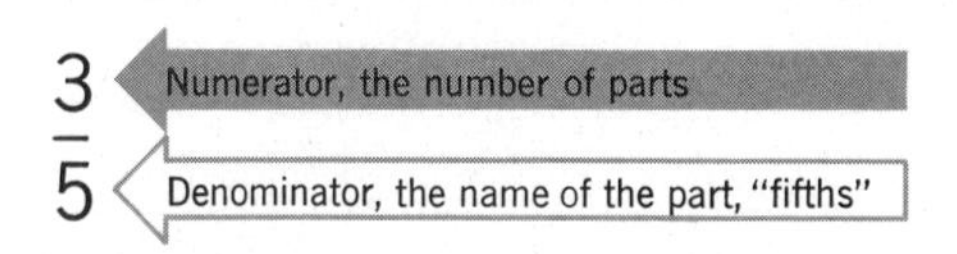

A textbook costs $6 and I have $5. What fraction of its cost do I have? Write the answer as a fraction.

______, numerator = ______, denominator = ______

Check your answer in 5.

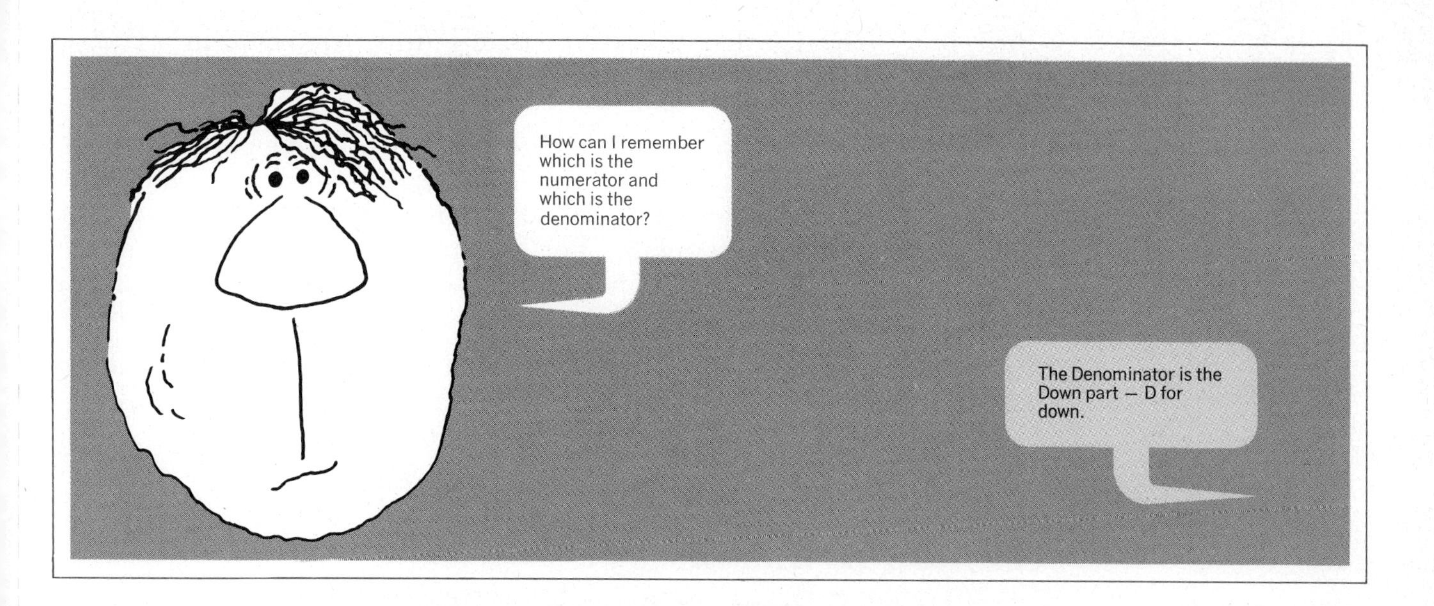

---

5 $\underbrace{\$\$\$\$\$}_{5}\ \$$ $\quad$ \$5 is $\frac{5}{6}$ of the total cost.

numerator = 5, denominator = 6

Complete these sentences by writing in the correct fraction.

(a) If we divide a length into eight equal parts, each part will be 1/8 of the total length.

(b) Then three of these parts will represent ______ of the total length.

(c) Eight of these parts will be ______ of the total length.

(d) Ten of these parts will be ______ of the total length.

Check your answers in **6**.

---

6 (a) $\frac{1}{8}$ (b) $\frac{3}{8}$ (c) $\frac{8}{8}$ (d) $\frac{10}{8}$

The original length is used as a standard and any other length—smaller or larger—can be expressed as a fraction of it. A *proper fraction* is a number less than 1, as you would suppose a fraction should be. It represents a quantity less than the standard. For example, $\frac{1}{2}$, $\frac{2}{3}$, and $\frac{17}{20}$ are all proper fractions. Notice that for a proper fraction, the numerator is less than the denominator—the top number is less than the bottom number in the fraction.

An *improper fraction* is a number greater than 1 and represents a quantity greater than the standard. If a standard length is 8 inches, a length of 11 inches would be $\frac{11}{8}$ of the standard. Notice that for an improper fraction the numerator is greater than the denominator—top number greater than the bottom number in the fraction.

Circle the proper fractions in the following list.

$$\frac{3}{2}, \frac{3}{4}, \frac{7}{8}, \frac{5}{4}, \frac{15}{12}, \frac{1}{16}, \frac{35}{32}, \frac{7}{50}, \frac{65}{64}, \frac{105}{100}$$

Go to **7** when you have finished.

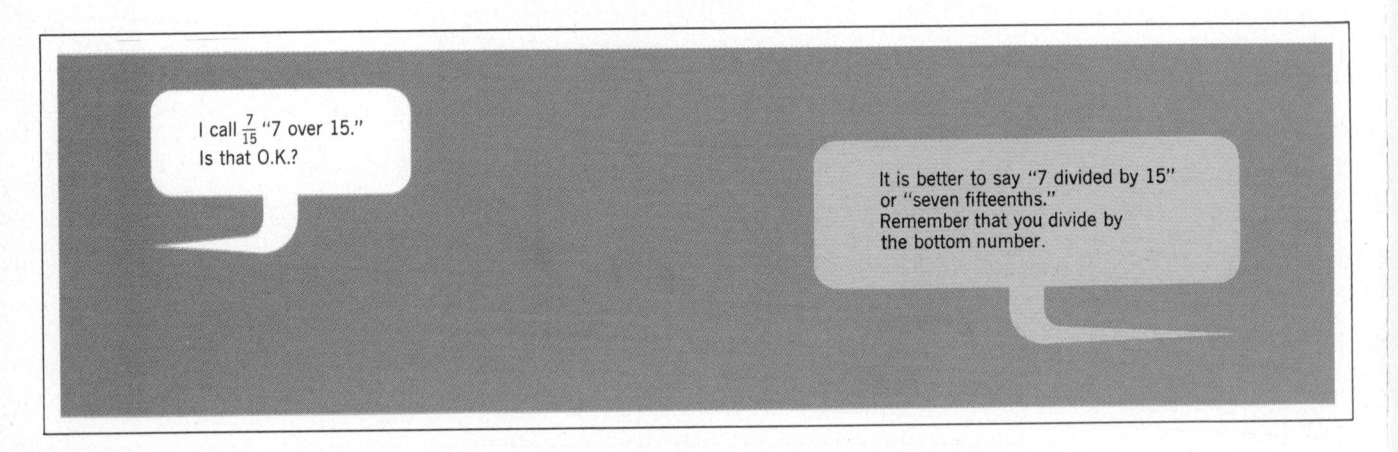

---

7 You should have circled the following proper fractions: $\frac{3}{4}$, $\frac{7}{8}$, $\frac{1}{16}$, and $\frac{7}{50}$. All are numbers less than 1. In each the numerator is less than the denominator.

The improper fraction $\frac{7}{3}$ can be shown graphically as follows:

Unit standard = ☐☐☐

$\frac{1}{3}$ = ■

then,

$\frac{7}{3}$ = ☐☐☐☐☐☐☐ (seven, count 'em)

We can rename this number by regrouping.

☐☐☐

☐☐☐ $2 + \frac{1}{3}$ or $2\frac{1}{3}$ standard units

☐

A *mixed number* is an improper fraction written as the sum of a whole number and a proper fraction.

$\frac{7}{3} = 2 + \frac{1}{3}$ or $2\frac{1}{3}$

We usually omit the + sign and write $2 + \frac{1}{3}$ as $2\frac{1}{3}$, and read it as "two and one-third." The numbers $1\frac{1}{2}$, $2\frac{3}{5}$, and $16\frac{2}{3}$ are all written as mixed numbers.

To write an improper fraction as a mixed number, divide numerator by denominator and form a new fraction as shown:

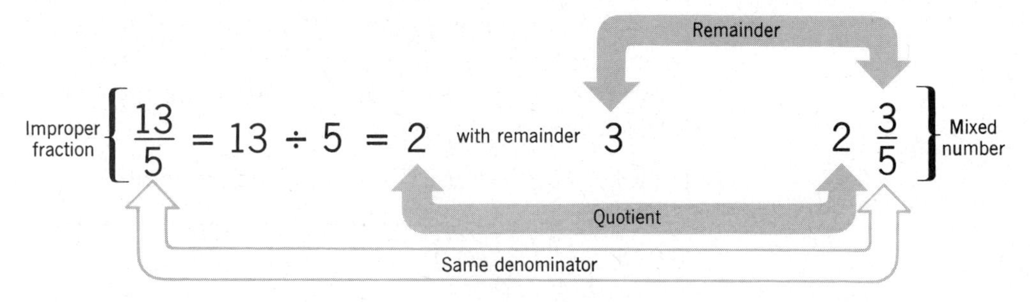

Now you try it. Rename $\frac{23}{4}$ as a mixed number. $\frac{23}{4}$ = ______

Follow the procedure shown above, then turn to **9**.

---

8 (a) $\frac{9}{5} = 1\frac{4}{5}$ (b) $\frac{13}{4} = 3\frac{1}{4}$ (c) $\frac{27}{8} = 3\frac{3}{8}$

(d) $\frac{31}{5} = 6\frac{1}{5}$ (e) $\frac{41}{12} = 3\frac{5}{12}$ (f) $\frac{17}{2} = 8\frac{1}{2}$

The reverse process, rewriting a mixed number as an improper fraction, is equally simple.

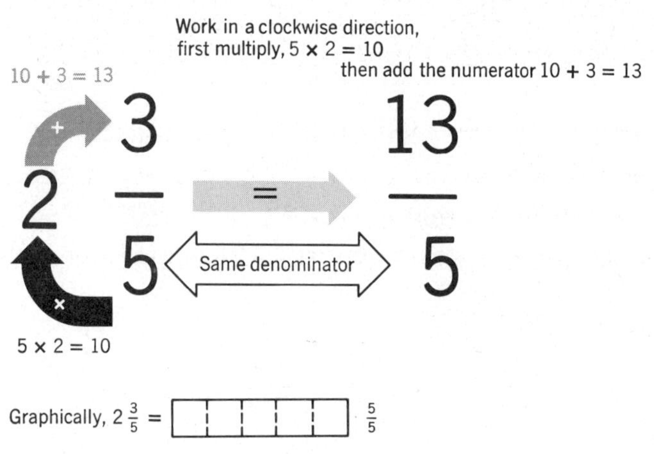

Graphically, $2\frac{3}{5}$ = $\frac{5}{5}$

$\frac{5}{5}$

or $\frac{13}{5}$, count them. $\frac{3}{5}$

Now you try it. Rewrite these mixed numbers as improper fractions.

(a) $3\frac{1}{6}$ (b) $4\frac{3}{5}$ (c) $1\frac{1}{2}$ (d) $8\frac{2}{3}$ (e) $15\frac{3}{8}$ (f) $9\frac{3}{4}$

Check your answers in **10**.

---

9 $\frac{23}{4} = 23 \div 4 = 5$ with remainder 3 ⟶ $5\frac{3}{4}$

If in doubt, check your work with a diagram like this

■■■■■■■■■■■■■■■■■■■■■■■ = ■■■■
23 ■■■■
■■■■ 5 rows of 4
■■■■
■■■■
■■■ 3 remaining

Now try these for practice. Write each improper fraction as a mixed number.

(a) $\frac{9}{5}$ (b) $\frac{13}{4}$ (c) $\frac{27}{8}$ (d) $\frac{31}{5}$ (e) $\frac{41}{12}$ (f) $\frac{17}{2}$

The answers are in **8**.

---

10 (a) $3\frac{1}{6} = \frac{19}{6}$ (b) $4\frac{3}{5} = \frac{23}{5}$ (c) $1\frac{1}{2} = \frac{3}{2}$

(d) $8\frac{2}{3} = \frac{26}{3}$ (e) $15\frac{3}{8} = \frac{123}{8}$ (f) $9\frac{3}{4} = \frac{39}{4}$

Two fractions are said to be *equivalent* if they are numerals or names for the same number. For example, $\frac{1}{2} = \frac{2}{4}$ since both fractions represent the same portion of some standard amount

$\frac{1}{2}$

$\frac{2}{4}$

There is a very large set of fractions equivalent to $\frac{1}{2}$

$$\frac{1}{2} = \frac{2}{4} = \frac{3}{6} = \frac{4}{8} = \frac{5}{10} = \cdots = \frac{46}{92} = \frac{61}{122} = \frac{1437}{2874} \text{ and so on.}$$

Each fraction is a name for the same number, and we can use these fractions interchangeably.

To obtain a fraction equivalent to any given fraction, multiply the original numerator and denominator by the same non-zero number. For example,

$$\frac{1}{2} = \frac{1 \times 3}{2 \times 3} = \frac{3}{6}$$

or

$$\frac{2}{3} = \frac{2 \times 5}{3 \times 5} = \frac{10}{15}$$

Rename as shown: $\frac{3}{4} = \frac{?}{20}$

Check your work in **11**.

---

11 $\frac{3}{4} = \frac{3 \times ?}{4 \times ?} = \frac{3 \times 5}{4 \times 5} = \frac{15}{20}$

$4 \times ? = 20$

? must be 5

The number value of the fraction has not changed, we have simply renamed it.

Practice with these.

(a) $\frac{5}{6} = \frac{?}{42}$ (b) $\frac{7}{16} = \frac{?}{48}$ (c) $\frac{3}{7} = \frac{?}{56}$ (d) $1\frac{2}{3} = \frac{?}{12}$

Look in **13** for the answers.

---

12 $\frac{90}{105} = \frac{2 \times \not{3} \times 3 \times \not{5}}{\not{3} \times \not{5} \times 7} = \frac{2 \times 3}{7} = \frac{6}{7}$

This process of eliminating common factors is usually called *cancelling*. When you cancel a factor you *divide* both top and bottom of the fraction by that factor. In the cancellation above we actually divided top and bottom by 3×5 or 15.

$\frac{90}{105} = \frac{90 \div 15}{105 \div 15} = \frac{6}{7}$

Be Careful

Cancel the same factor.

Cancel only multiplying factors.

This is **illegal:** $\frac{2 \times \not{3} \times 5}{\not{3} \times \not{3} \times 7}$ — Cancel only one 3. Since we divide the top by 3, we must divide the bottom by 3.

This is **illegal:** $\frac{2 + \not{5}}{3 + \not{5}}$ — 5 is not a multiplier. Cancelling means *dividing* the top and bottom of the fraction by the same number. It is illegal to subtract a number from top and bottom.

Reduce the following fractions to lowest terms:

(a) $\frac{15}{84}$ (b) $\frac{21}{35}$ (c) $\frac{4}{12}$ (d) $\frac{154}{1078}$ (e) $\frac{256}{208}$ (f) $\frac{378}{405}$

The answers are in **14**.

---

13 (a) $\frac{5}{6} = \frac{5 \times 7}{6 \times 7} = \frac{35}{42}$ (b) $\frac{7}{16} = \frac{7 \times 3}{16 \times 3} = \frac{21}{48}$

(c) $\frac{3}{7} = \frac{3 \times 8}{7 \times 8} = \frac{24}{56}$ (d) $1\frac{2}{3} = \frac{5}{3} = \frac{5 \times 4}{3 \times 4} = \frac{20}{12}$

Very often in working with fractions you will be asked to *reduce a fraction to lowest terms.* This means to replace it with the most simple fraction in its set of equivalent fractions. To reduce $\frac{15}{30}$ to its lowest terms means to replace it by $\frac{1}{2}$.

$$\frac{15}{30} = \frac{1 \times 15}{2 \times 15} = \frac{1}{2}$$

and $\frac{1}{2}$ is the simplest equivalent fraction to $\frac{15}{30}$ because its numerator and denominator are the smallest whole numbers of any in the set $\frac{1}{2}$, $\frac{2}{4}$, $\frac{3}{6}$, $\frac{4}{8}$, . . . $\frac{15}{30}$ . . .

How can you find the simplest equivalent fraction? For example, how would you reduce $\frac{30}{42}$ to lowest terms?

First, identify and factor numerator and denominator.

$$\frac{30}{42} = \frac{2 \times 3 \times 5}{2 \times 3 \times 7}$$

Second, eliminate common factors

$$\frac{30}{42} = \frac{2 \times 3 \times 5}{2 \times 3 \times 7}$$

2 is a common factor, cancel the 2s:

$$\frac{30}{42} = \frac{\not{2} \times 3 \times 5}{\not{2} \times 3 \times 7}$$

3 is a common factor, cancel the 3s:

$$\frac{30}{42} = \frac{2 \times \not{3} \times 5}{2 \times \not{3} \times 7}$$

and we see that

$$\frac{30}{42} = \frac{5}{7}$$

In effect, we have divided both top and bottom of the fraction by $2 \times 3 = 6$, the common factor.

$$\frac{30}{42} = \frac{30 \div 6}{42 \div 6} = \frac{5}{7}$$

Your turn. Reduce $\frac{90}{105}$ to lowest terms.

Look in **12** for the answer.

14 (a) $\frac{15}{84} = \frac{\cancel{3} \times 5}{2 \times 2 \times \cancel{3} \times 7} = \frac{5}{28}$ (b) $\frac{21}{35} = \frac{3 \times \cancel{7}}{5 \times \cancel{7}} = \frac{3}{5}$

(c) $\frac{4}{12} = \frac{\cancel{4}}{\cancel{4} \times 3} = \frac{1}{3}$ (d) $\frac{154}{1078} = \frac{\cancel{2} \times \cancel{7} \times \cancel{11}}{\cancel{2} \times \cancel{7} \times 7 \times \cancel{11}} = \frac{1}{7}$

(e) $\frac{256}{208} = \frac{\cancel{2} \times \cancel{2} \times \cancel{2} \times \cancel{2} \times 2 \times 2 \times 2 \times 2}{\cancel{2} \times \cancel{2} \times \cancel{2} \times \cancel{2} \times 13} = \frac{16}{13}$

(f) $\frac{378}{405} = \frac{2 \times \cancel{3} \times \cancel{3} \times \cancel{3} \times 7}{\cancel{3} \times \cancel{3} \times \cancel{3} \times 3 \times 5} = \frac{14}{15}$

Cancellation is division by a common factor.

Reduce the fraction $\frac{6}{3}$ to lowest terms.

Check your answer in **15**.

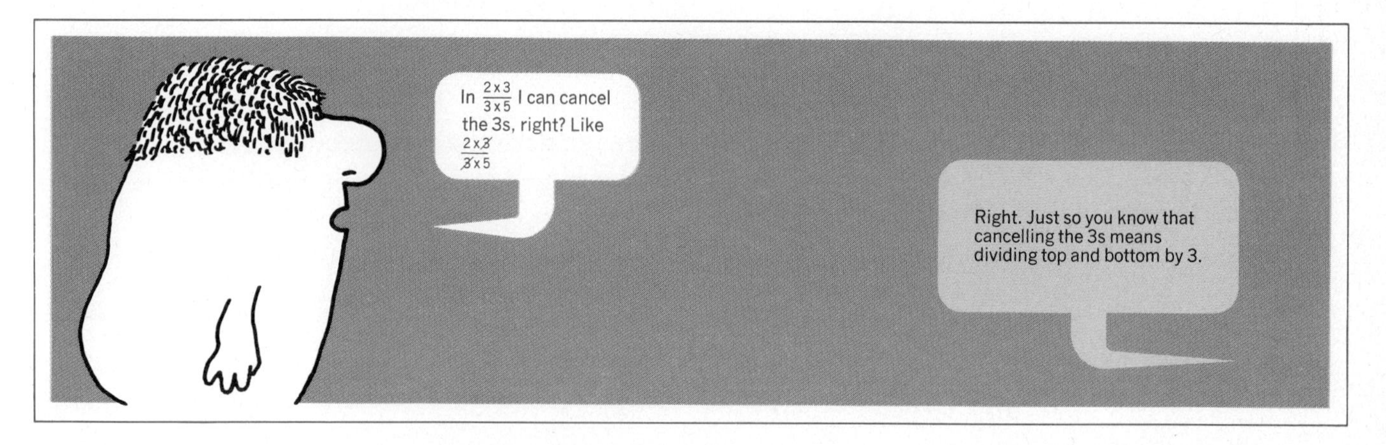

---

15 $\frac{6}{3} = \frac{2 \times 3}{1 \times 3} = \frac{2}{1}$ or simply 2.

Any whole number may be written as a fraction by using a denominator equal to 1.

$3 = \frac{3}{1}$ $4 = \frac{4}{1}$ and so on

Of course, the number 1 can be written as any fraction whose numerator and denominator are equal.

$1 = \frac{2}{2} = \frac{3}{3} = \frac{4}{4} = \cdots = \frac{72}{72} = \frac{1257}{1257} \cdots$ and so on

If you were offered your choice between $\frac{2}{3}$ of a certain amount of money and $\frac{5}{8}$ of it, which would you choose? Which is the larger fraction, $\frac{2}{3}$ or $\frac{5}{8}$?

Can you decide? Try. Renaming the fractions would help. The answer is in **16**.

---

16 To compare two fractions rename each by changing them to equivalent fractions with the same denominator.

$$\frac{2}{3} = \frac{2 \times 8}{3 \times 8} = \frac{16}{24} \quad \text{and} \quad \frac{5}{8} = \frac{5 \times 3}{8 \times 3} = \frac{15}{24}$$

Now compare the new fractions: $\frac{16}{24}$ is greater than $\frac{15}{24}$.

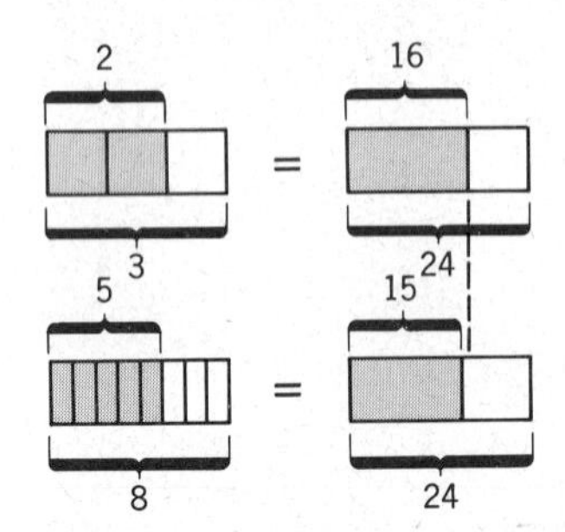

Notice: 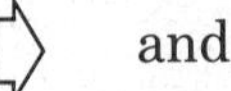 (1) The new denominator is the product of the original ones ($24 = 8 \times 3$).

and

(2) Once both fractions are written with the same denominator, the one with the larger numerator is the larger. (16 of the fractional parts is more than 15 of them.)

Which of the following pairs of fractions is the larger?

(a) $\frac{3}{4}$ and $\frac{5}{7}$ (b) $\frac{7}{8}$ and $\frac{19}{21}$ (c) 3 and $\frac{40}{13}$

(d) $1\frac{7}{8}$ and $\frac{5}{3}$ (e) $2\frac{1}{4}$ and $\frac{11}{5}$ (f) $\frac{5}{16}$ and $\frac{11}{35}$

Check your answer in **17**.

---

17 (a) $\frac{3}{4} = \frac{21}{28}, \frac{5}{7} = \frac{20}{28}, \frac{21}{28}$ is larger than $\frac{20}{28}$, so $\frac{3}{4}$ is larger than $\frac{5}{7}$

(b) $\frac{7}{8} = \frac{147}{168}, \frac{19}{21} = \frac{152}{168}, \frac{152}{168}$ is larger than $\frac{147}{168}$, so $\frac{19}{21}$ is larger than $\frac{7}{8}$

(c) $3 = \frac{39}{13}, \frac{40}{13}$ is larger than $\frac{39}{13}$, so $\frac{40}{13}$ is larger than 3

(d) $1\frac{7}{8} = \frac{15}{8} = \frac{45}{24}, \frac{5}{3} = \frac{40}{24}, \frac{45}{24}$ is larger than $\frac{40}{24}$, so $1\frac{7}{8}$ is larger than $\frac{5}{3}$

(e) $2\frac{1}{4} = \frac{9}{4} = \frac{45}{20}, \frac{11}{5} = \frac{44}{20}, \frac{45}{20}$ is larger than $\frac{44}{20}$, so $2\frac{1}{4}$ is larger than $\frac{11}{5}$

(f) $\frac{5}{16} = \frac{175}{560}, \frac{11}{35} = \frac{176}{560}, \frac{176}{560}$ is larger than $\frac{175}{560}$, so $\frac{11}{35}$ is larger than $\frac{5}{16}$

Now turn to **18** for some practice on renaming fractions.

MULTIPLICATION AND DIVISION OF FRACTIONS

19 The simplest arithmetic operation with fractions is multiplication and, happily, it is easy to show graphically. The multiplication of a whole number and a fraction may be illustrated this way:

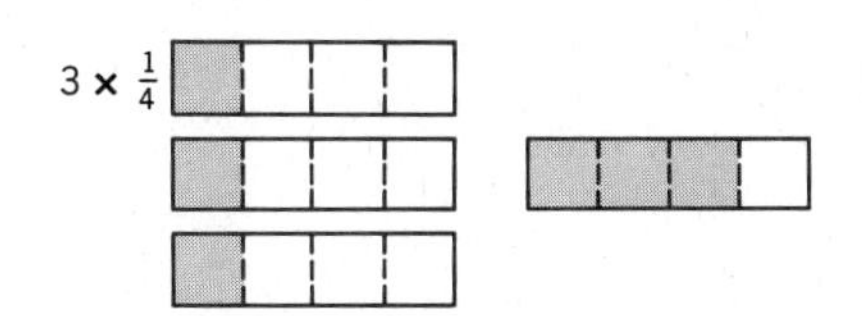

$$3 \times \frac{1}{4} = \frac{1}{4} + \frac{1}{4} + \frac{1}{4} = \frac{3}{4}$$

which is three segments each $\frac{1}{4}$ unit long.

Any fraction such as $\frac{3}{4}$ can be thought of as a product

$$3 \times \frac{1}{4}$$

The product of two fractions can also be shown graphically.

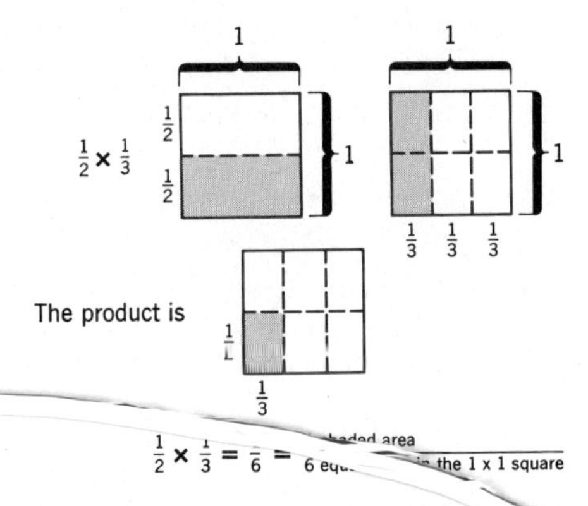

Another way to solve this is

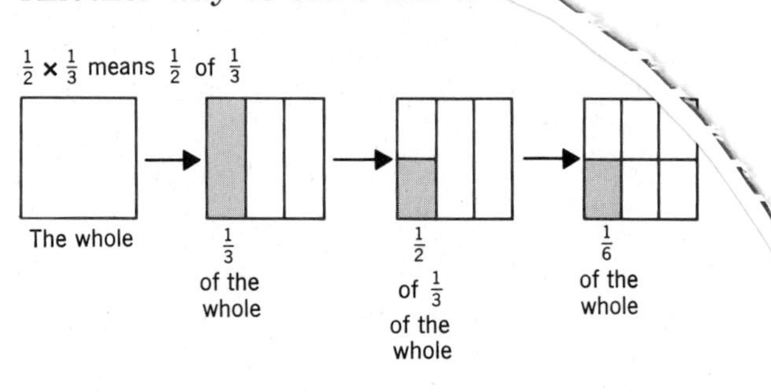

In general, we calculate this product as

$$\frac{1}{2} \times \frac{1}{3} = \frac{1 \times 1}{2 \times 3} = \frac{1}{6}$$

(*continued on the next page*)

The product of two fractions is a fraction whose numerator is the product of their numerators and whose denominator is the product of their denominators.

Multiply $\frac{5}{6} \times \frac{2}{3}$

(a) $\frac{5}{6} \times \frac{2}{3} = \frac{10}{18}$ — Go to **20**

(b) $\frac{5}{6} \times \frac{2}{3} = \frac{5}{9}$ — Go to **21**

(c) I don't know how to do it and I can't figure out how to draw the little boxes — Go to **22**

---

20 Right.

$$\frac{5}{6} \times \frac{2}{3} = \frac{5 \times 2}{6 \times 3} = \frac{10}{18}$$

Now reduce this answer to lowest terms and turn to **21**.

---

21 Excellent.

$$\frac{5}{6} \times \frac{2}{3} = \frac{5 \times 2}{6 \times 3} = \frac{5 \times \cancel{2}}{3 \times \cancel{2} \times 3} = \frac{5}{9}$$

Always reduce your answer to lowest terms. In this problem you probably recognized that 6 was evenly divisible by 2 and did it like this:

$$\frac{5}{\underset{3}{\cancel{6}}} \times \frac{\overset{1}{\cancel{2}}}{3} = \frac{5}{9}$$

Divide top and bottom of the fraction by 2.

It will save you time and effort if you eliminate common factors, such as the 2 above, *before* you multiply.

Do you see that $3 \times \frac{1}{4}$ is really the same sort of problem?

$$3 = \frac{3}{1}, \qquad \text{so} \qquad 3 \times \frac{1}{4} = \frac{3}{1} \times \frac{1}{4} = \frac{3 \times 1}{1 \times 4} = \frac{3}{4}$$

(continued)

Test your understanding with these problems. Multiply as shown:

(a) $\frac{7}{8}\times\frac{2}{3}=$ _____ (b) $\frac{8}{12}\times\frac{3}{16}=$ _____ (c) $\frac{3}{32}\times\frac{4}{15}=$ _____

(d) $\frac{15}{4}\times\frac{9}{10}=$ _____ (e) $\frac{3}{2}\times\frac{2}{3}=$ _____ (f) $1\frac{1}{2}\times\frac{2}{5}=$ _____

(g) $4\times\frac{7}{8}=$ _____ (h) $3\frac{5}{6}\times\frac{3}{10}=$ _____ (i) $\frac{4}{3}\times\frac{3}{4}=$ _____

(*Hint:* Change mixed numbers such as $1\frac{1}{2}$ and $3\frac{5}{6}$ into improper fractions, then solve as usual.)

Have you reduced all answers to lowest terms?

The correct answers are in **23**.

---

22 Now, now, don't panic. You needn't draw the little boxes to do the calculation. Try it this way:

$$\frac{5}{6}\times\frac{2}{3}=\frac{5\times 2}{6\times 3}$$

Multiply the numerators

Multiply the denominators

Finish the calculation and then return to **19** and choose an answer.

---

23 (a) $\frac{7}{8}\times\frac{2}{3}=\frac{7\times\cancel{2}}{(4\times\cancel{2})\times 3}=\frac{7}{4\times 3}=\frac{7}{12}$

Eliminate common factors before you multiply.

Your work will look like this when you learn to do these operations mentally

$$\frac{7}{\underset{4}{\cancel{8}}}\times\frac{\overset{1}{\cancel{2}}}{3}=\frac{7}{12}$$

(b) $\frac{8}{12}\times\frac{3}{16}=\frac{\overset{1}{\cancel{8}}\times\overset{1}{\cancel{3}}}{(4\times\cancel{3})\times(\cancel{8}\times 2)}=\frac{1}{4\times 2}=\frac{1}{8}$

or $\frac{\overset{1}{\cancel{8}}}{\underset{4}{\cancel{12}}}\times\frac{\overset{1}{\cancel{3}}}{\underset{2}{\cancel{16}}}=\frac{1}{8}$

(c) $\frac{\overset{1}{\cancel{3}}}{\underset{8}{\cancel{32}}}\times\frac{\overset{1}{\cancel{4}}}{\underset{5}{\cancel{15}}}=\frac{1}{40}$

(d) $\frac{\overset{3}{\cancel{15}}}{4}\times\frac{9}{\underset{2}{\cancel{10}}}=\frac{27}{8}=3\frac{3}{8}$

(*continued on the next page*)

(e) $\dfrac{\overset{1}{\cancel{3}}}{\underset{1}{\cancel{2}}} \times \dfrac{\overset{1}{\cancel{2}}}{\underset{1}{\cancel{3}}} = 1$

(f) $1\dfrac{1}{2} \times \dfrac{2}{5} = \dfrac{3}{\underset{1}{\cancel{2}}} \times \dfrac{\overset{1}{\cancel{2}}}{5} = \dfrac{3}{5}$

(g) $4 \times \dfrac{7}{8} = \dfrac{\overset{1}{\cancel{4}}}{1} \times \dfrac{7}{\underset{2}{\cancel{8}}} = \dfrac{7}{2} = 3\dfrac{1}{2}$

(h) $3\dfrac{5}{6} \times \dfrac{3}{10} = \dfrac{23}{\underset{2}{\cancel{6}}} \times \dfrac{\overset{1}{\cancel{3}}}{10} = \dfrac{23}{20} = 1\dfrac{3}{20}$

(i) $\dfrac{\overset{1}{\cancel{4}}}{\underset{1}{\cancel{3}}} \times \dfrac{\overset{1}{\cancel{3}}}{\underset{1}{\cancel{4}}} = 1$

Can you extend your new skills to solve these problems?

(a) $1\dfrac{4}{5} \times \dfrac{2}{3} \times \dfrac{1}{4} =$ ______ (b) $\left(\dfrac{1}{2}\right)^2 =$ ______

(c) $\left(1\dfrac{2}{3}\right)^3 =$ ______ (d) $\sqrt{\dfrac{16}{81}} =$ ______ $\left(\textit{Hint.}\ \sqrt{\dfrac{a}{b}} = \dfrac{\sqrt{a}}{\sqrt{b}}\right)$

Try them. An explanation is waiting in **25**.

25 (a) $1\frac{4}{5} \times \frac{2}{3} \times \frac{1}{4} = \frac{9}{5} \times \frac{2}{3} \times \frac{1}{4} = \frac{\overset{3}{\cancel{9}} \times \overset{1}{\cancel{2}} \times 1}{5 \times \underset{1}{\cancel{3}} \times \underset{2}{\cancel{4}}} = \frac{3}{10}$

Multiplication of three or more fractions involves nothing new. Be certain you change all mixed numbers to improper fractions before multiplying.

(b) $\left(\frac{1}{2}\right)^2 = \frac{1}{2} \times \frac{1}{2} = \frac{1 \times 1}{2 \times 2} = \frac{1}{4}$ Easy.

(c) $\left(1\frac{2}{3}\right)^3 = \left(\frac{5}{3}\right)^3 = \frac{5}{3} \times \frac{5}{3} \times \frac{5}{3} = \frac{5 \times 5 \times 5}{3 \times 3 \times 3} = \frac{125}{27}$

Again, you must change any mixed numbers to improper fractions *before* you multiply.

(d) $\sqrt{\frac{16}{81}} = \frac{\sqrt{16}}{\sqrt{81}} = \frac{4}{9}$ **Check:** $\frac{4}{9} \times \frac{4}{9} = \frac{4 \times 4}{9 \times 9} = \frac{16}{81}$

Square roots are not difficult if both numerator and denominator are perfect squares. Otherwise you must use the method explained in Unit 3.

Now turn to **24** for a set of practice problems on multiplication.

26

Division of Fractions

Addition and multiplication are both reversible arithmetic operations. For example,

2×3	and	3×2	both equal 6
$4 + 5$	and	$5 + 4$	both equal 9

The order in which you add addends or multiply factors is not important. This fact is called the commutative property of addition and multiplication.

In subtraction and division this kind of exchange is not allowed and because it is not allowed, many people find these operations very troublesome.

$7 - 5 = 2$

but $5 - 7$ is not equal to 2 and is not even a counting number.

$8 \div 4 = 2$

but $4 \div 8$ is not equal to 2 and is a very different kind of number.

In the division of fractions it is particularly important that you set up the process correctly.

Are these four numbers equal?

"8 divided by 4" $\quad 8\overline{)4} \quad 8 \div 4 \quad \frac{4}{8}$

Choose an answer: (a) Yes Go to **28**
(b) No Go to **29**

---

27 The divisor is $\frac{1}{2}$.

The division $5 \div \frac{1}{2}$ is read "5 divided by $\frac{1}{2}$" and it asks how many $\frac{1}{2}$ unit lengths are included in a length of 5 units.

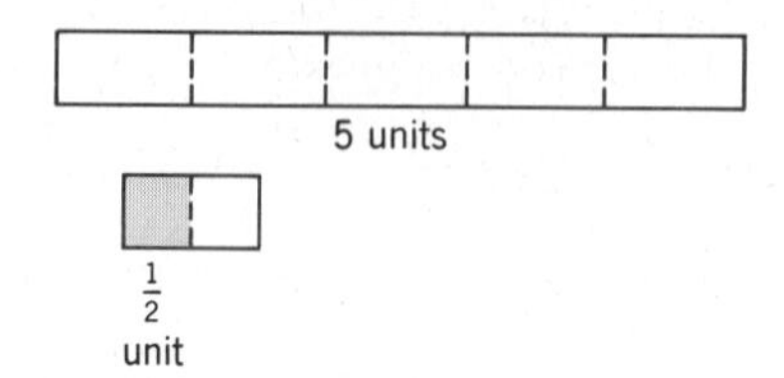

Division is defined in terms of multiplication.

$8 \div 4 = \square$ asks that you find a number $\square$ such that $8 = \square \times 4$. It is easy to see that $\square = 2$.

$5 \div \frac{1}{2} = \square$ asks that you find a number $\square$ such that $5 = \square \times \frac{1}{2}$.

The number $5 \div \frac{1}{2}$ answers the question "How many $\frac{1}{2}$s are there in 5?"

Working backward from this multiplication and the diagram above, find

$5 \div \frac{1}{2} =$ ______

Hop ahead to **30** to continue.

---

28 Your answer is incorrect. Be very careful about this.

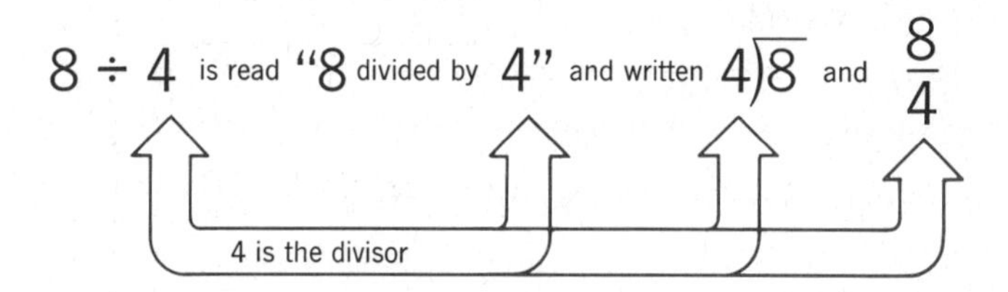

In the above problem you are being asked to divide a set of 8 objects into sets of 4 objects. The divisor 4 is the denominator or bottom number of the fraction. $8 \div 4$ or "8 divided by 4" are *not* equal to $\frac{4}{8}$ or $8\overline{)4}$.

In the division $5 \div \frac{1}{2}$, which number is the divisor?

Check your answer in **27**.

---

29 Right you are.

The phrase "8 divided by 4" is written $8 \div 4$. We can also write this as $4\overline{)8}$ or $\frac{8}{4}$. In all of these the divisor is 4, and you are being asked to divide a set of 8 objects into sets of 4 objects.

In the division $5 \div \frac{1}{2}$ which number is the divisor?

Check your answer in **27**.

30

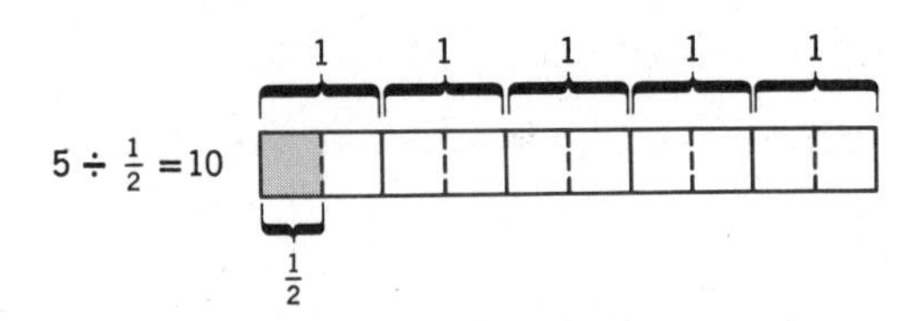

There are ten $\frac{1}{2}$ unit lengths contained in the 5 unit length.

Using a drawing of this sort to solve a division problem is difficult and clumsy. We need a simple rule. Here it is.

To divide by a fraction, invert the divisor and multiply.

For example,

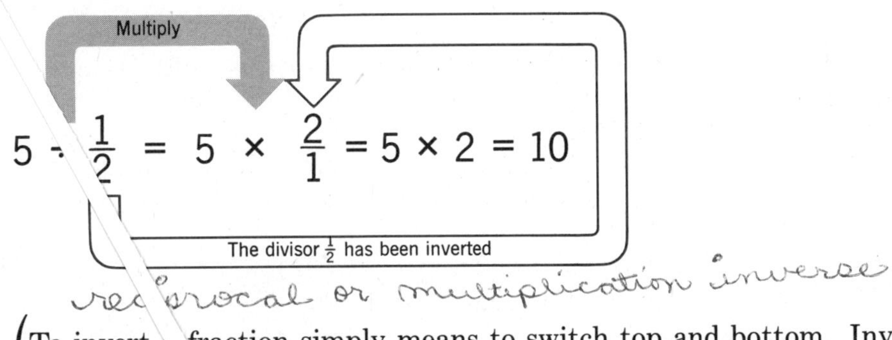

reciprocal or multiplication inverse

(To invert a fraction simply means to switch top and bottom. Inverting $\frac{2}{3}$ gives $\frac{3}{2}$. Inverting $\frac{1}{5}$ gives $\frac{5}{1}$ or 5. Inverting 7 gives $\frac{1}{7}$.)

Here is another example.

$$\frac{3}{5} \div \frac{2}{3} = \frac{3}{5} \times \frac{3}{2} = \frac{3 \times 3}{5 \times 2} = \frac{9}{10}$$

We have converted a division problem that is difficult to picture into a simple multiplication.

The final, and very important, step in every division is checking the answer.

(continued on the next page)

If $\frac{3}{5} \div \frac{2}{3} = \frac{9}{10}$ then $\frac{3}{5} = \frac{2}{3} \times \frac{9}{10}$

Check: $\frac{\overset{1}{\cancel{2}}}{\cancel{3}} \times \frac{\overset{3}{\cancel{9}}}{\underset{5}{\cancel{10}}} = \frac{3}{5}$

Why does this "invert and multiply" process work?

The division $\frac{3}{5} \div \frac{2}{3} = \square$ means that there is some number $\square$ such that $\frac{3}{5} = \square \times \frac{2}{3}$

Multiply both sides of the last equation by $\frac{3}{2}$.

$$\left(\frac{3}{5}\right) \times \frac{3}{2} = \left(\square \times \frac{2}{3}\right) \times \frac{3}{2}$$

But we can do multiplications in any order we wish and we know that

$$\frac{2}{3} \times \frac{3}{2} = \frac{6}{6} = 1$$

so $\frac{3}{5} \times \frac{3}{2} = \square$

Our answer, the unknown number we have labeled $\square$, is simply the product of the dividend $\frac{3}{5}$ and the inverted divisor $\frac{3}{2}$.

Try this one:

$\frac{7}{8} \div \frac{3}{2} =$ ______

Solve it by inverting the divisor and multiplying. Check your answer in **31**.

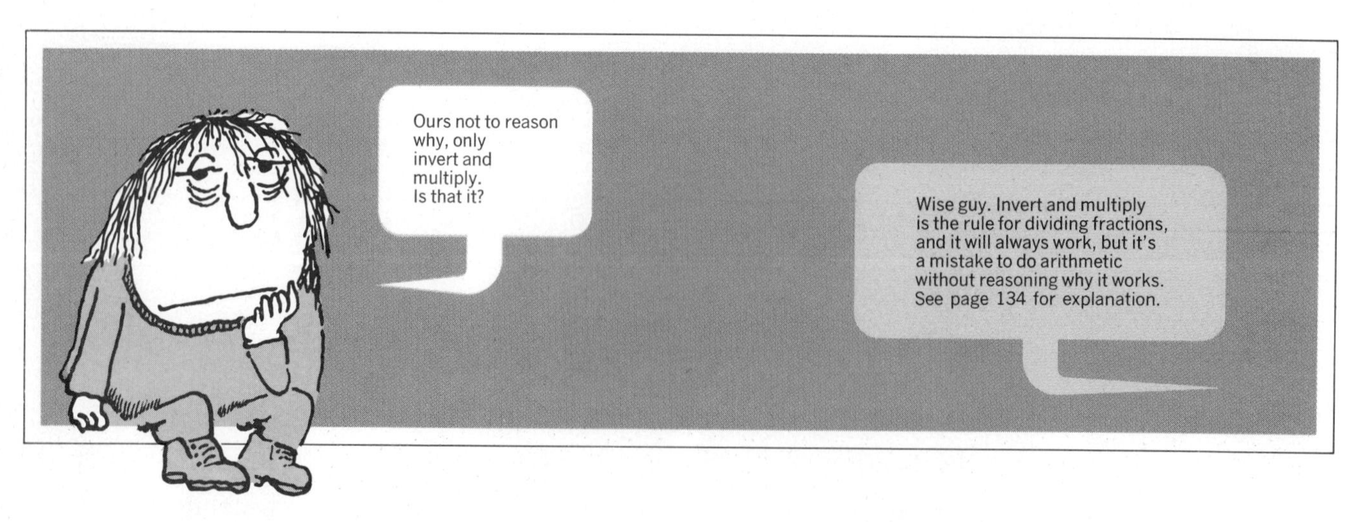

Ours not to reason why, only invert and multiply. Is that it?
Wise guy. Invert and multiply is the rule for dividing fractions, and it will always work, but it's a mistake to do arithmetic without reasoning why it works. See page 134 for explanation.

$$\frac{7}{8} \div \frac{3}{2} = \frac{7}{8} \times \frac{2}{3} = \frac{7 \times \overset{1}{\cancel{2}}}{\underset{4}{\cancel{8}} \times 3} = \frac{7}{12}$$

Check: $\dfrac{\overset{1}{\cancel{3}}}{2} \times \dfrac{7}{\underset{4}{\cancel{12}}} = \dfrac{7}{8}$

The chief source of confusion for many people in dividing fractions is deciding which fraction to invert. It will help if you

(1) Put every division problem in the form (dividend) ÷ (divisor) first, then invert the divisor, and finally multiply to obtain the quotient.
(2) Check your answer by multiplying. The product (divisor) × (quotient or answer) should equal the dividend.

Here are a few problems to test your understanding.

(a) $\dfrac{2}{5} \div \dfrac{3}{8} =$ ______ (b) $\dfrac{7}{40} \div \dfrac{21}{25} =$ ______

(c) $3\dfrac{3}{4} \div \dfrac{5}{2} =$ ______ (d) $4\dfrac{1}{5} \div 1\dfrac{4}{10} =$ ______

(e) $3\dfrac{2}{3} \div 3 =$ ______ (f) Divide $\dfrac{3}{4}$ by $2\dfrac{5}{8}$ ______

(g) Divide 8 by $\dfrac{1}{2}$ ______ (h) Divide $1\dfrac{1}{4}$ by $1\dfrac{7}{8}$ ______

Work carefully, check each answer, then turn to **32** for our worked solutions.

WHY DO WE INVERT AND MULTIPLY WHEN WE DIVIDE FRACTIONS?

The division $8 \div 4$ can be written $\frac{8}{4}$. Similarly,

$$\frac{1}{2} \div \frac{2}{3}$$

can be written

$$\frac{\frac{1}{2}}{\frac{2}{3}}$$

To simplify this fraction, multiply by $\frac{\frac{3}{2}}{\frac{3}{2}}$

$$\frac{\frac{1}{2}}{\frac{2}{3}} = \frac{\frac{1}{2} \times \frac{3}{2}}{\frac{2}{3} \times \frac{3}{2}} = \frac{\frac{1}{2} \times \frac{3}{2}}{1} = \frac{1}{2} \times \frac{3}{2}$$

$$\frac{2}{3} \times \frac{3}{2} = \frac{2 \times 3}{3 \times 2} = \frac{6}{6} = 1$$

Therefore

$$\frac{1}{2} \div \frac{2}{3} = \frac{1}{2} \times \frac{3}{2}$$

We have inverted the fraction $\frac{2}{3}$ and multiplied by it.

32 (a) $\frac{2}{5} \div \frac{3}{8} = \frac{2}{5} \times \frac{8}{3} = \frac{16}{15} = 1\frac{1}{15}$

Check: $\frac{\overset{1}{\cancel{3}}}{\underset{1}{\cancel{8}}} \times \frac{\overset{2}{\cancel{16}}}{\underset{5}{\cancel{15}}} = \frac{2}{5}$ $\left(\text{Remember } \frac{3}{8} \times \frac{16}{15} = \frac{3 \times 16}{8 \times 15} = \frac{\cancel{3} \times 2 \times \cancel{8}}{\cancel{8} \times \cancel{3} \times 5} = \frac{2}{5}\right)$

(b) $\frac{7}{40} \div \frac{21}{25} = \frac{7}{40} \times \frac{25}{21} = \frac{5}{24}$

Check: $\frac{\overset{7}{\cancel{21}}}{\underset{5}{\cancel{25}}} \times \frac{\overset{1}{\cancel{5}}}{\underset{8}{\cancel{24}}} = \frac{7}{40}$

(c) $3\frac{3}{4} \div \frac{5}{2} = \frac{15}{4} \div \frac{5}{2} = \frac{\overset{3}{\cancel{15}}}{\underset{2}{\cancel{4}}} \times \frac{\overset{1}{\cancel{2}}}{\underset{1}{\cancel{5}}} = \frac{3}{2} = 1\frac{1}{2}$

Check: $\frac{5}{2} \times \frac{3}{2} = \frac{15}{4} = 3\frac{3}{4}$

(d) $4\frac{1}{5} \div 1\frac{4}{10} = \frac{21}{5} \div \frac{14}{10} = \frac{\overset{3}{\cancel{21}}}{\underset{1}{\cancel{5}}} \times \frac{\overset{2}{\cancel{10}}}{\underset{2}{\cancel{14}}} = 3$

Check: $1\frac{4}{10} \times 3 = \frac{14}{10} \times \frac{3}{1} = \frac{42}{10} = 4\frac{2}{10} = 4\frac{1}{5}$

(e) $3\frac{2}{3} \div 3 = \frac{11}{3} \div \frac{3}{1} = \frac{11}{3} \times \frac{1}{3} = \frac{11}{9} = 1\frac{2}{9}$

Check: $3 \times \frac{11}{9} = \frac{3}{1} \times \frac{11}{9} = \frac{11}{3} = 3\frac{2}{3}$

(f) $\frac{3}{4} \div 2\frac{5}{8} = \frac{3}{4} \div \frac{21}{8} = \frac{\overset{1}{\cancel{3}}}{\underset{1}{\cancel{4}}} \times \frac{\overset{2}{\cancel{8}}}{\underset{7}{\cancel{21}}} = \frac{2}{7}$

Check: $2\frac{5}{8} \times \frac{2}{7} = \frac{21}{8} \times \frac{2}{7} = \frac{3}{4}$

(g) $8 \div \frac{1}{2} = \frac{8}{1} \times \frac{2}{1} = 16$

Check: $\frac{1}{2} \times 16 = 8$

(h) $1\frac{1}{4} \div 1\frac{7}{8} = \frac{5}{4} \div \frac{15}{8} = \frac{\overset{1}{\cancel{5}}}{\underset{1}{\cancel{4}}} \times \frac{\overset{2}{\cancel{8}}}{\underset{3}{\cancel{15}}} = \frac{2}{3}$

Check: $1\frac{7}{8} \times \frac{2}{3} = \frac{15}{8} \times \frac{2}{3} = \frac{5}{4} = 1\frac{1}{4}$

Turn to **33** for a set of practice problems on dividing fractions.

When I divide fractions, it doesn't matter which I turn upside down, does it?
Oh yes it does! Invert the divisor, the number you are dividing by.

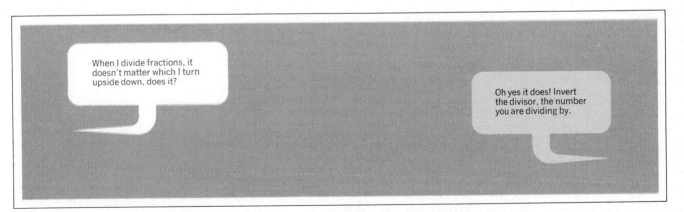

7. A box of modeling clay weighs $45\frac{1}{2}$ pounds. How many $1\frac{3}{4}$-lb packages will one such box fill?

____

8. How many pieces of wood each $2\frac{1}{2}$ inches long can be cut from a strip $142\frac{1}{2}$ inches long?

____

9. A large size can of fruit weighs $3\frac{1}{4}$ pounds. Into how many $\frac{1}{8}$ lb servings can this be divided?

____

10. Find: (a) $\frac{5}{2} \div 2$ (b) $\frac{5}{2} \div \frac{1}{2}$ (c) $2 \div \frac{5}{2}$ (d) $\frac{1}{2} \div \frac{5}{2}$

= ____ = ____ = ____ = ____

When you have had the practice you need either return to the preview test on page 105 or continue in **34** with the addition and subtraction of fractions.

Date

Name

Course/Section

34 At heart, adding fractions is a matter of counting:

$$\frac{1}{5} + \frac{3}{5} = \frac{1+3}{5} = \frac{4}{5}$$

$\frac{1}{5}$ 1 fifth

+

$\frac{3}{5}$ +3 fifths

=

$\frac{4}{5}$ =4 fifths, count them.

Add $\frac{2}{7} + \frac{3}{7} =$ ______

Check your answer in 35.

35

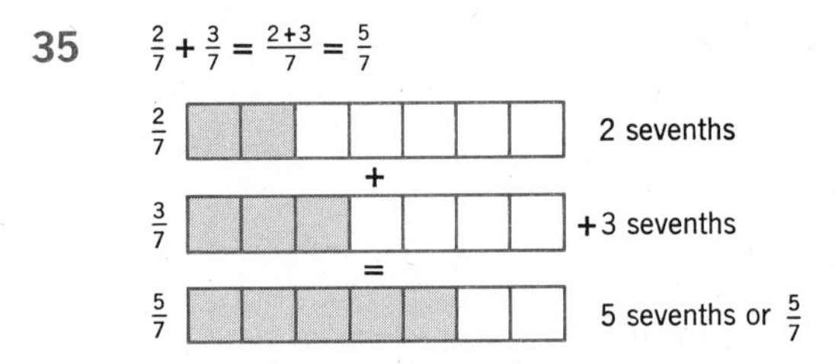

Fractions having the same denominator are called *like* fractions. In the problem above, $\frac{2}{7}$ and $\frac{3}{7}$ both have denominator 7 and are like fractions. Adding like fractions is easy: *first,* add the numerators to find the numerator of the sum and, *second,* use the denominator the fractions have in common as the denominator of the sum.

$$\frac{2}{9} + \frac{5}{9} = \frac{2+5}{9} = \frac{7}{9} \quad \begin{array}{l}\text{(add numerators)}\\ \text{(same denominator)}\end{array}$$

Adding three or more fractions presents no special problems:

$$\frac{3}{12} + \frac{1}{12} + \frac{5}{12} = \underline{\qquad}$$

Add the fractions as shown above, then turn to 36.

---

36 $\dfrac{3}{12} + \dfrac{1}{12} + \dfrac{5}{12} = \dfrac{3+1+5}{12} = \dfrac{9}{12} = \dfrac{3}{4}$

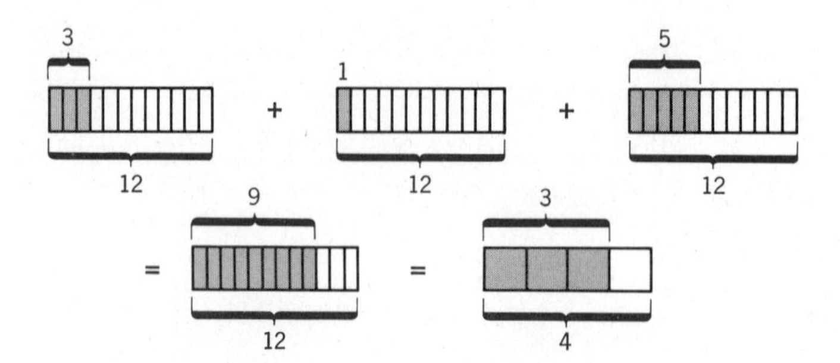

Notice that we reduce the sum to lowest terms.

Try these problems for exercise:

(a) $\dfrac{1}{8} + \dfrac{3}{8} =$ ______ (b) $\dfrac{7}{9} + \dfrac{5}{9} =$ ______

(c) $2\dfrac{1}{2} + 3\dfrac{3}{2} =$ ______ (d) $\dfrac{1}{7} + \dfrac{4}{7} + \dfrac{5}{7} + 1\dfrac{2}{7} + \dfrac{8}{7} =$ ______

(e) $2 + 3\dfrac{1}{2} =$ ______ (f) $\dfrac{3}{5} + 1\dfrac{1}{5} + 3 =$ ______

Go to **38** to check your work.

---

37 $\dfrac{3}{4} + \dfrac{2}{3} = \dfrac{9}{12} + \dfrac{8}{12} = \dfrac{17}{12} = 1\dfrac{5}{12}$

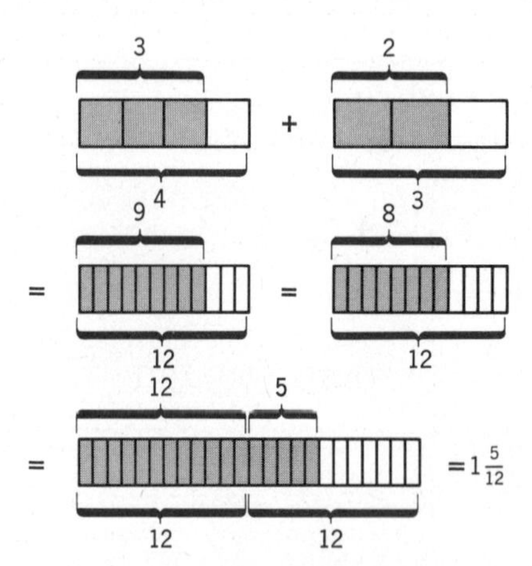

We change the original fractions to equivalent fractions with the same denominator and then add as before.

How do you know what number to use as the new denominator? In general, you cannot simply guess at the best new denominator. We need a method for finding it from the denominators of the fractions to be added. The new denominator we want is called the Least Common Multiple or LCM.

Every whole number has a set of *multiples* associated with it. We can find the multiples of any given number by multiplying it by each of the whole numbers in turn. For example,

The multiples of 2 are 2, 4, 6, 8, 10, 12, ...
(Multiply each integer 1, 2, 3, 4, . . . , by 2.)

The multiples of 3 are 3, 6, 9, 12, 15, 18, ...
(Multiply each integer 1, 2, 3, 4, . . . , by 3.)

The multiples of 4 are 4, 8, 12, 16, 20, 24, ...

Find the first few multiples of the following:

(a) 5 (b) 7 (c) 12 (d) 8

Our answers are in **39**.

---

38 (a) $\frac{1}{8} + \frac{3}{8} = \frac{1+3}{8} = \frac{4}{8} = \frac{1}{2}$, reduced to lowest terms

(b) $\frac{7}{9} + \frac{5}{9} = \frac{7+5}{9} = \frac{12}{9} = \frac{4}{3} = 1\frac{1}{3}$

(c) $2\frac{1}{2} + 3\frac{3}{2} = 2 + 3 + \frac{1}{2} + \frac{3}{2} = 2 + 3 + \frac{4}{2} = 2 + 3 + 2 = 7$

Add the whole number parts first.

(d) $\frac{1}{7} + \frac{4}{7} + \frac{5}{7} + 1\frac{2}{7} + \frac{8}{7} = \frac{1}{7} + \frac{4}{7} + \frac{5}{7} + \frac{9}{7} + \frac{8}{7} = \frac{1+4+5+9+8}{7}$

$= \frac{27}{7} = 3\frac{6}{7}$

(e) $2 + 3\frac{1}{2} = 2 + 3 + \frac{1}{2} = 5\frac{1}{2}$ Remember, $3\frac{1}{2}$ means $3 + \frac{1}{2}$.

(f) $\frac{3}{5} + 1\frac{1}{5} + 3 = 1 + 3 + \frac{3}{5} + \frac{1}{5} = 4 + \frac{4}{5} = 4\frac{4}{5}$

Add the whole number parts first.

(continued on the next page)

How do we add fractions whose denominators are not the same?

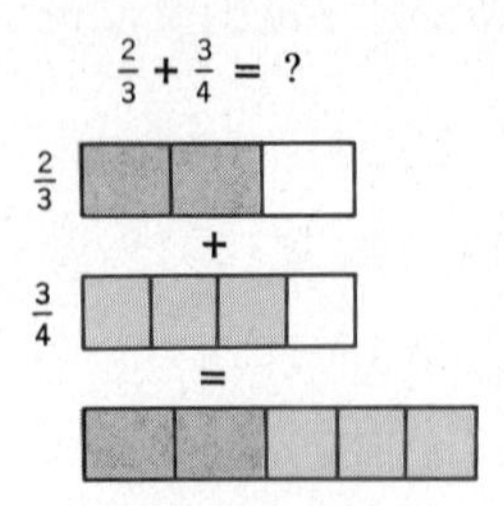

The problem is to find a simple numeral that names this new number. One way to find it is to change these fractions to equivalent fractions with the same denominator.

$$\frac{3}{4} = \frac{3 \times 3}{4 \times 3} = \frac{9}{12}$$

(Equivalent fractions are discussed on page 112 in frame **10**.)

$$\frac{2}{3} = \frac{2 \times 4}{3 \times 4} = \frac{8}{12}$$

Now add these fractions.

Continue in **37**.

---

39 (a) The multiples of 5 are 5, 10, 15, 20, 25, . . .

(b) The multiples of 7 are 7, 14, 21, 28, 35, . . .

(c) The multiples of 12 are 12, 24, 36, 48, 60, 72, . . .

(d) The multiples of 8 are 8, 16, 24, 32, 40, 48, 56, 64, 72, . . .

Notice that 24, 48, and 72 are multiples of both 8 and 12. They are the first three *common multiples* of 8 and 12.

List a few of the multiples common to 3 and 5.

Hop to **40** to check your answer.

---

40 The multiples of 3 are 3, 6, 9, 12, 15, 18, 21, 24, 27, 30, . . .

The multiples of 5 are 5, 10, 15, 20, 25, 30, 35, . . .

The multiples 3 and 5 have in common are 15, 30, 45, 60, and so on.

List several of the common multiples of the following pairs of numbers:

(a) 4 and 6 (b) 12 and 16 (c) 8 and 10

For each pair make two lists of multiples and find the numbers which appear on both lists.

Look in **41** for the correct answers.

---

41 (a) The multiples of 4 and 6 are

4: 4, 8, 12, 16, 20, 24, 28, 32, 36, 40, 44, 48, ...

6: 6, 12, 18, 24, 30, 36, 42, 48, 54, ...

The common multiples are 12, 24, 36, 48, and so on.

(b) The common multiples of 12 and 16 are
48, 96, 144, 192, and so on.

(c) The common multiples of 8 and 10 are
40, 80, 120, 160, and so on.

The *Least Common Multiple* or LCM is the smallest common multiple in the set of common multiples for a pair of numbers. For example, the Least Common Multiple of 4 and 6 in (a) above is 12. The LCM is the smallest number that both 4 and 6 will divide evenly.

The LCM of 12 and 16 is 48.

The LCM of 8 and 10 is 40.

What is the LCM of 8 and 12?

The answer is in **42**.

---

42 The LCM or Least Common Multiple of 8 and 12 is 24. It is the smallest number that both 8 and 12 will divide with zero remainder.

Making lists of multipliers is the most direct way of finding the LCM, but it is slow and time-consuming. Here is a short-cut method we call the *LCM Finder*.

Example

Step 1. Write each denominator as a product of primes.

To add $\frac{7}{60} + \frac{11}{72}$ find the LCM of 60 and 72

$60 = 2^2 \times 3^1 \times 5^1$
$72 = 2^3 \times 3^2$

Step 2. Write each base prime that appears in either number.

2 3 5

Step 3. Attach to each prime the largest exponent that appears on it in either number.

2^3 ↑ from 2^3 in 72; 3^2 ↑ from 3^2 in 72; 5^1 ↑ from 5^1 in 60

Step 4. Multiply to find the LCM.

$\text{LCM} = 2^3 \times 3^2 \times 5^1$
$= 8 \times 9 \times 5 = 360$

Use this as the new denominator.

Try it on this pair of numbers. Find the LCM of 12 and 18.

Check your work in **43**.

---

43 Step 1: Factor 12 and 18. $12 = 2^2 \times 3$

$18 = 2 \times 3^2$

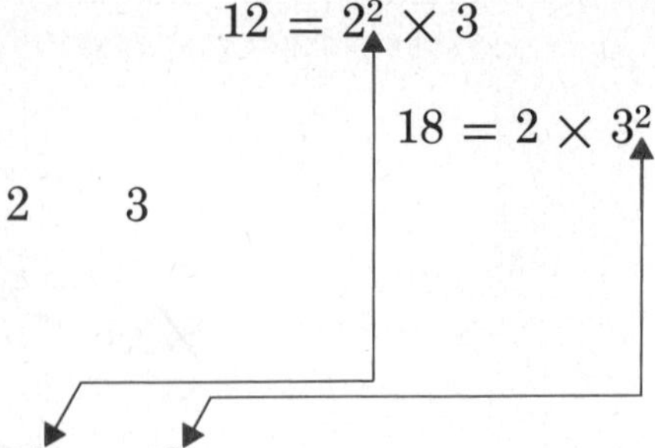

Step 2: Write the primes that appear as factors of 12 and 18. 2 3

Step 3: Attach the highest powers of each that appear in either number (the "power" is the exponent number . . . remember?) 2^2 3^2

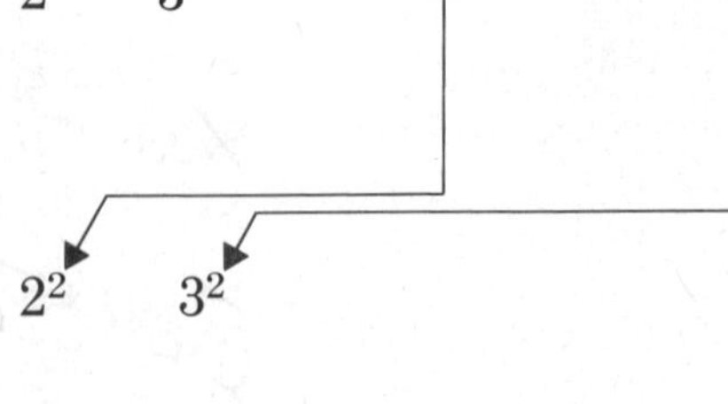

Step 4: Multiply. $2^2 \times 3^2 = 4 \times 9 = 36$

The LCM = 36.

A bit of practice will groove the idea into place. Find the Least Common Multiple of each of the following pairs of numbers.

(a) 15 and 6 (b) 20 and 24 (c) 54 and 180

(d) 525 and 90 (e) 8 and 18 (f) 140 and 50

Answers in **44**.

---

44 (a) $15 = 3^1 \times 5^1$ $6 = 2^1 \times 3^1$
LCM $= 2^1 \times 3^1 \times 5^1 = 30$

(b) $20 = 2^2 \times 5^1$ $24 = 2^3 \times 3^1$
LCM $= 2^3 \times 3^1 \times 5^1 = 8 \times 3 \times 5 = 120$

(c) $54 = 2^1 \times 3^3$ $180 = 2^2 \times 3^2 \times 5^1$
LCM $= 2^2 \times 3^3 \times 5^1 = 4 \times 27 \times 5 = 540$

(d) $525 = 3^1 \times 5^2 \times 7$ $90 = 2^1 \times 3^2 \times 5^1$
LCM $= 2^1 \times 3^2 \times 5^2 \times 7 = 2 \times 9 \times 25 \times 7 = 3150$

(e) $8 = 2^3$ $18 = 2^1 \times 3^2$
LCM $= 2^3 \times 3^2 = 8 \times 9 = 72$

(f) $140 = 2^2 \times 5^1 \times 7^1$ $50 = 2^1 \times 5^2$
LCM $= 2^2 \times 5^2 \times 7^1 = 4 \times 25 \times 7 = 700$

The process is exactly the same with three or more numbers. For example, the LCM of 24, 90, and 75 is found this way:

$24 = 2^3 \times 3^1$ $90 = 2^1 \times 3^2 \times 5^1$ $75 = 3^1 \times 5^2$

LCM $= 2^3 \times 3^2 \times 5^2 = 8 \times 9 \times 25 = 1800$

Find the sum of $\frac{3}{18} + \frac{5}{12}$. Write both numbers as equivalent fractions with the LCM of 18 and 12 as the denominator, then add.

Our solution is in **45**.

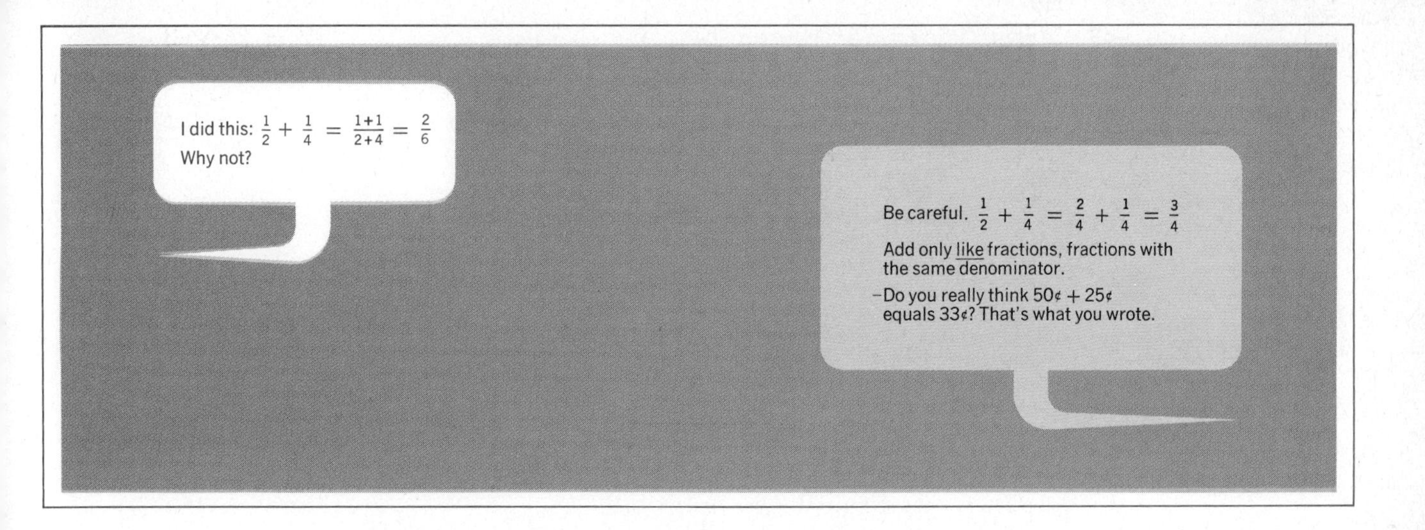

45 $18 = 2^1 \times 3^2$ $\quad$ $12 = 2^2 \times 3$ $\quad$ $\text{LCM} = 2^2 \times 3^2 = 4 \times 9 = 36$

$$\frac{3}{18} = \frac{?}{36} = \frac{3 \times 2}{18 \times 2} = \frac{6}{36}$$

$$\frac{5}{12} = \frac{?}{36} = \frac{5 \times 3}{12 \times 3} = \frac{15}{36}$$

Add: $\frac{3}{18} + \frac{5}{12} = \frac{6}{36} + \frac{15}{36} = \frac{6 + 15}{36} = \frac{21}{36} = \frac{7}{12}$ in lowest terms

Remember, to add fractions:

1. Add the whole number parts of the fractions.
2. Find the LCM of the denominators of the fractions to be added.
3. Write the fractions so that they are equivalent fractions with the LCM as denominator.
4. The answer numerator is the sum of the numerators and the denominator of the answer is the LCM.
5. Reduce to lowest terms.

If this way of adding fractions seems rather long and involved, here is why:

1. It *is* involved, but it is the only sure way to arrive at the answer.
2. It is new to you and you will need lots of practice at it before it comes quickly. Take each problem step by step, work slowly at first, and gradually you will become very quick at adding fractions.

(continued on the next page)

Practice by adding the following:

(a) $\frac{5}{18} + \frac{5}{16} =$ ______ (b) $\frac{7}{12} + \frac{3}{16} =$ ______

(c) $\frac{10}{32} + \frac{4}{30} =$ ______ (d) $\frac{3}{7} + \frac{4}{5} =$ ______

(e) $1\frac{2}{15} + \frac{5}{9} =$ ______ (f) $\frac{1}{2} + \frac{5}{6} + \frac{3}{4} =$ ______

(g) $2\frac{5}{12} + 1\frac{1}{9} + 2\frac{3}{8} =$ ______ (h) $\frac{13}{16} + 2\frac{1}{18} + 1\frac{11}{12} =$ ______

The worked solutions are in **48**.

---

46 $\frac{3}{8} - \frac{1}{8} = \frac{3-1}{8} = \frac{2}{8} = \frac{1}{4}$

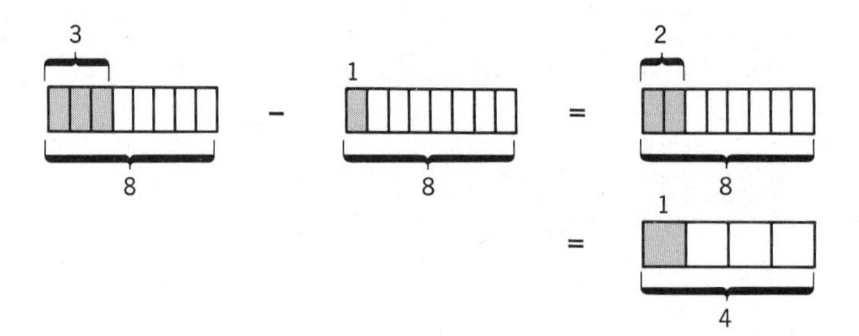

Easy enough? If the denominators are the same we subtract numerators and write this difference over the common denominator.

Subtract $\frac{3}{4} - \frac{1}{5} =$ ______

Our solution is in **47**.

---

47 The LCM of 4 and 5 is 20.

$$\frac{3}{4} = \frac{?}{20} = \frac{3 \times 5}{4 \times 5} = \frac{15}{20}$$

$$\frac{1}{5} = \frac{?}{20} = \frac{1 \times 4}{5 \times 4} = \frac{4}{20}$$

Then $\frac{3}{4} - \frac{1}{5} = \frac{15}{20} - \frac{4}{20} = \frac{11}{20}$

When the two fractions have different denominators we must change them to equivalent fractions with the LCM as denominator before subtracting.

If you have not yet learned how to find the Least Common Multiple (LCM) turn to **41**. Otherwise, continue by finding the following differences.

(a) $3\frac{1}{4} - 1\frac{1}{12} =$ ______ (b) $10 - 3\frac{5}{16} =$ ______

(c) $4\frac{7}{18} - 2\frac{11}{12} =$ ______ (d) $2\frac{6}{32} - 1\frac{1}{6} =$ ______

Check your answers in **49**.

WHAT IS THE LCM?

An LCM is not a *L*ittle *C*rooked *M*artian, but a *L*east *C*ommon *M*ultiple.

A *multiple* of a whole number is a number evenly divisible by it.

3, 6, 9, 12, 15, 18, 21, . . . are all multiples of 3.

A pair of numbers will have *common multiples.*

The multiples of 3 are 3, 6, 9, 12, 15, 18, 21, 24, 27, 30, 33, . . .

The multiples of 5 are 5, 10, 15, 20, 25, 30, 35, . . .

The common multiples of 3 and 5 are 15, 30, 45, and so on.

The smallest common multiple or *least common multiple* or LCM is the smallest of all the common multiples of a pair of whole numbers.

The LCM of 3 and 5 is 15. The LCM 15 is the smallest whole number that is evenly divisible by both 3 and 5.

48 (a) $18 = 2^1 \times 3^2 \qquad 16 = 2^4 \qquad$ The LCM of 18 and 16 is $2^4 \times 3^2 = 144$

$$\frac{5}{18} = \frac{?}{144} = \frac{5 \times 8}{18 \times 8} = \frac{40}{144}$$

$$\frac{5}{16} = \frac{?}{144} = \frac{5 \times 9}{16 \times 9} = \frac{45}{144}$$

$$\frac{5}{18} + \frac{5}{16} = \frac{40}{144} + \frac{45}{144} = \frac{40 + 45}{144} = \frac{85}{144}$$

(b) The LCM of 12 and 16 is 48.

$$\frac{7}{12} = \frac{7 \times 4}{12 \times 4} = \frac{28}{48} \qquad \frac{3}{16} = \frac{3 \times 3}{16 \times 3} = \frac{9}{48}$$

$$\frac{7}{12} + \frac{3}{16} = \frac{28}{48} + \frac{9}{48} = \frac{37}{48}$$

(c) The LCM of 32 and 30 is 480.

$$\frac{10}{32} = \frac{10 \times 15}{32 \times 15} = \frac{150}{480} \qquad \frac{4}{30} = \frac{4 \times 16}{30 \times 16} = \frac{64}{480}$$

$$\frac{10}{32} + \frac{4}{30} = \frac{150}{480} + \frac{64}{480} = \frac{214}{480} = \frac{107}{240}$$

(d) The LCM of 7 and 5 is 35.

$$\frac{3}{7} = \frac{3 \times 5}{7 \times 5} = \frac{15}{35} \qquad \frac{4}{5} = \frac{4 \times 7}{5 \times 7} = \frac{28}{35}$$

$$\frac{3}{7} + \frac{4}{5} = \frac{15}{35} + \frac{28}{35} = \frac{43}{35} = 1\frac{8}{35}$$

(e) The LCM of 15 and 9 is 45.

$$\frac{2}{15} = \frac{2 \times 3}{15 \times 3} = \frac{6}{45} \qquad \frac{5}{9} = \frac{5 \times 5}{9 \times 5} = \frac{25}{45}$$

$$1\frac{2}{15} + \frac{5}{9} = 1 + \frac{2}{15} + \frac{5}{9} = 1 + \frac{6}{45} + \frac{25}{45} = 1 + \frac{31}{45} = 1\frac{31}{45}$$

(f) The LCM of 2, 6, and 4 is 12.

$$\frac{1}{2} = \frac{6}{12} \qquad \frac{5}{6} = \frac{10}{12} \qquad \frac{3}{4} = \frac{9}{12}$$

$$\frac{1}{2} + \frac{5}{6} + \frac{3}{4} = \frac{6}{12} + \frac{10}{12} + \frac{9}{12} = \frac{25}{12} = 2\frac{1}{12}$$

(g) The LCM of 12, 9, and 8 is 72.

$$\frac{5}{12} = \frac{5 \times 6}{12 \times 6} = \frac{30}{72} \qquad \frac{1}{9} = \frac{1 \times 8}{9 \times 8} = \frac{8}{72} \qquad \frac{3}{8} = \frac{3 \times 9}{8 \times 9} = \frac{27}{72}$$

$$2\frac{5}{12} + 1\frac{1}{9} + 2\frac{3}{8} = 2 + 1 + 2 + \frac{5}{12} + \frac{1}{9} + \frac{3}{8}$$

$$= 5 + \frac{30}{72} + \frac{8}{72} + \frac{27}{72}$$

$$= 5 + \frac{65}{72} = 5\frac{65}{72}$$

(h) The LCM of 16, 18, and 12 is 144.

$$\frac{13}{16} = \frac{117}{144} \qquad \frac{1}{18} = \frac{8}{144} \qquad \frac{11}{12} = \frac{132}{144}$$

$$\frac{13}{16} + 2\frac{1}{18} + 1\frac{11}{12} = 2 + 1 + \frac{117}{144} + \frac{8}{144} + \frac{132}{144} = 3 + \frac{257}{144} = 4 + \frac{113}{144}$$

$$= 4\frac{113}{144}$$

Again, add the whole number part of the mixed numbers first.

Once you have mastered the process of adding fractions, subtraction is very simple indeed.

Calculate $\frac{3}{8} - \frac{1}{8} =$ ______

Turn to **46**.

---

49 (a) The LCM of 4 and 12 is 12.

$$3\frac{1}{4} = \frac{13}{4} = \frac{13 \times 3}{4 \times 3} = \frac{39}{12} \qquad 1\frac{1}{12} = \frac{13}{12}$$

$$3\frac{1}{4} - 1\frac{1}{12} = \frac{39}{12} - \frac{13}{12} = \frac{26}{12} = \frac{13}{6} = 2\frac{1}{6}$$

(b) $$10 = \frac{10}{1} = \frac{10 \times 16}{1 \times 16} = \frac{160}{16} \qquad 3\frac{5}{16} = \frac{53}{16}$$

$$10 - 3\frac{5}{16} = \frac{160}{16} - \frac{53}{16} = \frac{107}{16} = 6\frac{11}{16}$$

(c) The LCM of 18 and 12 is 36.

$$4\frac{7}{18} = \frac{79}{18} = \frac{79 \times 2}{18 \times 2} = \frac{158}{36}$$

$$2\frac{11}{12} = \frac{35}{12} = \frac{35 \times 3}{12 \times 3} = \frac{105}{36}$$

$$4\frac{7}{18} - 2\frac{11}{12} = \frac{158}{36} - \frac{105}{36} = \frac{53}{36} = 1\frac{17}{36}$$

(d) The LCM of 32 and 6 is 96.

$$2\frac{6}{32} = \frac{70}{32} = \frac{70 \times 3}{32 \times 3} = \frac{210}{96}$$

$$1\frac{1}{6} = \frac{7}{6} = \frac{7 \times 16}{6 \times 16} = \frac{112}{96}$$

$$2\frac{6}{32} - 1\frac{1}{6} = \frac{210}{96} - \frac{112}{96} = \frac{98}{96} = \frac{49}{48} = 1\frac{1}{48}$$

Now turn to **50** for a set of practice problems on addition and subtraction of fractions.

WORD PROBLEMS WITH FRACTIONS

51 The mathematics you use every day in your work or at play usually appears wrapped in words and hidden in sentences. Neat little sets of directions seldom are attached; no "Divide and reduce to lowest terms" or "Write as equivalent fractions and add." The difficulty with real problems is that they must be translated from words to mathematics. You need to learn to *talk* arithmetic, not just juggle numbers.

Certain words and phrases appear again and again in arithmetic. They are signals alerting you to the mathematical operations to be done. Here is a list of such *signal words:*

SIGNAL WORDS	TRANSLATE AS
Is, is equal to, equals, the same as	$=$
Of, the product of, multiply, times, multiplied by	$\times$
Add, in addition, plus, more, more than, sum, and, increased by, added to	$+$
Subtract, subtract from, less, less than, difference, diminished by, decreased by	$-$
Divide, divided by	$\div$
Twice, double, twice as much	$2\times$
half of, half	$\frac{1}{2}\times$

Translate the phrase "six times some number."

First, make a word equation by using parentheses:

(six) (times) (some number)

Second, substitute symbols:

(6)(×)(□)

Translate signal words to math symbols and use □ or a letter of the alphabet to represent any unknown quantity. (The little box □ is handy because you can write in it if you ever learn its numerical value.)

Translate the phrase "a number divided by seven."

Check your answer in **52**

52 Do it this way:

"a number divided by seven" ⟶ (a number)(divided by)(seven)

$\longrightarrow \square \div 7$ or $\dfrac{\square}{7}$ or $7\overline{)\square}$

Translate the following phrases to mathematical expressions.

(a) "Three more than some number"
(b) "One-half of some quantity"
(c) "The sum of $\frac{2}{3}$ and a number"
(d) "Six less than a number"
(e) "Five more than twice a number"
(f) "A number divided by $\frac{1}{2}$"
(g) "One-third of a quantity is equal to $\frac{3}{4}$"
(h) "A number diminished by two-thirds."

Our translations are in **53**.

53

(a) $3 + \square$ (e) $5 + (2 \times \square)$

(b) $\frac{1}{2} \times \square$ (f) $\square \div \frac{1}{2}$

(c) $\frac{2}{3} + \square$ (g) $\frac{1}{3} \times \square = \frac{3}{4}$

(d) $\square - 6$ (h) $\square - \frac{2}{3}$

Several kinds of word problems involving fractions appear again and again. Translate these:

(a) What fraction of $3\frac{3}{4}$ is $1\frac{1}{5}$?
(b) $\frac{3}{5}$ of what number is $1\frac{2}{3}$?
(c) $\frac{4}{9}$ of $1\frac{7}{8}$ is what number?

Translate as before, then turn to **54** for our answers.

Study for final

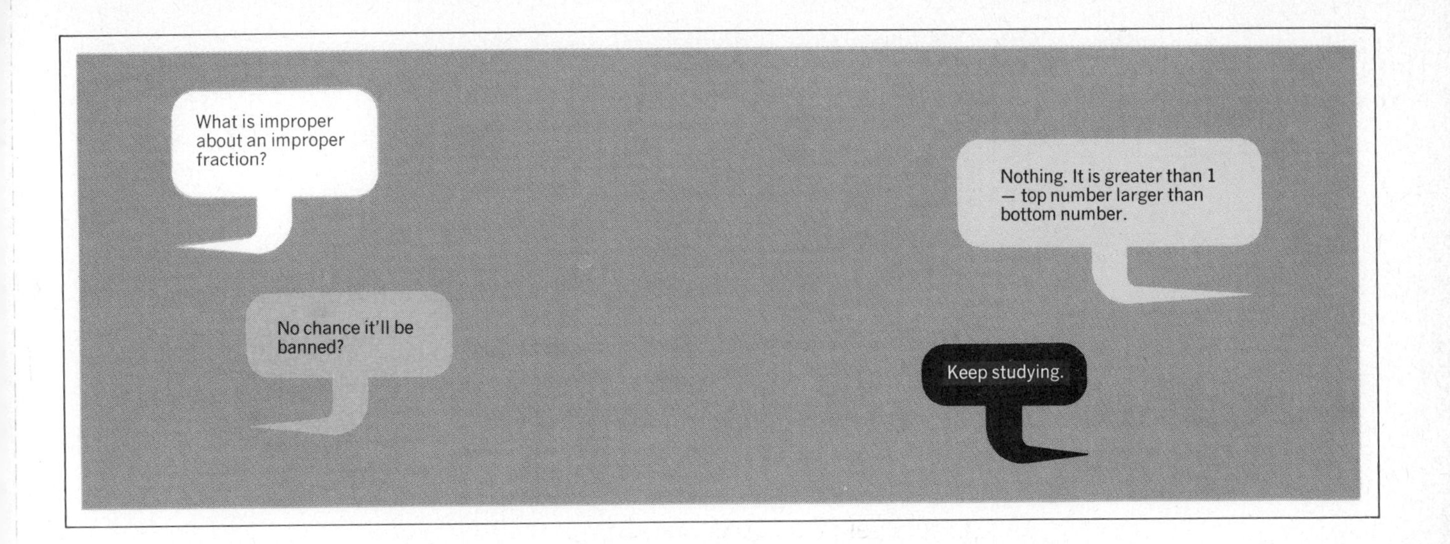

54 (a) (What fraction) (of) $\left(3\frac{3}{4}\right)$ (is) $\left(1\frac{1}{5}\right)$?

$$\square \times 3\frac{3}{4} = 1\frac{1}{5}$$

or $\square \times 3\frac{3}{4} = 1\frac{1}{5}$

(b) $\left(\frac{3}{5}\right)$ (of) (what number) (is) $\left(1\frac{2}{3}\right)$?

$$\frac{3}{5} \times \square = 1\frac{2}{3}$$

or $\frac{3}{5} \times \square = 1\frac{2}{3}$

(c) $\left(\frac{4}{9}\right)$ (of) $\left(1\frac{7}{8}\right)$ (is) (what number)?

$$\frac{4}{9} \times 1\frac{7}{8} = \square$$

or $\frac{4}{9} \times 1\frac{7}{8} = \square$

Problem (c) is easy to solve. If you do not remember how to do it, you should return to **21** and refresh your memory. Let's concentrate on the other two.

Can you solve problem (a)?

It will help if you make up a similar problem, one that you can solve easily, and compare them. For example, "what number multiplied by 3 equals 6?"

$\square \times 3 = 6$

Of course the answer is 2. How did you get that answer? It didn't simply pop into your head. Your brain did a quick calculation that went something like this:

"What times 3 gives 6?"

(continued on the next page)

"Hmm. I know that 6 divided by 3 is 2. Is 2 the answer?"

"Try it: does 2 times 3 give 6?"

"Yep. The answer must be 2."

In math language your brain did this

$$\square \times 3 = 6$$

so $\square = 6 \div 3 = 2$

Check: $\boxed{2} \times 3 = 6$ Ok.

Now solve $\square \times 3\frac{3}{4} = 1\frac{1}{5}$ in exactly the same way you solved $\square \times 3 = 6$.

Check your work in 55.

55

If $\square \times 3\dfrac{3}{4} = 1\dfrac{1}{5}$

then $\square = 1\dfrac{1}{5} \div 3\dfrac{3}{4}$

or $\square = \dfrac{6}{5} \div \dfrac{15}{4}$

$\square = \dfrac{6}{5} \times \dfrac{4}{15}$

$\square = \dfrac{8}{25}$

Check:

$\boxed{\dfrac{8}{25}} \times 3\dfrac{3}{4} = 1\dfrac{1}{5}$

$\dfrac{8}{25} \times \dfrac{15}{4} = \dfrac{6}{5} = 1\dfrac{1}{5}$

Similar Problem

$\square \times 3 = 6$

$\square = 6 \div 3$

$\square = 2$

Check: $\boxed{2} \times 3 = 6$

$6 = 6$

Here is the very best way of solving problems of this kind:

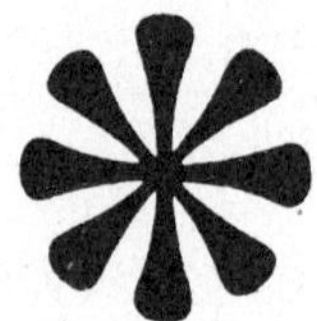

(1) Make up a similar problem using small whole numbers—an easy problem that you can solve immediately.
(2) Solve the difficult problem in an exactly parallel way.
(3) Check your answer by substituting it for the unknown quantity in the original equation.

The similar problem you make up allows you to use your feel for the logic of the problem to help you decide how to do it. You get to "think it out" without letting numbers get in the way. Don't hesitate on this, learn to trust yourself—but check every answer.

Now try to solve

$$\square \times 1\frac{7}{8} = \frac{5}{16}$$

The solution is in 57.

---

56 $\left(\frac{3}{4}\right)$ (of) (what number) (is) $\left(1\frac{2}{3}\right)$

$$\frac{3}{4} \times \square = 1\frac{2}{3}$$

or $\frac{3}{4} \times \square = 1\frac{2}{3}$

Make up a similar, but very easy, problem:

"Two times what number equals six?"

The answer is 3. How did you get it?

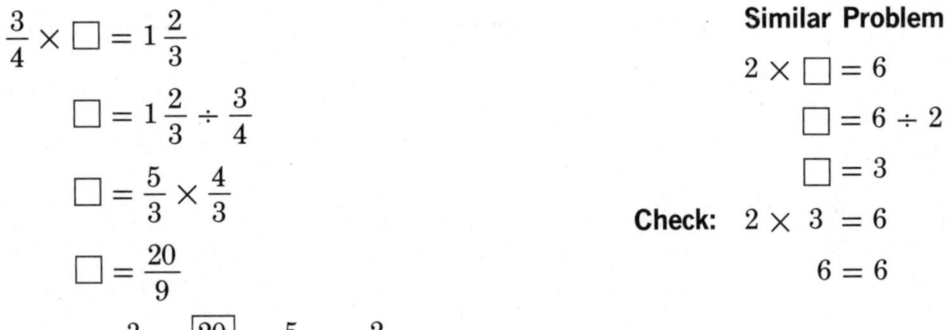

$$\frac{3}{4} \times \square = 1\frac{2}{3}$$
$$\square = 1\frac{2}{3} \div \frac{3}{4}$$
$$\square = \frac{5}{3} \times \frac{4}{3}$$
$$\square = \frac{20}{9}$$

Check: $\frac{3}{4} \times \boxed{\frac{20}{9}} = \frac{5}{3} = 1\frac{2}{3}$

Similar Problem

$$2 \times \square = 6$$
$$\square = 6 \div 2$$
$$\square = 3$$

Check: $2 \times 3 = 6$

$6 = 6$

If you are one of those people who like to memorize things, it may help you to remember that

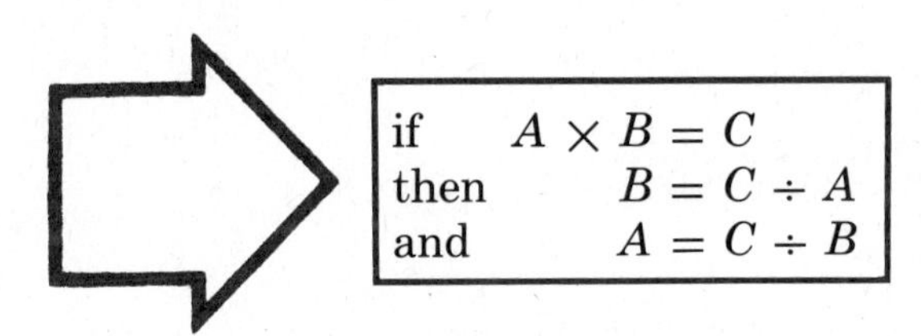

if $A \times B = C$
then $B = C \div A$
and $A = C \div B$

for any numbers A, B, and C not equal to zero.

As a general rule, it is better not to try to memorize a rule for every variety of math problem and far better to develop your ability to use basic principles, but if you need to memorize helpful rules this is a very useful one.

Use these problems for practice. Translate each into an equation and solve it.

(a) $\frac{3}{8}$ of what number is equal to $1\frac{5}{16}$?
(b) What fraction of $4\frac{1}{2}$ is $6\frac{3}{4}$?
(c) Find a number such that $\frac{3}{11}$ of it is $2\frac{2}{5}$.
(d) The product of $1\frac{7}{8}$ and $2\frac{1}{3}$ is what number?
(e) What fraction of $8\frac{3}{4}$ is $\frac{7}{12}$?

The answers are in **58**.

57 This is similar to the problem "What number times 4 equals 8?"

$$\square \times 1\frac{7}{8} = \frac{5}{16}$$
$$\square = \frac{5}{16} \div 1\frac{7}{8}$$
$$\square = \frac{5}{16} \div \frac{15}{8}$$
$$\square = \frac{5}{16} \times \frac{8}{15}$$
$$\square = \frac{1}{6}$$

Check: $\frac{1}{6} \times 1\frac{7}{8} = \frac{5}{16}$

$\frac{1}{6} \times \frac{15}{8} = \frac{5}{16}$ Ok.

Similar Problem

$$\square \times 4 = 8$$
$$\square = 8 \div 4$$
$$\square = 2$$

Check: $\boxed{2} \times 4 = 8$ Ok.

Isn't it easier when you have the simple problem as a guideline?

Try another problem:

"$\frac{3}{4}$ of what number is $1\frac{2}{3}$?"

Translate to an equation in symbols and solve.

Check your work in **56**.

A MEMORY GIMMICK

If $A \times B = C$ then $A = \dfrac{C}{B}$ and $B = \dfrac{C}{A}$ for any numbers A, B, and C that are not zero.

Do you need a memory jogger? Try this one.

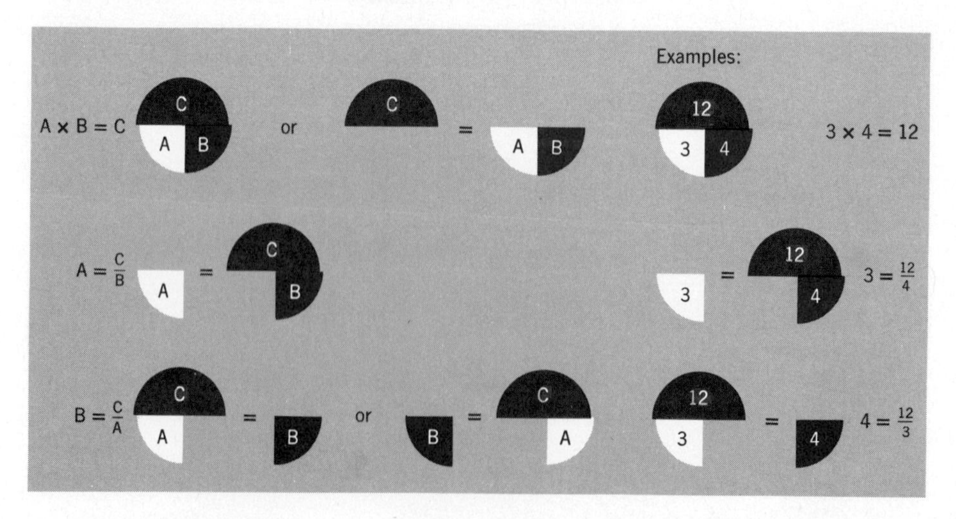

58 (a) $\left(\frac{3}{8}\right)$ (of) (what number) (is equal to) $\left(1\frac{5}{16}\right)$

$$\frac{3}{8} \times \square = 1\frac{5}{16}$$

$$\square = 1\frac{5}{16} \div \frac{3}{8}$$

$$\square = \frac{21}{16} \times \frac{8}{3}$$

$$\square = \frac{7}{2} = 3\frac{1}{2}$$

Check: $\frac{3}{8} \times \boxed{\frac{7}{2}} = \frac{21}{16} = 1\frac{5}{16}$

(b) (What fraction) (of) $\left(4\frac{1}{2}\right)$ (is) $\left(6\frac{3}{4}\right)$?

$$\square \times 4\frac{1}{2} = 6\frac{3}{4}$$

$$\square = 6\frac{3}{4} \div 4\frac{1}{2}$$

$$\square = \frac{27}{4} \times \frac{2}{9}$$

$$\square = \frac{3}{2} = 1\frac{1}{2}$$

Check: $\boxed{\frac{3}{2}} \times 4\frac{1}{2} = \frac{3}{2} \times \frac{9}{2} = \frac{27}{4} = 6\frac{3}{4}$

(c) Find (a number) such that $\left(\frac{3}{11}\right)$ (of) (it) (is) $\left(2\frac{2}{5}\right)$.

$$\frac{3}{11} \times \square = 2\frac{2}{5}$$

$$\square = 2\frac{2}{5} \div \frac{3}{11}$$

$$\square = \frac{12}{5} \times \frac{11}{3}$$

$$\square = \frac{44}{5} = 8\frac{4}{5}$$

Check: $\frac{3}{11} \times \boxed{\frac{44}{5}} = \frac{12}{5} = 2\frac{2}{5}$

(continued on the next page)

(d) The (product of) $\left(1\frac{7}{8}\right)$ and $\left(2\frac{1}{3}\right)$ (is) (what number)?

$$1\frac{7}{8} \times 2\frac{1}{3} = \square$$

$$\frac{15}{8} \times \frac{7}{3} = \square$$

$$\frac{35}{8} = \square$$

$$\square = 4\frac{3}{8}$$

(e) (What fraction) (of) $\left(8\frac{3}{4}\right)$ (is) $\left(\frac{7}{12}\right)$?

$$\square \times 8\frac{3}{4} = \frac{7}{12}$$

$$\square = \frac{7}{12} \div 8\frac{3}{4}$$

$$\square = \frac{7}{12} \div \frac{35}{4}$$

$$\square = \frac{7}{12} \times \frac{4}{35}$$

$$\square = \frac{1}{15}$$

Check: $\boxed{\frac{1}{15}} \times 8\frac{3}{4} = \frac{1}{15} \times \frac{35}{4} = \frac{7}{12}$

Another type of problem involving fractions is of this sort:

"If $6\frac{2}{3}$ dozen doorknobs cost \$100, what will 10 dozen doorknobs cost?"

First, make up a similar and much easier problem in order to get the feel of it. For example,

"If 2 apples cost 10¢, what will 6 apples cost?"

Solution: 2 cost 10¢
1 costs 10¢ ÷ 2 = 5¢
then 6 cost 6 × 5¢ = 30¢

In other words we solve the problem by dividing it into two parts. *First* find the unit cost, the cost of one item, then *second* find the cost of any number of items.

Apply this method to the doorknob problem above. Work exactly as you did in the simpler problem.

Check your work in **59**.

---

59 $6\frac{2}{3}$ cost \$100

$$1 \text{ costs } \left(\$100 \div 6\frac{2}{3}\right)$$

$$= \$100 \div \frac{20}{3}$$

$$= \$100 \times \frac{3}{20} = \$15$$

Then 10 cost $(10 \times \$15) = \$150$.

One way to check your answer is to compare it with what you might have guessed the answer to be before you did the arithmetic. For example, if $6\frac{2}{3}$ cost \$100, then 10 will cost almost double that, and the answer will be between \$100 and \$200. The actual answer \$150 is reasonable.

Here are a few practice problems for you.

(a) If $4\frac{1}{2}$ dozen pencils cost $2\frac{1}{4}$ dollars, what will $8\frac{1}{2}$ dozen cost?
(b) If you walk 24 miles in $7\frac{1}{2}$ hours, how many miles will you walk in 10 hours, assuming you go at the same rate?
(c) If you can swim $\frac{7}{8}$ of one mile in $\frac{2}{3}$ hour, how far can you swim in $2\frac{1}{2}$ hours at the same rate?

Look in **60** for the answers.

Similar Problem

2 cost 10¢
1 costs 10¢ ÷ 2 = 5¢
6 cost 6 × 5¢ = 30¢

(a) $4\frac{1}{2}$ cost $\$2\frac{1}{4}$

$$1 \text{ costs } \$2\frac{1}{4} \div 4\frac{1}{2} = \$\frac{9}{4} \div \frac{9}{2}$$

$$= \$\frac{9}{4} \times \frac{2}{9} = \$\frac{1}{2}$$

Then $8\frac{1}{2}$ cost $8\frac{1}{2} \times \$\frac{1}{2} = \frac{17}{2} \times \$\frac{1}{2} = \$\frac{17}{4} = \$4\frac{1}{4}$ or $4.25

$8\frac{1}{2}$ dozen cost $4.25.

(b) $7\frac{1}{2}$ hours ⟶ 24 miles

$$1 \text{ hour} \longrightarrow 24 \div 7\frac{1}{2} = 24 \times \frac{2}{15}$$

$$= \frac{16}{5} = 3\frac{1}{5} \text{ miles}$$

Then 10 hours ⟶ $10 \times \frac{16}{5} = 32$ miles

(c) $\frac{2}{3}$ hour for $\frac{7}{8}$ mile

$$1 \text{ hour} \longrightarrow \frac{7}{8} \div \frac{2}{3} = \frac{7}{8} \times \frac{3}{2}$$

$$= \frac{21}{16} \text{ mile}$$

$$2\frac{1}{2} \text{ hours} \longrightarrow \frac{21}{16} \times 2\frac{1}{2} = \frac{21}{16} \times \frac{5}{2} = \frac{105}{32}$$

$$= 3\frac{9}{32} \text{ hours}$$

Now go to **61** for some practice on word problems involving fractions.

Unit 2
Fractions

62

Self-Test

1. Write $7\frac{3}{16}$ as an improper fraction. ________
2. Write $\frac{37}{11}$ as a mixed number. ________
3. $\frac{3}{8} = \frac{?}{40}$ ________
4. Reduce to lowest terms: $\frac{195}{255}$ ________
5. $\frac{3}{5} + \frac{2}{7} =$ ________
6. $\frac{1}{4} + \frac{2}{3} + \frac{2}{5} =$ ________
7. $1\frac{3}{8} + 2\frac{1}{4} + 2\frac{2}{3} =$ ________
8. $\frac{3}{4} - \frac{1}{3} =$ ________
9. $2\frac{2}{5} - 1\frac{1}{4} =$ ________
10. $6\frac{2}{3} - 3\frac{1}{4} =$ ________
11. $\frac{9}{15} \times \frac{5}{3} =$ ________
12. $2\frac{2}{7} \times 2\frac{1}{4} =$ ________
13. $\frac{2}{3} \div \frac{3}{5} =$ ________
14. $3\frac{2}{7} \div 7\frac{1}{3} =$ ________
15. $1\frac{3}{16} \div 4\frac{3}{4} =$ ________
16. $\left(2\frac{1}{3}\right)^2 =$ ________
17. $1\frac{3}{5} \times 4\frac{7}{8} \times 7\frac{1}{2} =$ ________
18. What fraction of 16 is 7? ________
19. What fraction of $7\frac{1}{2}$ is 3? ________
20. What fraction of $4\frac{2}{3}$ is $3\frac{1}{2}$? ________

Date ________

Name ________

Course/Section ________

(*continued on next page*)

21. $\frac{7}{8}$ of what number is $1\frac{3}{4}$? ________

22. $1\frac{2}{3}$ of what number is $\frac{7}{15}$? ________

23. Find a number such that $\frac{2}{7}$ of it is $3\frac{1}{2}$. ________

24. What fraction of 50 is $4\frac{1}{2}$? ________

25. If $16\frac{2}{3}$ pounds of peanuts cost 60¢, what will 20 pounds cost? ________

Answers are on page 513.

SUPPLEMENTARY PROBLEM SET 2A

Multiplication of fractions

A. Multiply and reduce the answer to lowest terms:

1. $\frac{1}{2} \times \frac{1}{3} =$ ______
2. $\frac{2}{3} \times \frac{3}{4} =$ ______
3. $\frac{1}{5} \times \frac{2}{3} =$ ______
4. $\frac{5}{8} \times \frac{1}{2} =$ ______
5. $\frac{4}{5} \times \frac{3}{5} =$ ______
6. $3 \times \frac{1}{4} =$ ______
7. $\frac{5}{7} \times \frac{4}{10} =$ ______
8. $\frac{2}{3} \times 12 =$ ______
9. $\frac{21}{7} \times \frac{5}{6} =$ ______
10. $\frac{5}{9} \times \frac{3}{10} =$ ______
11. $\frac{7}{12} \times \frac{3}{15} =$ ______
12. $\frac{4}{7} \times \frac{20}{16} =$ ______
13. $\frac{2}{5} \times \frac{35}{20} =$ ______
14. $\frac{24}{9} \times \frac{15}{28} =$ ______
15. $\frac{3}{8} \times \frac{4}{9} =$ ______
16. $\frac{24}{35} \times \frac{5}{12} =$ ______

B. Multiply and reduce the answer to lowest terms:

1. $1\frac{1}{2} \times \frac{2}{3} =$ ______
2. $2\frac{1}{3} \times \frac{1}{4} =$ ______
3. $3\frac{2}{3} \times \frac{6}{5} =$ ______
4. $5\frac{1}{2} \times 4 =$ ______
5. $1\frac{2}{7} \times 1\frac{1}{3} =$ ______
6. $4\frac{3}{8} \times \frac{2}{5} =$ ______
7. $6\frac{2}{3} \times \frac{6}{5} =$ ______
8. $9\frac{4}{5} \times \frac{3}{7} =$ ______
9. $3\frac{1}{5} \times 1\frac{5}{8} =$ ______
10. $5\frac{3}{5} \times 4\frac{2}{7} =$ ______
11. $12 \times 1\frac{2}{3} =$ ______
12. $7\frac{4}{10} \times 2\frac{1}{2} =$ ______
13. $6\frac{2}{5} \times 3\frac{1}{8} =$ ______
14. $4\frac{1}{8} \times \frac{4}{11} =$ ______
15. $\frac{3}{7} \times \frac{1}{4} \times \frac{2}{3} =$ ______
16. $5\frac{1}{4} \times 2\frac{2}{5} \times 1\frac{2}{3} =$ ______
17. $\frac{1}{4} \times \frac{3}{5} \times \frac{1}{3} =$ ______
18. $6 \times 1\frac{1}{2} \times \frac{1}{9} =$ ______
19. $\left(4\frac{1}{2}\right)^2 =$ ______
20. $\left(1\frac{1}{2}\right)^3 =$ ______

C. Word Problems:

1. Water has a density of $62\frac{1}{2}$ pounds per cubic foot. Find the weight of $\frac{2}{5}$ cubic foot of water.

____

March 14, 1978
Date

Stephanie Catanzaro
Name

Math 50
Course/Section

2. How many miles can you travel on $8\frac{2}{5}$ gallons of gas if your car uses gas at the rate of $25\frac{1}{2}$ miles per gallon?

____

3. What is the area in square miles of a plot of land $1\frac{3}{8}$ miles wide by $3\frac{1}{5}$ miles long?

____

4. What is the total cost of $5\frac{3}{4}$ yards of fabric selling for $3 per yard?

____

5. Stock in the Golden Gadget Co. sells for $34\frac{5}{8}$ per share. What is the cost of eight shares?

____

SUPPLEMENTARY PROBLEM SET 3A

Dividing Fractions

A. Divide and reduce the answer to lowest terms:

1. $\frac{1}{2} \div \frac{3}{4} =$ ____
2. $\frac{2}{3} \div \frac{2}{5} =$ ____
3. $\frac{5}{8} \div \frac{2}{5} =$ ____
4. $2 \div \frac{1}{2} =$ ____
5. $\frac{1}{2} \div 2 =$ ____
6. $\frac{3}{5} \div \frac{3}{8} =$ ____
7. $\frac{3}{7} \div 3 =$ ____
8. $\frac{3}{11} \div \frac{3}{5} =$ ____
9. $1 \div \frac{1}{3} =$ ____
10. $1\frac{1}{3} \div \frac{1}{6} =$ ____
11. $2\frac{1}{2} \div \frac{2}{3} =$ ____
12. $2\frac{2}{3} \div 3 =$ ____
13. $3 \div 2\frac{2}{3} =$ ____
14. $3\frac{3}{5} \div 1\frac{4}{5} =$ ____
15. $2\frac{1}{7} \div 2\frac{1}{2} =$ ____
16. $8\frac{1}{4} \div 5\frac{1}{2} =$ ____
17. $16\frac{1}{4} \div 6\frac{1}{2} =$ ____
18. $7\frac{1}{7} \div 12\frac{1}{2} =$ ____
19. $4\frac{7}{8} \div 3\frac{1}{4} =$ ____
20. $9\frac{1}{7} \div 2\frac{2}{7} =$ ____

B. Divide:

1. $\dfrac{4}{\frac{1}{3}} =$ ____
2. $\dfrac{15}{\frac{3}{5}} =$ ____
3. $\dfrac{\frac{3}{4}}{12} =$ ____
4. $\dfrac{1\frac{2}{3}}{\frac{3}{5}} =$ ____
5. $\dfrac{1\frac{1}{2}}{2\frac{1}{2}} =$ ____
6. $\dfrac{2\frac{2}{3}}{\frac{2}{3}} =$ ____
7. $\frac{3}{5}$ divided by $\frac{1}{3} =$ ____
8. $1\frac{4}{5}$ divided by $1\frac{1}{2} =$ ____
9. Divide 8 by $1\frac{3}{5} =$ ____
10. Divide $3\frac{1}{3}$ by $5 =$ ____

C. Word Problems:

1. How many strips of wood $6\frac{3}{4}$ inch long can be cut from a strip 9 feet long?

____

2. What is the average speed of an automobile traveling $12\frac{1}{2}$ miles in $\frac{5}{12}$ hr?

____

March 14, 1978
Date

Stephanie Catanzaro
Name

Math 50
Course/Section

3. What is the volume of $12\frac{1}{2}$ pounds of water if the density of water is $62\frac{1}{2}$ pounds per cubic foot? (*Hint.* Divide $12\frac{1}{2}$ by $62\frac{1}{2}$.)

____

4. What is the diameter of a circle whose circumference is $5\frac{1}{2}$ feet? (*Hint.* Use $\pi = \frac{22}{7}$ and divide $5\frac{1}{2}$ by $\frac{22}{7}$)

____

5. The total weight of three apples is $17\frac{3}{4}$ oz. What is the average weight of an apple?

____

SUPPLEMENTARY PROBLEM SET 4A

Adding and Subtracting Fractions

A. Add or subtract as shown:

1. $\frac{1}{2} + \frac{1}{3} =$ ______
2. $\frac{2}{3} + \frac{3}{4} =$ ______
3. $\frac{3}{7} + \frac{1}{2} =$ ______
4. $\frac{1}{2} - \frac{1}{5} =$ ______
5. $\frac{7}{8} - \frac{1}{2} =$ ______
6. $\frac{5}{9} - \frac{2}{3} =$ ______
7. $\frac{7}{16} - \frac{1}{4} =$ ______
8. $\frac{3}{5} + \frac{2}{3} =$ ______
9. $\frac{7}{24} + \frac{1}{3} =$ ______
10. $\frac{1}{6} + \frac{3}{4} =$ ______
11. $\frac{2}{3}$ subtracted from $\frac{7}{8} =$ ______
12. $\frac{1}{2}$ reduced by $\frac{1}{3} =$ ______
13. $\frac{3}{5}$ less than $\frac{7}{8} =$ ______
14. subtract $\frac{2}{5}$ from $\frac{2}{3} =$ ______

B. Add or subtract as shown:

1. $1\frac{1}{2} + \frac{2}{3} =$ ______
2. $1\frac{2}{3} + 1\frac{1}{5} =$ ______
3. $2\frac{1}{4} - 1\frac{7}{8} =$ ______
4. $3\frac{3}{4} - 2\frac{1}{5} =$ ______
5. $3 - 1\frac{7}{8} =$ ______
6. $3\frac{4}{5} + 1\frac{1}{2} =$ ______
7. $4\frac{1}{7} - 3\frac{2}{3} =$ ______
8. $5\frac{1}{2} - 2\frac{1}{5} =$ ______
9. $4\frac{3}{7} + 1\frac{3}{4} =$ ______
10. $\frac{1}{2} + 1\frac{1}{3} + 2\frac{1}{4} =$ ______
11. $2\frac{5}{8} + 1\frac{2}{3} + 2\frac{3}{4} =$ ______
12. $4\frac{5}{8} + 2\frac{1}{4} + 1\frac{2}{5} =$ ______

C. Word Problems:

1. A dress pattern requires $3\frac{1}{4}$ yards of material. How much will be left if you start with $4\frac{3}{8}$ yards?

______

2. What is the total weight of three cartons whose gross weights are $14\frac{1}{4}$ pounds, $23\frac{1}{2}$ pounds, and $9\frac{3}{4}$ pounds?

______

Date ______

Stephanie Catanzaro
Name

Math 50
Course/Section

3. If you start a trip with $14\frac{1}{4}$ gallons of gas in the tank and end it with $3\frac{5}{8}$ gallons, how much gas did you use?

____

4. Two pieces of wood are glued together. What is the total thickness of the combination if one is $\frac{7}{8}$ inches thick and the other is $2\frac{9}{16}$ inches thick?

____

5. The opening market price of stock from the Golden Gadget Company was $34\frac{7}{8}$ and it closed at $32\frac{5}{16}$. What was the net drop in price?

____

6. A metal pipe is $16\frac{3}{4}$ inches long. How much must be cut off to leave a pipe $11\frac{3}{16}$ inches long?

____

SUPPLEMENTARY PROBLEM SET 5A

Word Problems

A. Write the following as mathematical expressions:

1. A number increased by $\frac{2}{7}$ ______
2. A number decreased by $\frac{2}{3}$ ______
3. The product of $\frac{3}{5}$ and a number ______
4. $\frac{1}{3}$ more than half of a number ______
5. A number divided by $\frac{2}{5}$ ______
6. $\frac{2}{3}$ of a number is equal to $1\frac{1}{4}$ ______
7. Some number divided by $\frac{3}{5}$ equals 6 ______
8. Some number is equal to the sum of 4 and $\frac{1}{2}$ of the number ______
9. A number increased by $\frac{1}{2}$ is the same as 2 less than double the number ______
10. Four times a certain number is $\frac{7}{8}$ ______
11. 8 decreased by a number ______
12. 8 less than a number ______
13. 8 less than a certain number ______
14. 8 subtracted from a number ______

B. Solve:

1. What fraction of $\frac{1}{2}$ is $\frac{1}{3}$? ______
2. What fraction of $\frac{2}{5}$ is $1\frac{1}{2}$? ______
3. What part of $2\frac{3}{4}$ is $1\frac{1}{2}$? ______
4. $\frac{2}{3}$ of what number is $\frac{3}{4}$? ______
5. $1\frac{2}{5}$ of what number is $2\frac{1}{2}$? ______
6. $3\frac{3}{8}$ of $\frac{8}{27}$ is what number? ______
7. $\frac{1}{4}$ of $1\frac{1}{4}$ is what number? ______

Date: Stephanie

Name: stephanie Catanzaro

Course/Section: Math 50

Study for final

8. Find a number such that $\frac{2}{3}$ of it is 1. ______

9. Find a number such that $\frac{3}{5}$ of it is $\frac{2}{3}$. ______

10. Find a number such that $1\frac{1}{3}$ of it is $2\frac{1}{4}$ ______

11. What fraction of 10 is $2\frac{1}{3}$? ______

12. What fraction of $2\frac{1}{2}$ is 4? ______

13. What part of 5 is $1\frac{1}{3}$? ______

14. What part of $3\frac{1}{5}$ is 5? ______

Decimals

PREVIEW 3

Objective	Sample Problems			Where To Go for Help	
Upon successful completion of this program you will be able to:				Page	Frame
1. Add and subtract decimal numbers.	(a)	$4.1672 + 17.009 + 2.9$	= ______	183	1
	(b)	$81.62 - 79.627$	= ______	183	1
2. Multiply and divide decimal numbers.	(a)	13.05×4.6	= ______	195	10
	(b)	210.84×3.4	= ______	195	10
	(c)	$3 \div 0.064$	= ______	201	16
	(d)	round 4.1563 to two decimal places	= ______	204	20
	(e)	$(10.3)^2$	= ______	195	10
3. Convert fractions to decimals and decimals to fractions.	(a)	Change $\frac{13}{35}$ to a decimal	= ______	213	25
	(b)	Change 7.325 to a fraction in lowest terms	= ______	213	25
4. Work with decimal fractions.	(a)	What decimal part of 16.2 is 4.131? (round to three decimal places)	= ______	213	25
	(b)	If 0.7 of a number is 12.67, find the number.	= ______	213	25
5. Add, subtract, multiply, and divide signed numbers.	(a)	$4 - (-7)$	= ______	225	35
	(b)	$8 - 14$	= ______	225	35
	(c)	$(-6.4) \times (-3.1)$	= ______	225	35
	(d)	$(-12.2) \div (-4.0)$	= ______	225	35
6. Calculate square roots.	(a)	$\sqrt{34.6}$ (to two decimal places)	= ______	235	44
	(b)	$\sqrt{315.4}$ (to two decimal places)	= ______	235	44

(Answers to these sample problems are on the back of this sheet.)

If you are certain you can work all of these problems correctly, turn to page 247 for a self test. If you want help with any of these objectives or if you cannot work one of the sample problems, turn to the page indicated. Super-students, who are eager to learn everything in this unit, will turn to frame **1** and begin work there.

PREVIEW 3

Answers to Sample Problems

1. (a) 24.0762
   (b) 1.993
2. (a) 60.03
   (b) 716.856
   (c) 46.875
   (d) 4.16
   (e) 106.09
3. (a) $0.3\overline{714285}$
   (b) $7\frac{13}{40}$
4. (a) 0.255
   (b) 18.1
5. (a) 11
   (b) -6
   (c) 19.84
   (d) 3.05
6. (a) 5.88
   (b) 17.76

Date ________________

Name ________________

Course/Section ________________

3 Decimals

PEANUTS
"SETS."....
"ONE TO ONE MATCHING."
10-5

"EQUIVALENT SETS."...
"NON-EQUIVALENT SETS."...
"SETS OF ONE."..."SETS OF TWO".

"RENAMING TWO."..."SUBSETS."..
"JOINING SETS."..."NUMBER SENTENCES."..."PLACEHOLDERS."..

ALL I WANT TO KNOW IS, HOW MUCH IS TWO AND TWO?
SCHULZ

1

ADDITION AND SUBTRACTION OF DECIMALS

Little Sally has allowed the words to get in the way of what she wants to learn. She may not be able to add 2 and 2 in class or recognize equivalent sets, but you can be certain she makes no mistakes checking her change at the local candy store or adding the money in her piggy bank. Words never interfere with the really important things.

In Unit 1 you learned that whole numbers are written in a place value system based on powers of ten. A number such as

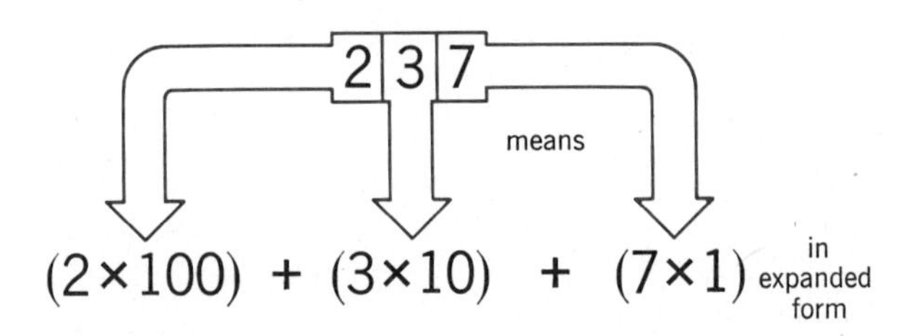
2 3 7
means
(2×100) + (3×10) + (7×1)
in expanded form

This way of writing numbers can be extended to fractions. A *decimal* number is a fraction whose denominator is a power of 10. For example,

Decimal form				Fraction form
.6	=	6 tenths	=	$\frac{6}{10}$
.05	=	5 hundredths	=	$\frac{5}{100}$
.32	=	32 hundredths	=	$\frac{32}{100}$
.004	=	4 thousandths	=	$\frac{4}{1000}$
.267	=	267 thousandths	=	$\frac{267}{1000}$

(Remember that a "power of 10" is simply a multiple of 10, that is, 10, 100, 1000, and so on.)

(*continued on the next page*)

We may also write the decimal number .267 in expanded form

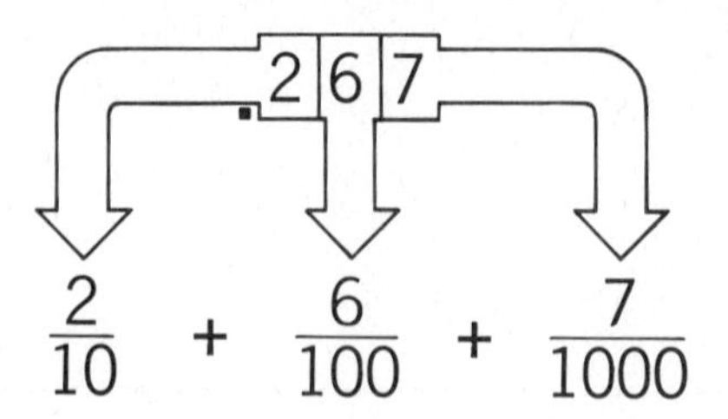

Write the decimal number .526 in expanded form. Check your answer in **2**.

What is a power of ten?
It is a product of 10s — 10, 10 x 10 = 100, 10 x 10 x 10 = 1000, and so on.

---

2 $.526 = \frac{5}{10} + \frac{2}{100} + \frac{6}{1000}$

Decimal notation enables us to extend the idea of place value to numbers less than one. A decimal number often has both a whole number part and a fraction part. For example, the number 324.576 is

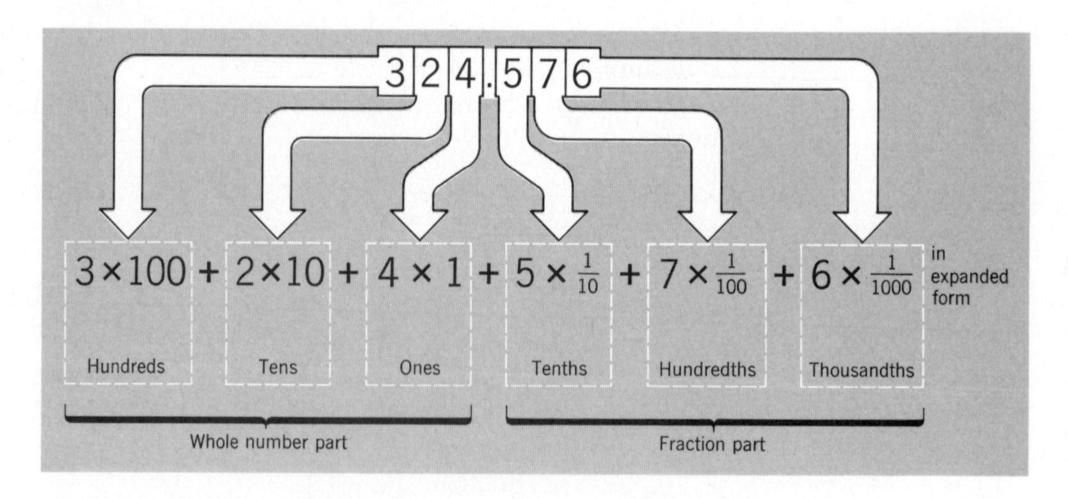

You are already familiar with this way of interpreting decimal numbers from working with money.

$243.78

means $2 \times \$100 + 4 \times \$10 + 3 \times \$1 + 7 \times 10¢ + 8 \times 1¢$

Write the following in expanded form:

(a) $86.42
(b) 43.607
(c) 14.5060
(d) 235.22267

Compare your answer with ours in **3**.

---

3 (a) $\$86.42 = 8 \times 10 + 6 \times 1 + 4 \times \frac{1}{10} + 2 \times \frac{1}{100}$
$= 80 + 6 + \frac{4}{10} + \frac{2}{100}$

(b) $43.607 = 4 \times 10 + 3 \times 1 + 6 \times \frac{1}{10} + 0 \times \frac{1}{100} + 7 \times \frac{1}{1000}$
$= 40 + 3 + \frac{6}{10} + \frac{0}{100} + \frac{7}{1000}$

(c) $14.5060 = 1 \times 10 + 4 \times 1 + \frac{5}{10} + \frac{0}{100} + \frac{6}{1000} + \frac{0}{10000}$

(d) $235.22267 = 2 \times 100 + 3 \times 10 + 5 \times 1 + \frac{2}{10} + \frac{2}{100} + \frac{2}{1000} + \frac{6}{10000} + \frac{7}{100000}$

Notice that the denominators in the decimal fractions increase by a factor of 10. For example,

3247 . 8956

3 × 1000	thousands	3000 . 0006	ten-thousandths	6 × 0.0001
2 × 100	hundreds	200 . 005	thousandths	5 × 0.001
4 × 10	tens	40 . 09	hundredths	9 × 0.01
7 × 1	ones	7 . 8	tenths	8 × 0.1

Each row changes by a factor of ten

$1 \times 10 = 10$ $\quad 0.01 \times 10 = 0.1$
$10 \times 10 = 100$ $\quad 0.001 \times 10 = 0.01$
$100 \times 10 = 1000$ $\quad 0.0001 \times 10 = 0.001$

In the decimal number 86.423 the digits 4, 2, and 3 are called *decimal digits.* The number 43.6708 has four decimal digits. All digits to the right of the decimal point, those that name the fractional part of the number, are decimal digits.

How many decimal digits are included in the numeral 324.0576?

Count them, then turn to **4**.

---

4 The number 324.0576 has four decimal digits, 0, 5, 7, and 6 the digits to the right of the decimal point.

Notice that the decimal point is simply a way of separating the whole number part from the fraction part; it is a place marker. In whole numbers the decimal point usually is not written, but its location should be clear to you.

2 = 2.

(The decimal point)

or 324 = 324.

(The decimal point)

Very often additional zeros are annexed to the decimal numbers without changing its value. For example,

8.5 = 8.50 = 8.5000 and so on
6 = 6. = 6.0 = 6.00 and so on

The value of the number is not changed but the additional zeros may be useful, as we shall see.

The decimal number .6 is often written 0.6. The zero added on the left is used to call attention to the decimal point. It is easy to mistake .6 for 6, but the decimal point in 0.6 cannot be overlooked.

Add the following decimal numbers:

$0.2 + 0.5 = \frac{2}{10} + \frac{5}{10} =$ ______

Try it, using what you know about adding fractions, then turn to **5**.

---

5 $0.2 + 0.5 = \frac{2}{10} + \frac{5}{10} = \frac{2+5}{10} = \frac{7}{10} = 0.7$

or, in other words,

$0.2 + 0.5 = 0.7$

Because decimal numbers represent fractions with denominators equal to powers of ten, addition is very simple.

$$\begin{array}{r} 2.34 = 2 + \frac{3}{10} + \frac{4}{100} \\ +5.23 = 5 + \frac{2}{10} + \frac{3}{100} \\ \hline 7 + \frac{5}{10} + \frac{7}{100} = 7.57 \end{array}$$

Add, using expanded form.

$$\begin{array}{r} 1.45 \\ +3.42 \\ \hline \end{array}$$

Check your answer in **6**.

6

$$
\begin{array}{r}
1.45 = 1 + \frac{4}{10} + \frac{5}{100} \\
+3.42 = 3 + \frac{4}{10} + \frac{2}{100} \\
\hline
4 + \frac{8}{10} + \frac{7}{100} = 4.87
\end{array}
$$

Of course we need not use expanded form in order to add decimal numbers. As with whole numbers, we may arrange the digits in vertical columns and add directly.

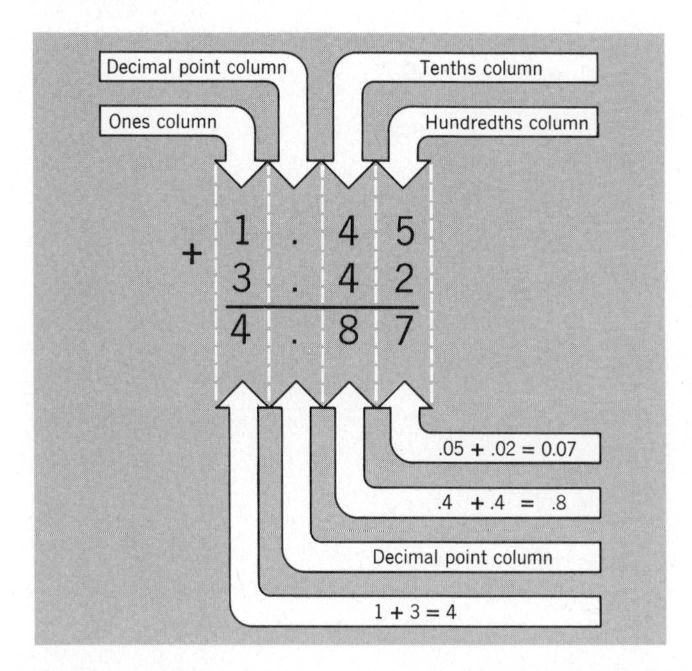

Digits of the same power of ten are placed in the same vertical column. Decimal points are always lined up vertically.

If one of the addends is written with fewer decimal digits than the other, annex as many zeros as needed to write both addends with the same number of decimal digits.

$$
\begin{array}{r} 2.345 \\ +1.5 \\ \hline \end{array} \quad \text{becomes} \quad \begin{array}{r} 2.345 \\ +1.500 \\ \hline \end{array}
$$

Except for the preliminary step of lining up decimal points, addition of decimal numbers is exactly the same process as addition of whole numbers.

Add the following decimal numbers.

(a) \$4.02 + \$3.67 = ______ (d) 14.6 + 1.2 + 3.15 = ______

(b) 13.2 + 1.57 = ______ (e) 5.7 + 3.4 = ______

(c) 23.007 + 1.12 = ______ (f) 42.768 + 9.37 = ______

Arrange each sum vertically, placing the decimal points in the same vertical column, then add as with whole numbers.

Check your work in **7**.

7

(a) Decimal points in line vertically

$$\begin{array}{r} \$4.02 \\ \$3.67 \\ \hline 7.69 \end{array}$$

$.02 + .07 = .09$ Add cents

$.0 + .6 = .6$ Add 10¢ units

$4 + 3 = 7$ Add dollars

As a check notice that the sum is roughly \$4 + \$3 or \$7, which agrees with the actual answer. Always check your answer by first estimating it, then comparing your estimate or rough guess with the final answer.

(b) Decimal points in line

$$\begin{array}{r} 13.20 \\ +\ 1.57 \\ \hline 14.77 \end{array}$$

Annex a zero to provide the same number of decimal digits as in the other addend.

Place answer decimal point in the same vertical line.

Check: $13 + 1 = 14$, which agrees roughly with the answer.

(c)
$$\begin{array}{r} 23.007 \\ +\ 1.120 \\ \hline 24.127 \end{array}$$

(d)
$$\begin{array}{r} 14.60 \\ 1.20 \\ +\ 3.15 \\ \hline 18.95 \end{array}$$

(e)
$$\begin{array}{r} \scriptstyle 1 \\ 5.7 \\ +3.4 \\ \hline 9.1 \end{array}$$

$.7 + .4 = 1.1$ Write .1 Carry 1

Carry 1 + 5 + 3 = 9

In expanded form

$$\begin{aligned} 5.7 &= 5 + \tfrac{7}{10} \\ 3.4 &= 3 + \tfrac{4}{10} \\ \hline &= 8 + \frac{7+4}{10} = 8 + \frac{11}{10} = 8 + \frac{10}{10} + \frac{1}{10} \\ &= 8 + 1 + \tfrac{1}{10} = 9 + \tfrac{1}{10} \\ &= 9.1 \end{aligned}$$

The carry 1 (pointing to $\frac{10}{10}$)

The carry 1 (pointing to the 1 in $8 + 1$)

(f)
$$\begin{array}{r} \scriptstyle 11\ 1\ \ \ \\ 42.768 \\ 9.370 \\ \hline 52.138 \end{array}$$

In expanded form

$$\begin{aligned} & 42 + \tfrac{7}{10} + \tfrac{6}{100} + \tfrac{8}{1000} \\ & + 9 + \tfrac{3}{10} + \tfrac{7}{100} + \tfrac{0}{1000} \\ \hline & 51 + \tfrac{10}{10} + \tfrac{13}{100} + \tfrac{8}{1000} \\ &= 51 + 1 + \tfrac{10}{100} + \tfrac{3}{100} + \tfrac{8}{1000} \\ &= 52 + \tfrac{1}{10} + \tfrac{3}{100} + \tfrac{8}{1000} \\ &= 52.138 \end{aligned}$$

$$\frac{10}{10} = 1$$

$$\frac{13}{100} = \frac{10+3}{100}$$

$$\frac{10}{100} = \frac{1}{10}$$

Beware! You must line up the decimal points carefully to be certain of getting a correct answer.

Subtraction is equally simple if you are careful to line up decimal points carefully and attach any needed zeros before you begin work.

For example, \$437.56 − \$41

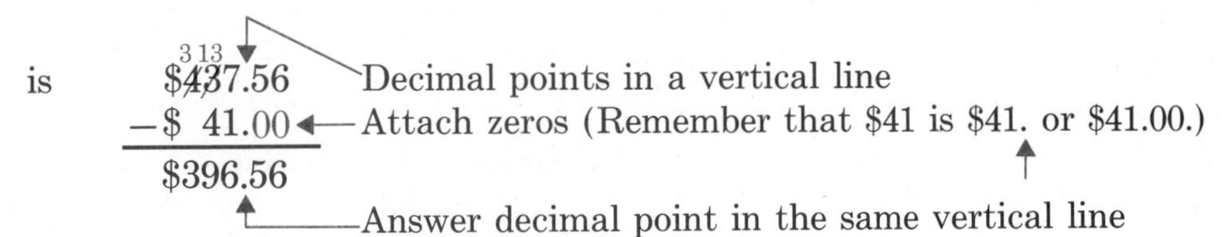

or again, 19.452 − 7.3615

```
   3 15 1 10
 19.4520      Decimal points in a vertical line
−  7.3615     Attach zero
 12.0905
Attach zeros (Remember that \$41 is \$41. or \$41.00.)
              Answer decimal point in the same vertical line
```

Try these problems to test yourself on subtraction of decimal numbers.

(a) \$37.66 − 14.57 = ______

(b) 248.3 − 135.921 = ______

(c) 6.4701 − 3.2 = ______

(d) 7.304 − 2.59 = ______

Work carefully. The answers are in **8**.

How do I do a problem like 90 −25.4?
Easy. Write 90 as 90.0, then line up the decimal points
90.0
−25.4 and subtract
90.0
−25.4
64.6 Then check your answer:
25.4 + 64.6 = 90.0

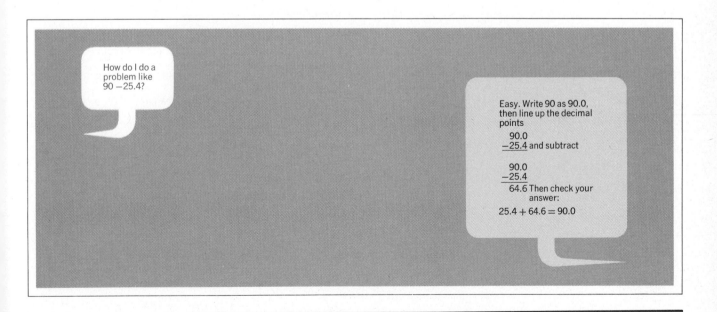

8

(a) $\begin{array}{r} \overset{5\ 16}{\$37.\cancel{6}\cancel{6}} \\ -\$14.57 \\ \hline 23.09 \end{array}$ ← Line up decimal points

Check: $14.57 + 23.09 = 37.66$

(b) $\begin{array}{r} 248.300 \\ -135.921 \\ \hline 112.379 \end{array}$ ← Line up decimal points; Attach zeros

Check: $135.921 + 112.379 = 248.300$

Answer decimal point in the same vertical line

(c) $\begin{array}{r} 6.4701 \\ -3.2000 \\ \hline 3.2701 \end{array}$ **Check:** $3.2000 + 3.2701 = 6.4701$

(d) $\begin{array}{r} 7.304 \\ -2.590 \\ \hline 4.714 \end{array}$ **Check:** $2.590 + 4.714 = 7.304$

Notice that each problem is checked by comparing the sum of the difference (answer) and subtrahend (number subtracted) with the minuend. Avoid careless mistakes by always checking your answer.

Now for a set of practice problems on addition and subtraction of decimal numbers turn to **9**.

---

4. The 400 meter race run in the Olympic games is 437.444 yards long. What is the difference between this distance and one-quarter of a mile (440 yards)?

____

5. A certain machine part is 2.345 inches thick. What is its thickness after 0.078 inches are ground off?

____

6. One minute is defined as exactly 60 seconds in sun time. One minute as measured by movement of the earth with respect to the stars is 59.836174 seconds. Find the difference between sun time and star time for one minute.

____

7. Can you balance a checkbook? At the start of a shopping spree your balance was $472.33. While shopping you wrote checks for $12.57, $8.95, $4, $7.48, and $23.98. What is your new balance?

____

8. Find the perimeter (distance around) the field shown. All distances are in meters.

____

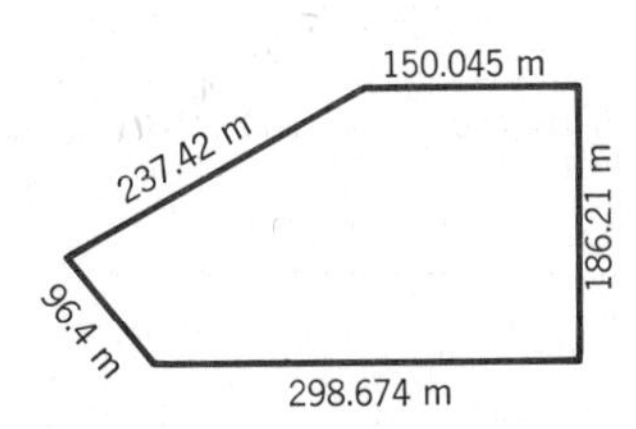

9. At the start of the semester Joe College bought a math textbook for $8.75, a history book for $3.49, a biology book for $12.50, and an English workbook for $7.40. What was the total cost of these four books?

____

10. How much change should you receive from a $20 bill if you bought groceries costing $7.93?

____

11. What is the difference between the following numbers?

(a) .05 and .005 ______

(b) 9.9 and 9.09 ______

(c) 4.32 and .432 ______

(d) 1 and 0.9994 ______

12. A bicycle normally selling for $139.45 is advertised at $15.50 off. What is the sale price? ______

When you have had the practice you need, either return to the preview on page 175 or continue in frame **10** with the study of multiplication and division of decimal numbers.

---

10 A decimal number is a fraction with a power of 10 as denominator. Multiplication of decimals should therefore be no more difficult than the multiplication of fractions.

Find $0.5 \times 0.3 =$ ______

Write out the two numbers as fractions and multiply, then choose an answer.

(a) 15 Go to **11**
(b) 1.5 Go to **12**
(c) .15 Go to **13**

---

11 You answered that $0.5 \times 0.3 = 15$ and that is incorrect. A wise first step would be to guess at the answer. Both 0.5 and 0.3 are less than 1; therefore their product is also less than 1, and 15 is not a reasonable answer.

Try calculating the sum this way:

$0.5 = \frac{5}{10}$ $\quad 0.3 = \frac{3}{10}$

$0.5 \times 0.3 = \frac{5}{10} \times \frac{3}{10}$

Complete this multiplication, then return to **10** and choose a better answer.

---

12 Your answer is incorrect. Don't get discouraged; we'll never tell.

The first step is to guess: since both 0.5 and 0.3 are less than 1, their product will be less than 1. Next, convert the decimals to fractions with 10 as denominator.

$0.5 = \frac{5}{10}$ $\quad 0.3 = \frac{3}{10}$

Finally, multiply:

$0.5 \times 0.3 = \frac{5}{10} \times \frac{3}{10} =$ ______

Complete this multiplication, then return to **10** and choose a better answer.

---

13 Excellent.

Notice that both 0.5 and 0.3 are less than 1; therefore their product will be less than 1. This provides a rough guess at the answer.

$0.5 = \frac{5}{10} \qquad 0.3 = \frac{3}{10}$

$$0.5 \times 0.3 = \frac{5}{10} \times \frac{3}{10} = \frac{5 \times 3}{10 \times 10} = \frac{15}{100} = 0.15$$

Of course it would be very, very clumsy and time-consuming to calculate every decimal multiplication in this way. We need a simpler method. Here is the procedure most often used.

Step 1. Multiply the two decimal numbers as if they were whole numbers. Pay no attention to the decimal points.

Step 2. The sum of the decimal digits in the factors will give you the number of decimal digits in the product.

For example,

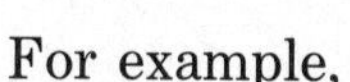

3.2 × .41 = ______

Step 1. Multiply without regard to the decimal points.

$$\begin{array}{r} 32 \\ \times 41 \\ \hline 1312 \end{array}$$

Step 2. Count decimal digits in the factors:

3.2 has *one* decimal digit (2)
.41 has *two* decimal digits (4 and 1)

The total number of decimal digits in the two factors is 3. The product will have *three* decimal digits. Count over *three* digits to the left in the product.

1.312

Three decimal digits

Check: 3.2 × .41 is roughly $3 \times \frac{1}{2}$ or about $1\frac{1}{2}$. The answer 1.3 agrees with our rough guess.

Try these simple decimal multiplications:

(a) 0.5 × 0.5 = ______ (b) 0.1 × 0.1 = ______

(c) 10 × 0.6 = ______ (d) 2 × 0.4 = ______

(e) 1 × 0.1 = ______ (f) 2 × 0.003 = ______

(g) 0.01 × 0.02 = ______ (h) 0.04 × 0.005 = ______

Follow the steps outlined above. Count decimal digits carefully. Check your answers in **14**.

HOW TO NAME DECIMAL NUMBERS

The decimal number 3,254,935.4728 should be interpreted as

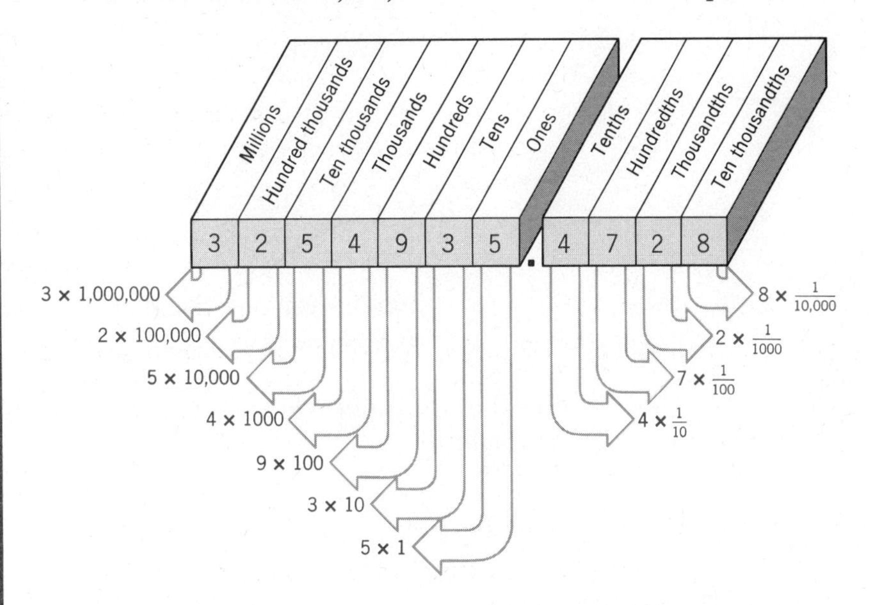

It may be read "three million, two hundred fifty-four thousand, nine hundred thirty-five, and four thousand one hundred twenty-eight ten thousandths."

Notice that the decimal point is read "and."

It is useful to recognize that, for example, the digit 8 represents 8 ten-thousandths or $\frac{8}{10,000}$ and the digit 7 represents 7 hundredths or $\frac{7}{100}$. Most often however, this number is read more simply as "three million, two hundred fifty-four thousand, nine hundred thirty-five, *point* four, seven, two, eight." This way of reading the number is easiest to write, to say, and to understand.

14 (a) $0.5 \times 0.5 =$ ______

First, multiply $5 \times 5 = 25$. Second, count decimal digits.

0.5×0.5

One decimal digit — One decimal digit = a total of *two* decimal digits.

Count over *two* decimal digits from the right

.25 The product is 0.25.

Two decimal digits

(*continued on the next page*)

Check: Both factors (0.5) are less than 1; therefore their product will be less than 1, and 0.25 seems reasonable.

(b) 0.1×0.1 $\quad 1 \times 1 = 1$

Count over *two* decimal digits from the right. Since there are not two decimal digits in the product attach a few on the left.

$1 \longrightarrow 0.01$

Two decimal digits

So $0.1 \times 0.1 = 0.01$

Check: $\frac{1}{10} \times \frac{1}{10} = \frac{1}{100}$ Ok.

(c) 10×0.6 $\quad 10 \times 6 = 60$

Count over *one* decimal digit from the right. 6.0 so that $10 \times 0.6 = 6.0$

Notice that multiplication by 10 simply shifts the decimal place one digit to the right.

$10 \times 6.2 = 62$
$10 \times 0.075 = 0.75$
$10 \times 8.123 = 81.23$ and so on.

(d) 2×0.4 $\quad 2 \times 4 = 8$

Count over *one* decimal digit.

.8 $\quad 2 \times 0.4 = 0.8$

(e) 1×0.1 (One decimal digit) $\quad 1 \times 1 = 1$

Count over *one* decimal digit

.1 $\quad 1 \times 0.1 = 0.1$

(f) 2×0.003 (Three decimal digits) $\quad 2 \times 3 = 6$

Count over *three* decimal digits

0.006 (Three decimal digits) $\quad 2 \times 0.003 = 0.006$

(g) 0.01×0.02 (Two decimal digits, Two decimal digits: Total of 4 decimal digits) $\quad 1 \times 2 = 2$

Count over *four* decimal digits

0.0002 (Four decimal digits) $\quad 0.01 \times 0.02 = 0.0002$

(h) 0.04×0.005 — Two decimal digits, Three decimal digits; Total of 5 decimal digits

$4 \times 5 = 20$

Count over *five* decimal digits

0.00020 — Five decimal digits

Remember...

Do not try to do this entire process mentally until you are certain you will not misplace zeros.

Always estimate before you begin the arithmetic, and finally check your answer against your estimate.

Multiplication of larger decimal numbers is performed in exactly the same manner. Try these:

(a) $4.302 \times 12.05 =$ ______

(b) $6.715 \times 2.002 =$ ______

(c) $3.144 \times 0.00125 =$ ______

Look in **15** for the answers.

15 (a) **Guess:** 4×12 is 48. The product will be about 48.

Multiply
```
    4302
   ×1205
 5183910
```
(If you cannot do this multiplication correctly turn to page 36 in Unit 1 for help with multiplication of whole numbers.)

The factors contain a total of five decimal digits (three in 4.302 and two in 12.05). Count over five decimal digits from the right in the product

51.83910

so that

$4.302 \times 12.05 = 51.8391$

Check: The answer 51.8391 is approximately equal to the guess 48.

(b) **Guess:** 6.7×2 is about 7×2 or 14.

Multiply
```
     6715
    ×2002
13.443430
```
Six decimal digits

6.715 has *three* decimal digits, 2.002 has *three* decimal digits, a total of *six* decimal digits.

$6.715 \times 2.002 = 13.44343$, which agrees with our guess.

Check: The answer and the guess are roughly equal.

(c) **Guess:** 3×0.001 is about 0.003.

Multiply
```
     3144
   ×  125
.00393000
```
Eight decimal digits

3.144 has *three* decimal digits, 0.00125 has *five* decimal digits, a total of *eight* decimal digits.

$3.144 \times 0.00125 = 0.00393$

Check: The answer and the guess are roughly equal.

Now go to **16** for a look at division of decimal numbers.

16 *Division* of decimal numbers is very similar to division of whole numbers. For example,

$6.8 \div 1.7$ can be written $\frac{6.8}{1.7}$

and if we multiply top and bottom of the fraction by 10,

$$\frac{6.8}{1.7} = \frac{6.8 \times 10}{1.7 \times 10} = \frac{68}{17}$$

$\frac{68}{17}$ is a normal whole number division.

$$\frac{68}{17} = 68 \div 17 = 4$$

Therefore $6.8 \div 1.7 = 4$ **Check:** $1.7 \times 4 = 6.8$

Rather than take the trouble to write the division as a fraction, we may use a short cut.

		Example
Step 1.	Write the divisor and dividend in standard long division form.	$6.8 \div 1.7$ $1.7\overline{)6.8}$
Step 2.	Shift the decimal point in the divisor to the right so as to make the divisor a whole number.	$1.7.\overline{)}$
Step 3.	Shift the decimal point in the dividend *the same amount.* (Add zeros if necessary.)	$1.7.\overline{)6.8.}$
Step 4.	Place the decimal point in the answer space directly above the new decimal position in the dividend.	$.$ $17.\overline{)68.}$
Step 5.	Complete the division exactly as you would with whole numbers. The decimal points in divisor and dividend may now be ignored.	$4.$ $17.\overline{)68.}$ $\underline{68}$

$6.8 \div 1.7 = 4$

Notice in steps 2 and 3 that we have simply multiplied both divisor and dividend by 10.

Repeat the process above with this division:

$1.38 \div 2.3$

Work carefully, then compare your work with ours in **17**.

---

17 Let's do it step by step.

$2.3\overline{)1.38}$ Shift the decimal point in the divisor 2.3 one digit to the right so that the divisor becomes a whole number. Then shift the decimal point in the dividend 1.38 the *same* number of digits. 2.3 becomes 23. 1.38 becomes 13.8. This is the same as multiplying both numbers by 10.

$2.3.\overline{)1.3.8}$

$23.\overline{)13.8}$ Place the answer decimal point directly above the decimal point in the dividend.

$$\begin{array}{r} .6 \\ 23.\overline{)13.8} \\ \underline{138} \end{array}$$

Divide as you would with whole numbers.

$6 \times 23 = 138$

$1.38 \div 2.3 = 0.6$ **Check:** $2.3 \times 0.6 = 1.38$

Never forget to check your answer.

How would you do this one?

$2.6 \div 0.052 =$ __?__

Look in **18** for the solution after you have tried it.

18

$.052.\overline{)2.6}$

To shift the decimal place three digits in the dividend, we must attach several zeros to its right.

$.052.\overline{)2.600.}$

Now place the decimal point in the answer space above that in the dividend.

$$\begin{array}{r} 50. \\ 52.\overline{)2600.} \\ \underline{260}\downarrow \\ 0 \\ \underline{0} \end{array}$$

$5 \times 52 = 260$

$2.6 \div 0.052 = 50$ **Check:** $0.052 \times 50 = 2.6$

Shifting the decimal point three digits and attaching zeros to the right of the decimal point in this way is equivalent to multiplying both divisor and dividend by 1000.

Try this problem set.

(a) $3.5 \div 0.001 =$ ______

(b) $9 \div 0.02 =$ ______

(c) $.365 \div 18.25 =$ ______

(d) $8.8 \div 3.2 =$ ______

The answers are in **19**.

19

(a) $0.001.\overline{)3.500.}$

$$1.\overline{)3500.} \quad \text{quotient } 3500.$$

$3.5 \div 0.001 = 3500$

Check: $0.001 \times 3500 = 3.5$

(b) $0.02.\overline{)9.00.}$

$$2.\overline{)900.} \quad \text{quotient } 450.$$

$9 \div 0.02 = 450$

Check: $0.02 \times 450 = 9$

(c) $18.25.\overline{).36.5}$

$$1825.\overline{)36.50} \quad \text{quotient } .02$$

$36\ 50 \qquad 2 \times 1825 = 3650$

$0.365 \div 18.25 = 0.2$

Check: $18.25 \times 0.2 = .365$

(continued on the next page)

(d) $3.2.\overline{)8.8.}$

$$
\begin{array}{r l}
2.75 & \\
32.\overline{)88.00} & \\
\underline{64} & 2 \times 32 = 64 \\
240 & \\
\underline{224} & 7 \times 32 = 224 \\
160 & \\
\underline{160} & 5 \times 32 = 160
\end{array}
$$

$8.8 \div 3.2 = 2.75$

Check: $3.2 \times 2.75 = 8.8$

If the dividend is not exactly divisible by the divisor, we must either stop the process after some preset number of decimal places in the answer or we must round the answer. We do not generally indicate a remainder in decimal division.

Turn to **20** for some rules for rounding.

---

20 *Rounding* is a process of approximating a number. To round a number means to find another number roughly equal to the given number but expressed less precisely. For example,

$432.57 = $400 rounded to the nearest hundred dollars,
= $430 rounded to the nearest ten dollars,
= $433 rounded to the nearest dollar,
and so on.

1.376521 is equal to 1.377 rounded to three decimal digits
or 1.4 rounded to the nearest tenth
or 1 rounded to the nearest whole number

There are exactly 5280 feet in 1 mile. To the nearest thousand feet how many feet are in one mile? To the nearest hundred feet?

Check your answers in **21**.

---

21 5280 ft = 5000 ft rounded to the nearest thousand feet. In other words 5280 is closer to 5000 than to 4000 or 6000.

5280 ft = 5300 ft rounded to the nearest hundred feet. In other words 5280 is closer to 5300 than to 5200 or 5400.

For most rounding follow this simple rule:

		Example
Step 1.	Determine the number of digits or the place to which the number is to be rounded. Mark it with a ‸ .	Round 3.462 to one decimal place 3.4‸62
Step 2.	If the digit to the right of the mark is less than 5, replace all digits to the right of the mark by zeros. If the zeros are decimal digits, you may discard them.	2.8‸32 becomes 2.800 or 2.8
Step 3.	If the digit to the right of the mark is equal to or larger than 5, increase the digit to the left by 1.	3.4‸62 becomes 3.5

Try it with these.

(a) Round 74.238 to two decimal places.
(b) Round 8.043 to two decimal places.
(c) Round 156 to the nearest hundred.
(d) Round 6.07 to the nearest tenth.

Follow the rules, round them, then check your work in 22.

The human brain is a fantastic machine, isn't it?
Sure. It starts working the moment you're born and never stops — until you pick up a math book.

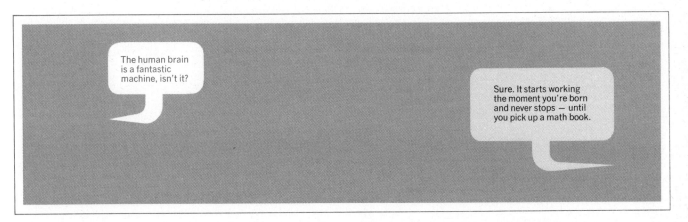

22 (a) 74.238 = 74.24 to two decimal places
(Write 74.23‸8; since 8 is larger than 5, increase the 3 to 4.)

(b) 8.043 = 8.04 to two decimal places
(Write 8.04‸3; since 3 is less than 5, drop the 3.)

(c) 156 = 200 to the nearest hundred
(Write 1‸56; since the digit to the right of the mark is 5, increase the 1 to 2.)

(d) 6.07 = 6.1 to the nearest tenth
(Write 6.0‸7; since 7 is greater than 5, increase the 0 to 1.)

There are a few very specialized situations where this rounding rule is *not* used:

1. Engineers use a more complex rule when rounding a number that ends in 5.
2. In business, fractions of a cent are usually rounded up in determining selling price. Three items for 25¢ or $8\frac{1}{3}$¢ each is rounded to 9¢ each.

Our rule will be quite satisfactory for most of your work in arithmetic.

Divide as shown and round your answer to two decimal places.

$6.84 \div 32.7 =$ ______

Careful now.

Check your work in **23**.

MULTIPLYING AND DIVIDING BY POWERS OF TEN

Many practical problems involve multiplying or dividing by 10, 100, or 1000. You will find it very useful to be able to multiply and divide by powers of 10 quickly and without using paper and pencil. The following rules will help.

1. To multiply a whole number by 10, 100, 1000, or a larger multiple of 10, attach as many zeros to the right of the number as there are in the multiple of 10.

   For example,

   $24 \times 10 = 240$

   Attach 1 zero

   $24 \times 100 = 2400$

   Attach 2 zeros

2. To multiply a decimal number by 10, 100, 1000, or a larger multiple of 10, move the decimal point as many places to the right as there are zeros in the multiplier.

   For example,

   $3.46 \times 10 = 34.6$

   Move the decimal point 1 place to the right

   $3.46 \times 100 = 346.$

   Move the decimal point 2 places to the right

   You may need to attach additional zeros before moving the decimal point.

   $2.4 \times 1000 = 2.400 \times 1000 = 2400.$

   Move the decimal point 3 places to the right

3. To divide a whole number or a decimal number by 10, 100, 1000, or a larger multiple of 10, move the decimal point as many places to the left as there are zeros in the divisor.

   $12.4 \div 10 = 1.24$

   Move the decimal point one place to the left

   $12.4 \div 100 = .124$

   Move the decimal point two places to the left

   You may need to attach additional zeros before moving the decimal point.

(*continued on the next page*)

$3.4 \div 100 = 03.4 \div 100 = .034$

Move the decimal point two places to the left

With a whole number the decimal point is usually not written and you must remember that it is understood to be after the units digit.

$4 = 4.$

$4 \div 100 = 4. \div 100 = .04$

Move the decimal point two places to the left

Here are a few problems for practice. Work quickly. No pencil or paper are needed. Do them in your head.

1. $4 \times 10 =$ ___
2. $64 \times 100 =$ ___
3. $16 \times 10{,}000 =$ ___
4. $3.5 \times 1000 =$ ___
5. $1.26 \times 10 =$ ___
6. $4.23 \times 100 =$ ___
7. $.4 \times 10 =$ ___
8. $.004 \times 100 =$ ___
9. $.075 \times 10 =$ ___
10. $.075 \times 1000 =$ ___
11. $1.257 \times 100 =$ ___
12. $2 \times 10{,}000 =$ ___
13. $45 \div 10 =$ ___
14. $376 \div 100 =$ ___
15. $82.1 \div 10 =$ ___
16. $82.1 \div 100 =$ ___
17. $82.1 \div 10{,}000 =$ ___
18. $4 \div 1000 =$ ___
19. $0.24 \div 10 =$ ___
20. $6 \div 100 =$ ___
21. $.035 \div 10 =$ ___

Answers are on page 503.

23 $32.7.\overline{)6.8.4}$

```
          .209
  327.)68.400
       65 4↓↓        2 × 327 = 654
        3 000
        2 943        9 × 327 = 2943
```

0.209 rounded to two decimal places is 0.21.

$6.84 \div 32.7 = 0.21$ rounded to two decimal places.

Check: $32.7 \times 0.21 = 6.867$, which is approximately equal to 6.84. (The check will not be exact because we have rounded.)

Go to **24** for a set of practice problems on multiplication and division of decimal numbers.

25 Since decimal numbers are fractions, they may be used, as fractions are used, to represent a part of some quantity. For example, recall that

"$\frac{1}{2}$ of 8 equals 4" means $\frac{1}{2} \times 8 = 4$

and therefore,

"0.5 of 8 equals 4" means $0.5 \times 8 = 4$

The word *of* used in this way indicates multiplication.

Find 0.35 of 8.4.

(If you need a review of problems of this kind, turn to frame **51**, page 157, in Unit 2 before continuing here.)

Go to **26** to check your answer to this problem.

---

26 $0.35 \times 8.4 =$ ______

```
   8.4 ←— 1 decimal digit
 ×0.35 ←— 2 decimal digits
  ----
   420
  252
  ----
 2.940 ←— A total of 3 decimal digits
```

Here is a second variety of problem:

What part of 16 is 4? ______

(What part → □; of → ×; 16 → 16; is → =; 4 → 4)

$\square \times 16 = 4$

$\square = 4 \div 16 = \frac{1}{4}$

What decimal part of 16 is 4?

(What decimal part → □; of → ×; 16 → 16; is → =; 4 → 4)

$\square \times 16 = 4$

$\square = 4 \div 16$

Dividing,

```
      .25
 16)4.00
    3 2
    ---
      80
      80
      --
```

$\square = 0.25$

Check: $0.25 \times 16 = 4.00$ or 4

What decimal part of 8 is 3?

Solve the problem as above, then hop to **28**.

---

27 (a) $\square \times 5 = 13$

$\square = 13 \div 5$

$\square = 2.6$

```
    2.6
 5)13.0
   10
    30
    30
```

Check: $2.6 \times 5 = 13.0$ or 13

(b) $0.8 \times \square = 10$

$\square = 10 \div 0.8$

$\square = 12.5$

```
        12.5
 0.8)10.00
      8
      20
      16
       40
       40
```

Check: $0.8 \times 12.5 = 10.00$ or 10

(c) $2.35 \times \square = 1.739$

$\square = 1.739 \div 2.35$

$\square = 0.74$

```
          .74
 2.35)1.7390
      1 645
        940
        940
```

Check: $2.35 \times 0.74 = 1.7390$ or 1.739

To convert a number from fraction form to decimal form, simply divide as indicated in the problems above. If the division has no remainder, the decimal number is called a *terminating* decimal.

For example, $\frac{5}{8} =$ ______

```
   .625
 8)5.000 ←— Attach as many zeros as needed
   4 8
     20
     16
      40
      40
         ←— Zero remainder; hence the decimal terminates or ends
```

$\frac{5}{8} = 0.625$

If a decimal does not terminate, you may round it to any desired number of decimal digits.

For example, $\frac{2}{13} =$ ______

Divide it out and round to three decimal digits.

Check your work in **29**.

28 What decimal part of 8 is 3?

$\square \times 8 = 3$

$\square = 3 \div 8$

$\square = 0.375$

$$\begin{array}{r} .375 \\ 8\overline{)3.000} \\ 2\,4 \\ \hline 60 \\ 56 \\ \hline 40 \end{array}$$

Check: $0.375 \times 8 = 3.000$ or 3.

Try these:

(a) What decimal part of 5 is 13?
(b) If 0.8 of a number is 10, find the number.
(c) If 2.35 of a number is 1.739, find the number.

Look in **27** for the answers.

29

$$\begin{array}{r} .1538 \\ 13\overline{)2.0000} \\ 1\,3 \\ \hline 70 \\ 65 \\ \hline 50 \\ 39 \\ \hline 110 \\ 104 \\ \hline 6 \end{array}$$

$\frac{2}{13} = 0.154$ rounded to three decimal digits.

Convert the following fractions to decimal form and round to two decimal digits.

(a) $\frac{2}{3} =$ ______ (b) $\frac{5}{6} =$ ______

(c) $\frac{17}{7} =$ ______ (d) $\frac{7}{16} =$ ______

Our work is in **30**.

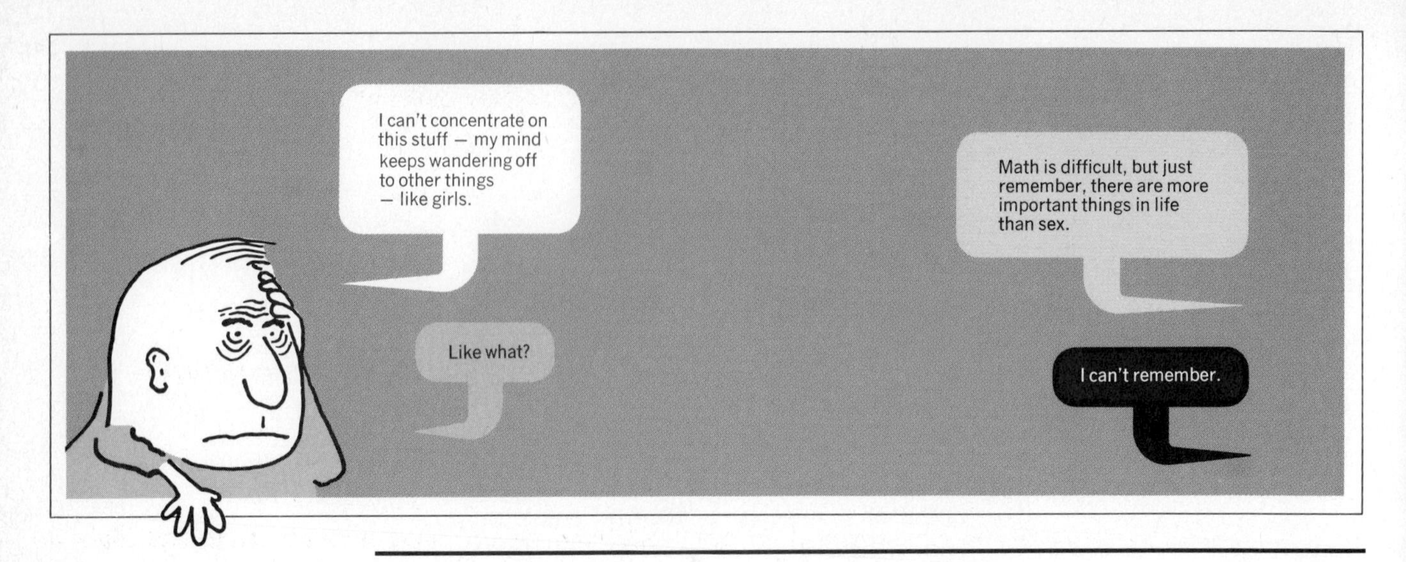

30

(a)
$$
\begin{array}{r}
.666 \\
3\overline{)2.000} \\
1\,8 \\
\hline
20 \\
18 \\
\hline
20 \\
18 \\
\hline
2
\end{array}
\qquad \frac{2}{3} = 0.67 \text{ rounded to two decimal digits}
$$

Notice that in order to round to two decimal digits, we must carry the division out to at least three decimal digits.

(b)
$$
\begin{array}{r}
.833 \\
6\overline{)5.000} \\
4\,8 \\
\hline
20 \\
18 \\
\hline
20 \\
18 \\
\hline
2
\end{array}
\qquad \frac{5}{6} = 0.83 \text{ rounded to two decimal digits}
$$

(c)
$$
\begin{array}{r}
2.428 \\
7\overline{)17.000} \\
14 \\
\hline
3\,0 \\
2\,8 \\
\hline
20 \\
14 \\
\hline
60 \\
56 \\
\hline
4
\end{array}
\qquad \frac{17}{7} = 2.43 \text{ rounded to two decimal digits}
$$

$$\begin{array}{r} .437 \\ 16\overline{)7.000} \\ \underline{6\,4} \\ 60 \\ \underline{48} \\ 120 \\ \underline{112} \\ 8 \end{array}$$

(d) $\frac{7}{16} = 0.44$ rounded to two decimal digits

Decimal numbers that do not terminate, repeat a sequence of digits. Such decimals are called *repeating decimals.* For example,

$\frac{1}{3} = 0.3333\ldots$

where the three dots are read "and so on" and indicate that the digit 3 continues without end.

Similarly, $\frac{2}{3} = 0.6666\ldots$ and $\frac{3}{11}$ is

$$\begin{array}{r} .2727 \\ 11\overline{)3.0000} \\ \underline{2\,2} \\ 80 \\ \underline{77} \\ 30 \\ \underline{22} \\ 80 \\ \underline{77} \\ 3 \end{array}$$

or $\frac{3}{11} = 0.272727\ldots$

Notice the remainder 3 is equal to the original dividend. This indicates that the decimal quotient repeats itself.

We may use a shorthand notation to show that the decimal repeats.

Write $\frac{1}{3} = 0.\overline{3}$ or $\frac{2}{3} = 0.\overline{6}$

where the bar means that the digits under the bar repeat endlessly.

$\frac{3}{11} = 0.\overline{27}$ means $0.272727\ldots$

$3\frac{1}{27} = 3.\overline{037}$ means $3.037037037\ldots$

Write $\frac{41}{33}$ as a repeating decimal using the "bar" notation. Check your answer in **31**.

31

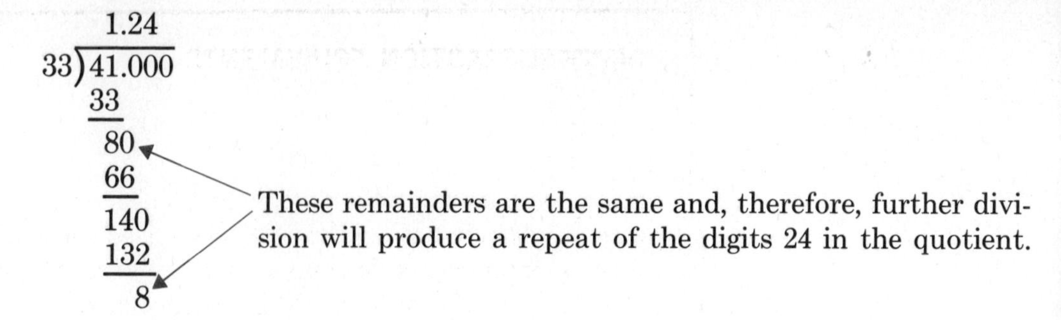

$\frac{41}{33} = 1.242424\ldots = 1.\overline{24}$

Converting decimal numbers to fractions is fairly easy.

$0.4 = \frac{4}{10}$ or $\frac{2}{5}$

$0.13 = \frac{13}{100}$

$0.275 = \frac{275}{1000} = \frac{13}{40}$ reduced to lowest terms

$0.035 = \frac{35}{1000} = \frac{7}{200}$ reduced to lowest terms

Follow this procedure:

Example

1. Write the digits to the right of the decimal point as the numerator in the fraction. — $0.00325 = ?$; $\frac{325}{?}$
2. In the denominator write 1 followed by as many zeros as there are decimal digits in the decimal number.

$0.00325 = \frac{325}{100000}$

5 digits ⟶ 5 zeros — $\frac{325}{100000}$

3. Reduce to lowest terms $\frac{325}{100000} = \frac{13 \times \cancel{25}}{4000 \times \cancel{25}} =$ — $\frac{13}{4000}$

Write 0.036 as a fraction in lowest terms.

Check your work in **32**.

DECIMAL-FRACTION EQUIVALENTS

Some fractions are used so often that it is worthwhile to list their decimal equivalents. Here are those used most often. Both rounded form and repeating decimal form are given. The items marked ★ are especially useful and should be memorized.

★ $\frac{1}{2} = 0.50$

★ $\frac{1}{3} = 0.\overline{3}$ or 0.33 rounded ★ $\frac{2}{3} = 0.\overline{6}$ or 0.67 rounded

★ $\frac{1}{4} = 0.25$ ★ $\frac{2}{4} = \frac{1}{2} = 0.50$ ★ $\frac{3}{4} = 0.75$

★ $\frac{1}{5} = 0.20$ ★ $\frac{2}{5} = 0.40$ ★ $\frac{3}{5} = 0.60$ ★ $\frac{4}{5} = 0.80$

$\frac{1}{6} = 0.1\overline{6}$ or 0.17 rounded $\frac{5}{6} = 0.8\overline{3}$ or 0.83 rounded

★ $\frac{1}{8} = 0.125$ $\frac{3}{8} = 0.375$ $\frac{5}{8} = 0.625$ $\frac{7}{8} = 0.875$

$\frac{1}{12} = 0.08\overline{3}$ $\frac{5}{12} = 0.41\overline{6}$ $\frac{7}{12} = 0.58\overline{3}$ $\frac{11}{12} = 0.91\overline{6}$

$\frac{1}{16} = 0.0625$ $\frac{3}{16} = 0.1875$ $\frac{5}{16} = 0.3121$ $\frac{7}{16} = 0.4375$

$\frac{9}{16} = .5625$ $\frac{11}{16} = 0.6875$ $\frac{13}{16} = 0.8125$

$\frac{1}{20} = 0.05$ $\frac{1}{25} = 0.04$ $\frac{1}{50} = 0.02$

32

$$0.036 = \frac{36}{1000} = \frac{9}{250}$$ reduced to lowest terms

(3 digits → 3 zeros)

If the decimal number has a whole number portion, convert the decimal part to a fraction first and then add the whole number part. For example,

$3.85 = 3 + 0.85$

and

$0.85 = \frac{85}{100} = \frac{17}{20}$ reduced to lowest terms

Therefore $3.85 = 3 + \frac{17}{20} = 3\frac{17}{20}$

Write the following decimal numbers as fractions in lowest terms.

(a) 0.0075 (b) 2.08 (c) 3.11

Check your work in **33**.

 study for final

HOW TO WRITE A REPEATING DECIMAL AS A FRACTION

A repeating decimal is one in which some sequence of digits is endlessly repeated. For example, $0.333\ldots = 0.\overline{3}$ and $0.272727\ldots = 0.\overline{27}$ are repeating decimals. The bar over the number is a shorthand way of showing that those digits are repeated.

What fraction is equal to $0.\overline{3}$? To answer this, form a fraction with numerator equal to the repeating digits and denominator equal to a number formed with the same number of 9s.

$0.\overline{3} = \frac{3}{9} = \frac{1}{3}$

$0.\overline{27} = \frac{27}{99} = \frac{3}{11}$ Two digits in $0.\overline{27}$; therefore use 99 as the denominator.

$0.\overline{123} = \frac{123}{999} = \frac{41}{333}$ Three digits in $0.\overline{123}$; therefore use 999 as the denominator.

33 (a) $0.0075 = \dfrac{75}{10000} = \dfrac{3}{400}$

(b) $2.08 = 2 + \dfrac{8}{100} = 2 + \dfrac{2}{25} = 2\dfrac{2}{25}$

(c) $3.11 = 3 + \dfrac{11}{100} = 3\dfrac{11}{100}$

Now turn to **34** for a set of practice problems on decimal fractions.

35 What kind of number would you use to name each of the following: a golf score two strokes below par, a loss of \$2 in a poker game, a debt of \$2, a loss of two yards on a football play, a temperature two degrees below zero, the year 2 B.C.? The answer is that they are all *negative numbers,* all equal to -2.

If we display the whole numbers as points along a line, we have:

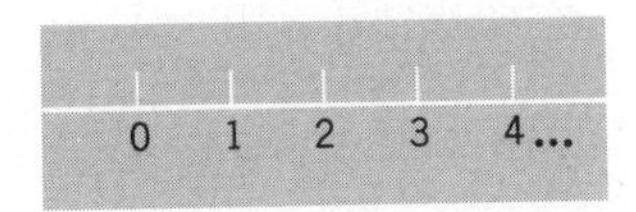

It is reasonable that we should be able to extend the line to the left and talk about numbers less than zero. Notice that every positive number is paired with a negative number.

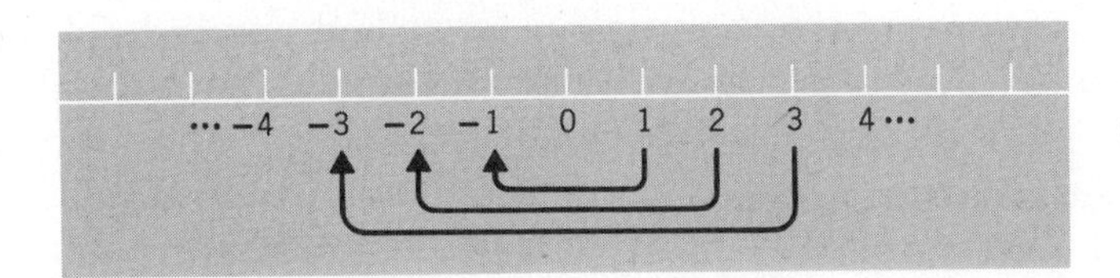

This number line can be drawn to any length (it should be infinitely long but we can draw only part of it), and we may choose any point on the line as a zero position. Using any convenient length unit, we mark off equal intervals to the right and left of the zero point. We call all numbers to the right of the zero *positive* numbers and all numbers to the left of the zero *negative* numbers. On the number line above we have labeled only the positive and negative integers . . . -3, -2, -1, 0, 1, 2, 3, . . . but of course other, non-integer, numbers such as $\frac{1}{2}$, -0.2, 4.2, or -3.17 may be located there as well.

Mark the following points on the number line above:

(a) $-2\frac{1}{2}$ (b) -0.4 (c) -1.2

Compare your answer with ours in **36**.

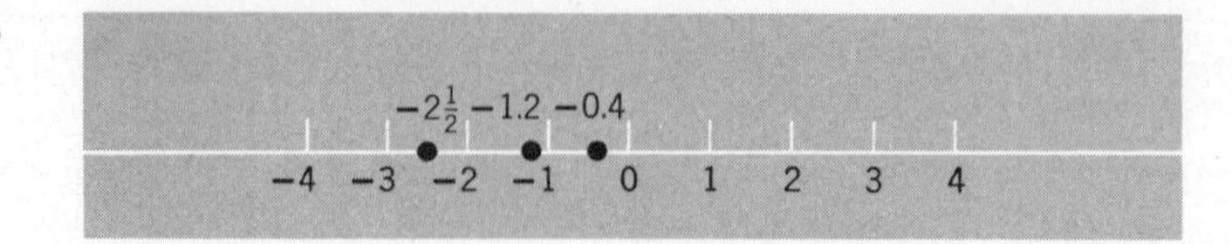

Signed numbers are a bookkeeping concept. They were used in bookkeeping by the ancient Greeks, Chinese, and Hindus more than 2000 years ago. Chinese merchants wrote positive numbers in black and negative numbers in red in their account books. (Being "in the red" still means losing money or having a negative income.) The Hindus used a dot or circle to show that a number was negative. We use a minus sign (−) to show that a number is negative, and because this symbol is also used for subtraction, some confusion results.

The + and − signs have a dual role in mathematics. With positive numbers + means "add" and − means "subtract." But these signs also are used to indicate position on a number line. The number +2 names a position on the line two units to the *right* of the zero. The number −2 names a position on the line two units to the *left* of the zero. A golf score of −2 is two strokes below par; the year 50 B.C. might be written as −50; Nitrogen freezes at a temperature of −210° C. If you receive a check for $8.00, your net gain is +8. If you send a check for $8.00, your net gain is −8.

> Read −8 as "negative 8" rather than "minus 8" and it may be less confusing for you.

Addition and subtraction of signed numbers may be pictured using the number line. A positive number may be represented on the line by an arrow to the right or positive direction. Think of a positive arrow as a "gain" of some quantity such as money. The sum of two numbers is the net "gain."

For example, the sum 4 + 3 or (+4) + (+3) is

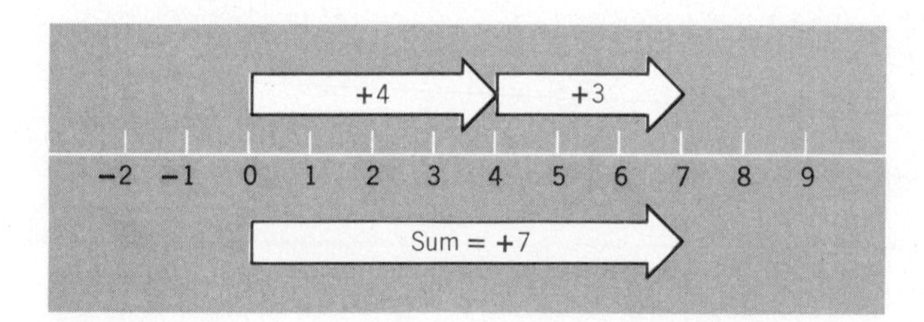

The numbers +4 and +3 are represented by arrows directed to the right. The +3 arrow begins where the +4 arrow ends. The sum 4 + 3 is the arrow that begins at the start of the +4 arrow and ends at the tip of the +3 arrow. In terms of money, a "gain" of $4 followed by a "gain" of $3 produces a net gain of $7.

Try setting up a number line to show the solution to this problem:

4 − 3 = ______

Check your answer in **37**.

Signed numbers don't make sense to me. How can you take 6 apples from 4 apples?

Suppose you are given 4 apples and owe someone 6 apples. Give them the 4 and now you owe 2.

+4 −6 = −2
Give 4 Owe 6 → Owe 2

37 $4 - 3$ is $(+4) - (+3)$ or $(+4) + (-3)$

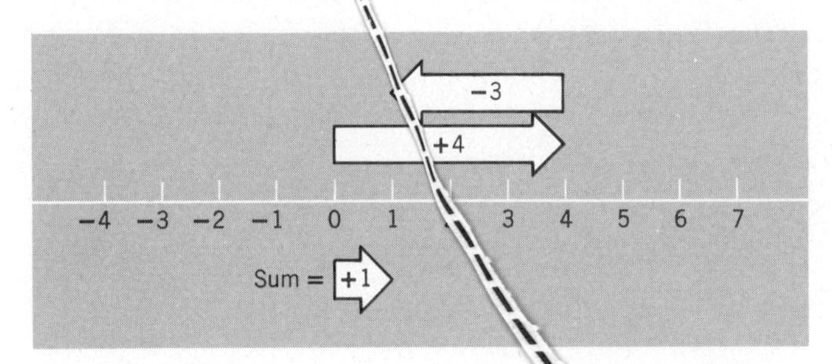

The number $+4$ is represented by an arrow directed to the right. The number -3 is represented by an arrow starting at the tip of $+4$ and directed to the left.

A "gain" of \$4 followed by a "loss" of \$3 produces a net "gain" of \$1.

Of course you did not need the number line to find that $4 - 3 = 1$, but this way of seeing negative numbers is helpful when the problems get more difficult.

Notice that we wrote the subtraction $4 - 3$ as a sum of signed numbers $(+4) + (-3)$. It is always true that

$$+3 = -(-3)$$

and $$-3 = -(+3)$$

Now set up this problem on a number line:

$3 - 8 =$ ______

Check your work in **39**.

38 (a) $11 - 14 = -3$

The signs are different $(+11 - 14)$. Subtract $(14 - 11 = 3)$ and attach the sign $(-)$ of the larger number (14).

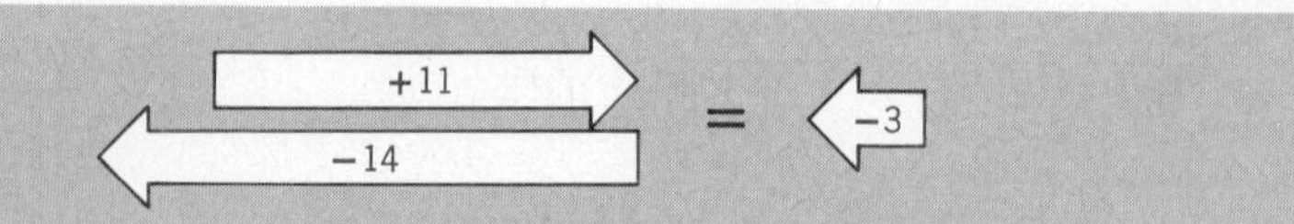

(b) $-6 - 5 = -11$

The signs are the same $(-6 - 5)$. Add $(6 + 5 = 11)$ and attach the sign $(-)$.

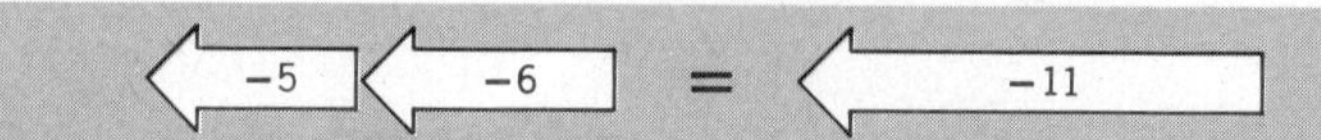

(c) $-10 - (-7) = -10 + 7 = -3$

Replace $-(-7)$ by $+7$. The signs are now different $(-10 + 7)$. Subtract $(10 - 7 = 3)$ and attach the sign $(-)$ of the larger number (10).

(d) $13 - (-6) = 13 + 6 = 19$

Replace $-(-6)$ by $+6$. The signs are the same $(+13 + 6)$. Add $(13 + 6 = 19)$ and attach the sign $(+)$.

 Notice that when no sign is present the + sign is implied.

$6 = +6$

Beware! Never use two signs together without parentheses. The phrase "+4 added to −2" is written $+4 + (-2)$ but *never* $+4 + -2$.

Now here are a few problems for practice:

(a) $3 - 7 =$ ______

(b) $-13 + 6 =$ ______

(c) $14 - (-18) =$ ______

(d) $3\frac{1}{2} - (-1\frac{1}{2}) =$ ______

(e) $-4 - (-12) =$ ______

(f) $\frac{1}{2} - 2\frac{1}{2} =$ ______

(g) $-9 - 4\frac{1}{2} =$ ______

(h) $17 - (+6) =$ ______

(i) $-13 - (3\frac{1}{4}) =$ ______

The answers are in **40**.

39 $3 - 8 = (+3) + (-8)$

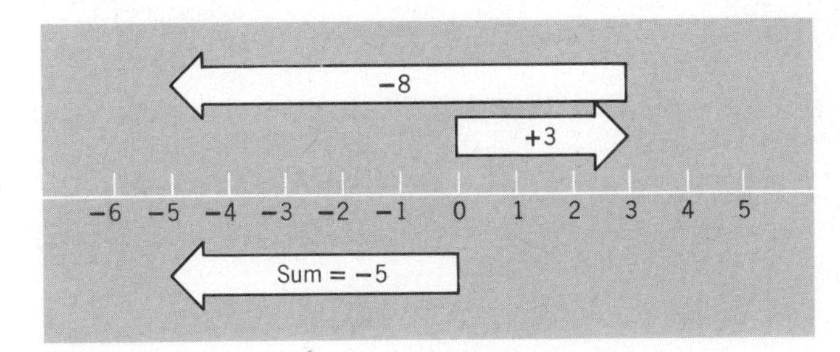

A "gain" of \$3 followed by a "loss" of \$8 produces a net loss of \$5. A loss of \$5 is equivalent to a "gain" of −\$5.

Number lines are a nice way of seeing what is happening when we add and subtract signed numbers, but we need a nice neat rule that involves no pictures or fancy imagination. Here it is:

Rule for adding or subtracting signed numbers:

★ 1. Replace any number of the form $-(-a)$ with $+a$. ★
★ 2. If both numbers have the *same* sign, add them and give the sum ★
★ the sign of the original numbers. ★
★ 3. If the two numbers have *opposite* signs, subtract them and give the ★
★ difference the sign of the number which is larger. ★

For example,

$8 + 2 = +10$	$+8$ and $+2$ have the same sign. Add and give the sum (10) the sign of the original numbers (+).
$8 - 2 = +6$	$+8$ and -2 have opposite signs. Take the difference (6) and give it the sign (+) of the larger number (8). (8 is larger than 2.)
$-7 + 3 = -4$	-7 and $+3$ have opposite signs. Take the difference (4) and give it the sign (−) of the larger number (7). (7 is larger than 3.)
$7 - (-3) = 7 + 3 = 10$	Replace $-(-3)$ by $+3$. Add.
$-7 - 12 = -19$	-7 and -12 have the same sign. Add and give the sum (19) the sign (−) of the original numbers.

Apply this rule to the following problems:

(a) $11 - 14 =$ ______ (b) $-6 - 5 =$ ______

(c) $-10 - (-7) =$ ______ (d) $13 - (-6) =$ ______

Check your work in **38**.

40

(a) -4	(b) -7	(c) 32
(d) $+5$	(e) $+8$	(f) -2
(g) $-13\frac{1}{2}$	(h) 11	(i) $-16\frac{1}{4}$

Multiplication of signed numbers

The product of two positive numbers presents nothing new. For example,

$(+3) \times (+4) = +12$

The product of numbers with opposite signs can be understood in terms of repeated addition. For example,

$(+3) \times (-4) = -12$

because

$3 \times (-4) = (-4) + (-4) + (-4)$

In the same way

$(-3) \times (+4) = -12$

because

$4 \times (-3) = (-3) + (-3) + (-3) + (-3)$

Finally, the product of two negative numbers is always positive. For example,

$(-3) \times (-4) = +12$ (For an explanation of this you may want to peek in the box on page 232.)

Following these examples, find:

(a) $(+5) \times (-7) =$ ______

(b) $(-2) \times (-8) =$ ______

(c) $(-6) \times (+5) =$ ______

Look in **41** for the answers.

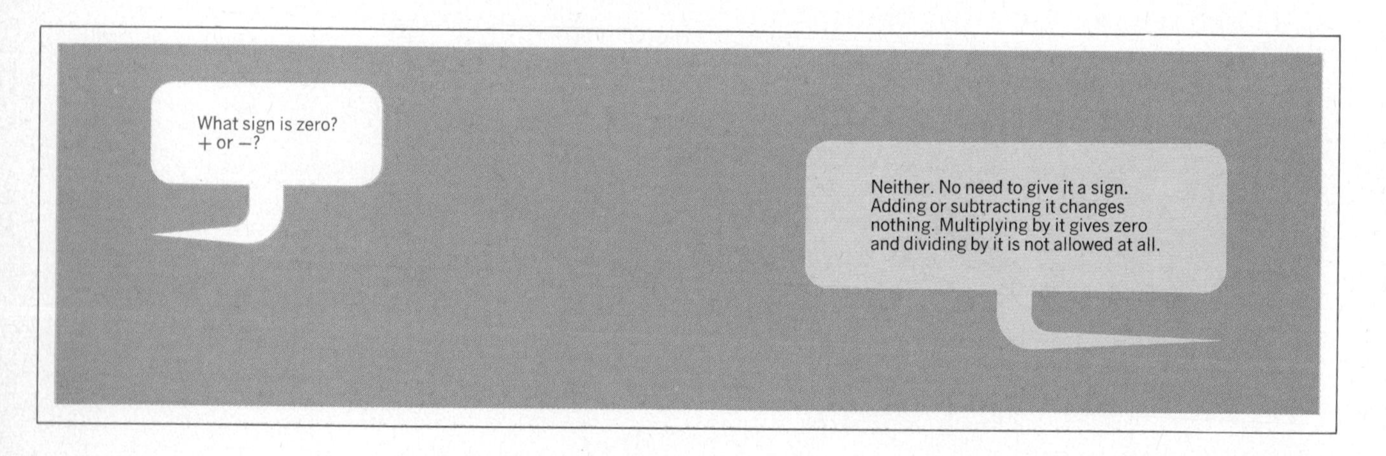

What sign is zero? + or −?
Neither. No need to give it a sign. Adding or subtracting it changes nothing. Multiplying by it gives zero and dividing by it is not allowed at all.

41 (a) -35 (b) $+16$ (c) -30

From these examples we may develop a general rule for the multiplication and division of signed numbers. Here it is:

> If both numbers have the *same* sign, the result of the operation will be a *positive* number; if both numbers have the *opposite* sign, the result of the operation will be a negative number.

We can summarize it this way:

Factors or Divisors	Product or Quotient
Same sign	+
Opposite sign	−

For multiplication,

$$\text{Same sign} \begin{cases} (4) \times (3) = +12 \\ (-4) \times (-3) = +12 \end{cases}$$

$$\text{Opposite sign} \begin{cases} (-4) \times (3) = -12 \\ (4) \times (-3) = -12 \end{cases}$$

For division:

$$\text{Same sign} \begin{cases} (8) \div (2) = +4 \\ (-8) \div (-2) = +4 \end{cases}$$

$$\text{Opposite sign} \begin{cases} (-8) \div (2) = -4 \\ (8) \div (-2) = -4 \end{cases}$$

Work these problems for practice in this very important rule.

(a) $(-1.2) \times (-3)$ (b) $(0.02) \times (-1.5)$ (c) $(-2.4) \times (3.5)$
(d) $(-1.5) \times (-2)$ (e) $(-3)^2$ (f) $(-2)^3$
(g) $(-4.5) \div (-5)$ (h) $(-20) \div (4)$ (i) $(3.6) \div (-2)$
(j) $(-7.2) \div (6)$ (k) $(-27) \div (-3)$ (l) $(4.2) \div (-0.07)$

Check your answers in **42**.

WHY DOES $(-1) \times (-1) = +1$?

Think of it this way:

Find the multiples of 3

$$\begin{array}{ccccccccc} \ldots(-4) & (-3) & (-2) & (-1) & 0 & 1 & 2 & 3 & 4\ldots \\ \times 3 & \times 3 & \times 3 & \times 3 & \times 3 & \times 3 & \times 3 & \times 3 & \times 3 \\ \hline \ldots -12 & -9 & -6 & -3 & 0 & 3 & 6 & 9 & 12 \end{array}$$

← The multiples of 3

Notice that each multiple is 3 more than its neighbor to the left.

Now find the multiples of -3

$$\begin{array}{cccc} \ldots\ (-4) & (-3) & (-2) & (-1) \\ \times(-3) & \times(-3) & \times(-3) & \times(-3) \\ \hline ? & ? & ? & ? \end{array}$$

$$\begin{array}{ccccc} 0 & 1 & 2 & 3 & 4\ \ldots \\ \times(-3) & \times(-3) & \times(-3) & \times(-3) & \times(-3) \\ \hline 0 & -3 & -6 & -9 & -12 \end{array}$$

← Multiples of (-3)

Everything is dandy for $0 \times (-3)$, $1 \times (-3)$, $2 \times (-3)$, and so on, but what about $(-1) \times (-3)$, $(-2) \times (-3)$, and so on? Because these are multiples of 3, we know that each multiple differs from its neighbors by 3. In this case the neighbor to the left must be larger by 3. Therefore,

$(-1) \times (-3) = +3$
$(-2) \times (-3) = +6$
$(-3) \times (-3) = +9$ and so on, in order to maintain the pattern.

Then $(-1) \times (-1) = +1$ or in general $(-a) \times (-a) = +a^2$ for any number a.

42 (a) $+3.6$ (b) -0.03 (c) -8.4
(d) $+3$ (e) $+9$ (f) -8
(g) $+0.9$ (h) -5 (i) -1.8
(j) -1.2 (k) $+9$ (l) -60

For more practice in the arithmetic of signed numbers turn to **43** for a problem set.

44 As we saw in Unit 1, a *perfect square* is a whole number that can be written as the product of equal whole numbers. For example, 25 is a perfect square because

$25 = 5 \times 5$

and 15,129 is a perfect square because

$15{,}129 = 123 \times 123$

Finding the square of any number is a simple process, but the reverse is difficult. The *square root* of a number is that number which when squared produces the original number. For example, the square root of 25, written $\sqrt{25}$, is 5 since

$5 \times 5 = 25$

$\sqrt{121} = 11$ since $11 \times 11 = 121$
$\sqrt{100} = 10$ since $10 \times 10 = 100$ and so on.

Notice that $(-5) \times (-5)$ also equals 25. Every positive number has two square roots, one positive and one negative. When we speak of *the* square root of a number, we are referring to its positive square root.

Find $\sqrt{36}$

Try it, then turn to **45**.

---

45 $\sqrt{36} = 6$ since $6 \times 6 = 36$

In general the square root of a number will not be a whole number or exact decimal. For most numbers we cannot write an exact value for the square root; instead we must approximate it.

For example, find an approximate value for $\sqrt{2}$.

Try to find a value for it, then turn to **46** and continue.

---

46 $1 \times 1 = 1$ or $\sqrt{1} = 1$

and $2 \times 2 = 4$ or $\sqrt{4} = 2$

Therefore $\sqrt{2}$ will be some number between 1 and 2. A guess of roughly 1.5 would be reasonable.

$1.5 \times 1.5 = 2.25$

A guess of 1.4 would be even better

$1.4 \times 1.4 = 1.96$ and 1.96 is quite close to 2

We can approximate $\sqrt{2}$ as closely as we wish.

$\sqrt{2} \cong 1.41$ ($\cong$ means "approximately equal")
since $1.41 \times 1.41 = 1.9881$

$\sqrt{2} \cong 1.414$ since $1.414 \times 1.414 = 1.999396$

There is no decimal number exactly equal to $\sqrt{2}$.

How do we find the square root of any given number? The easiest way is to consult a ready-made table of square roots. One such table is available on page 517 of this book. Use it to calculate the following:

(a) $\sqrt{86}$ (b) $\sqrt{130}$ (c) $\sqrt{174}$

Compare your answers with ours in **48**.

---

47 4.5 would be a good first guess.

$\sqrt{19}$ is a number between 4 and 5 since $4^2 = 16$ and $5^2 = 25$, and 19 is between 16 and 25.

First guess: $\sqrt{19} \cong 4.5$

First check:

```
        4.22
4.5.)19.0.00
     180
      100
       90
       100
        90
```

Notice that if 4.5 were exactly equal to $\sqrt{19}$, then the quotient would also be exactly 4.5.

To find a better second guess, take the average of the first guess 4.5 and the quotient 4.22.

Second guess: $\frac{1}{2}(4.5 + 4.22) = \frac{1}{2}(8.72) = 4.36$

Now check this new guess by dividing it into 19 and obtain a better third guess. Do it here . . .

. . . then turn to **49** to see if you have done it correctly.

48 (a) $\sqrt{86} \cong 9.2736$ **Check:** $(9.2736)^2 \cong 85.9996$
(b) $\sqrt{130} \cong 11.4018$ **Check:** $(11.4018)^2 \cong 130.0010$
(c) $\sqrt{174} \cong 13.1909$ **Check:** $(13.1909)^2 \cong 173.9998$

($\cong$ means "roughly equal to", . . . remember?)

Always check your answer by squaring when you have found a square root.

How do you determine the square root of a number not in the table of square roots or not a whole number? The answer is that you use the *guess n' check* method of finding square roots. The guess n' check method requires that you *guess* at the answer and then refine your guess by *check*ing it to get a closer second guess.

For example, find $\sqrt{5}$.

First guess: $\sqrt{5} \cong 2$

First check:

$$\begin{array}{r} 2.5 \\ 2\overline{)5.0} \\ \underline{4} \\ 10 \\ \underline{10} \end{array}$$

Divide the number whose square root you want by the guess. Take the quotient to one more place than the guess.

Second guess: $\frac{1}{2}(2 + 2.5) = \frac{1}{2}(4.5) = 2.25$

(The second guess is the average of the first guess and the quotient of the first check.)

2.25 is a much better approximation to $\sqrt{5}$ than is 2.

Second check:

$$\begin{array}{r} 2.222 \\ 2.25\overline{)5.00000} \\ \underline{4.50} \\ 500 \\ \underline{450} \\ 500 \\ \underline{450} \\ 500 \\ \underline{450} \\ 50 \end{array}$$

Divide again by the guess. Take the quotient to one more place than the guess.

Third guess: $\frac{1}{2}(2.25 + 2.222) = \frac{1}{2}(4.472) = 2.236$

2.236 is equal to $\sqrt{5}$ to three decimal places.

⇨ Continue the guess n' check process until you find that the digits in the quotient and the guess are the same to whatever number of places you desire.

Try another one. What would be a good first guess to $\sqrt{19}$?

Guess, then turn to **47**.

49 **First guess:** $\sqrt{19} \cong 4.5$

```
                4.22
First check:   4.5)19.000
```

Second guess: $\frac{1}{2}(4.5 + 4.22) = 4.36$

```
                  4.3578
Second check:  4.36)19.000000
                    17 44
                    ─────
                     1560
                     1308
                     ────
                      2520
                      2180
                      ────
                       3400
                       3052
                       ────
                        3480
```

Notice that the quotient 4.3578 when rounded to two decimal digits is 4.36, the same as the second guess. Therefore the answer is

$\sqrt{19} \cong 4.36$ to two decimal digits.

If a more accurate value for $\sqrt{19}$ is needed simply continue the guess n' check process through one or two more cycles.

Here is a summary of the *guess n' check* method for calculating the square root of a number.

Step 1. **Guess** at the square root.

Step 2. **Check** the guess by dividing the number by the guess.

Step 3. Take the quotient to one more place than the guess.

Step 4. Obtain a new and better guess by averaging the first guess and the quotient.

Step 5. Repeat the process using the new guess.

Stop the process when the quotient and the previous guess are the same to the desired number of decimal places.

Now apply this method to find $\sqrt{41}$ to two decimal digits.

Check your work in **50**.

THE MOST OFTEN USED SQUARE ROOTS

$$\sqrt{2} \simeq 1.4142 \simeq \tfrac{7}{5} \text{ or even closer } \tfrac{17}{12}$$
$$\sqrt{3} \simeq 1.7321 \simeq \tfrac{7}{4} \text{ or even closer } \tfrac{19}{11}$$
$$\sqrt{5} \simeq 2.2361 \simeq \tfrac{9}{4}$$
$$\sqrt{6} \simeq 2.4495 \simeq \tfrac{22}{9}$$
$$\sqrt{7} \simeq 2.6547 \simeq \tfrac{8}{3}$$
$$\sqrt{8} \simeq 2.8284 \simeq \tfrac{14}{5} \text{ or even closer } \tfrac{17}{6}$$
$$\sqrt{10} \simeq 3.1623 \simeq \tfrac{19}{6}$$

---

50 **First guess:** $\sqrt{41} \simeq 6.5$ (A good first guess would be any number between 6 and 7.)

First check:

```
        6.307
6.5 ) 41.00
      39 0
       2 00
       1 95
         500
```

Round the quotient to 6.31.

Second guess: $\frac{1}{2}(6.5 + 6.31) = \frac{1}{2}(12.81) = 6.405$

Second check:

```
          6.401
6.405 ) 41.000000
        38 430
         2 5700
         2 5620
             8000
             6405
```

Third guess: $\frac{1}{2}(6.405 + 6.401) = \frac{1}{2}(12.806) = 6.403$

To two decimal digits $\sqrt{41} \simeq 6.40$.

Even if you make a very bad first guess, the *guess n' check* method will quickly lead you to the correct answer. For example, suppose you wrote:

First guess: $\sqrt{41} \simeq 5$ (A very poor guess.)

First check:

```
    8.2
5 ) 41.0
```

Second guess: $\frac{1}{2}(5 + 8.2) = \frac{1}{2}(13.2) = 6.6$ (. . . getting closer.)

(continued on the next page)

Second check:

$$\begin{array}{r} 6.21 \\ 6.6\overline{)41.000} \\ \underline{39\,6} \\ 1\,40 \\ \underline{1\,32} \\ 80 \end{array}$$

Third guess: $\frac{1}{2}(6.6 + 6.21) = \frac{1}{2}(12.81) = 6.405$

This answer is very close to the correct value of $\sqrt{41}$.

$(6.405)^2 \cong 41.02$

The *guess n' check* method is goof-proof if you do all the steps in proper order.

Find $\sqrt{53}$. Start with a first guess of 7. What will be your second guess?

(a) 7.571 Go to **51**
(b) 7.6 Go to **52**
(c) 7.3 Go to **53**

51 Not quite.

First guess: $\sqrt{53} \cong 7$

$$
\begin{array}{r}
7.571 \\
7\overline{)53.00} \\
\underline{49} \\
4\,0 \\
\underline{3\,5} \\
50 \\
\underline{49} \\
10
\end{array}
$$

First check: Round 7.571 to 7.6.

To find the best second guess find the average of your first guess, 7, and the quotient of the division, 7.6 $\frac{1}{2}(7 + 7.6) = ?$

Try it, then return to **50** and choose a better answer.

---

52 Nope.

First guess: $\sqrt{53} \cong 7$

$$
\begin{array}{r}
7.57 \\
7\overline{)53.00} \\
\underline{49} \\
4\,0 \\
\underline{3\,5} \\
50
\end{array}
$$

First check: Round 7.57 to 7.6.

Now, to get the best second guess find the average of the first guess, 7, and the quotient of the division, 7.6. $\frac{1}{2}(7 + 7.6) = ?$

Try it, then return to **50** and choose the correct answer.

---

53 Very good!

First guess: $\sqrt{53} \cong 7$

First check: $7\overline{)53.00}$ with quotient 7.57. Round it to 7.6.

Second guess: $\frac{1}{2}(7 + 7.6) = \frac{1}{2}(14.6) = 7.3$

Now what would be the third guess?

Work it out, then look in **54**.

54

Second check:

$$\begin{array}{r} 7.260 \\ 7.3\overline{)53.000} \\ \underline{51\,1} \\ 1\,90 \\ \underline{1\,46} \\ 440 \\ \underline{438} \end{array}$$

Third guess: $\frac{1}{2}(7.3 + 7.260) = \frac{1}{2}(14.560) = 7.280$

Are you ready to apply the *guess n' check* method to a set of problems? If not, return to **44** and review. Otherwise, try these.

Use the *guess n' check* method to find the following square roots to two decimal digits.

(a) $\sqrt{13.6}$
(b) $\sqrt{8.25}$
(c) $\sqrt{130.5}$

The worked solutions are in **55**.

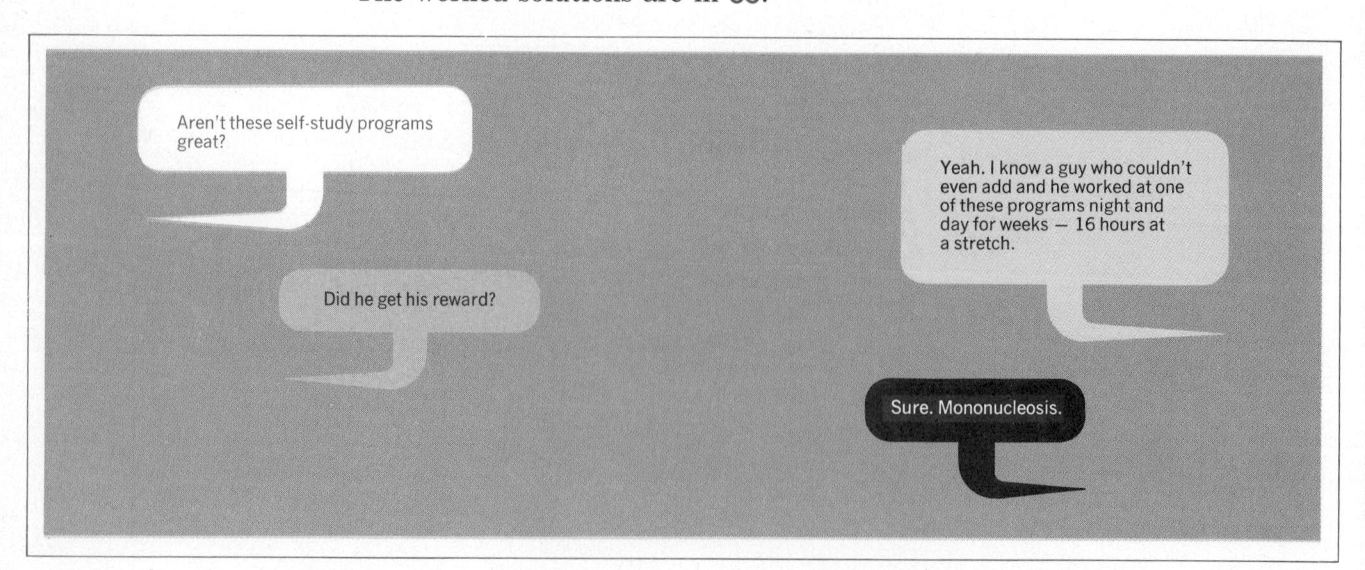

55 (a) **First guess:** $\sqrt{13.6} \cong 3.5$

First check:

$$\begin{array}{r} 3.886 \\ 3.5\overline{)13.6000} \\ \underline{10\,5} \\ 3\,10 \\ \underline{2\,80} \\ 300 \\ \underline{280} \\ 200 \end{array}$$

Round it to 3.89.

Second guess: $\frac{1}{2}(3.5 + 3.89) = \frac{1}{2}(7.39) = 3.69$

Second check: $3.69\overline{)13.6000}$ quotient 3.6856. Round it to 3.69.

Answer: $\sqrt{13.6} \cong 3.69$

(b) **First guess:** $\sqrt{8.25} \cong 3$

First check:

$$\begin{array}{r} 2.75 \\ 3\overline{)8.25} \\ \underline{6} \\ 2\,2 \\ \underline{2\,1} \\ 15 \end{array}$$

Round to 2.8.

Second guess: $\frac{1}{2}(3 + 2.8) = \frac{1}{2}(5.8) = 2.9$

Second check:

$$\begin{array}{r} 2.8448 \\ 2.9\overline{)8.250} \\ \underline{5\,8} \\ 2\,45 \\ \underline{2\,32} \\ 130 \end{array}$$

Round to 2.84.

Third guess: $\frac{1}{2}(2.9 + 2.84) = \frac{1}{2}(5.74) = 2.87$

Third check: $2.87\overline{)8.25000}$ quotient 2.874

Answer: $\sqrt{8.25} \cong 2.87$

(continued on the next page)

(c) **First guess:** $\sqrt{130.5} \cong 11.5 \qquad 11^2 = 121 \qquad 12^2 = 144$

First check:

$$
\begin{array}{r}
11.347 \\
11.5\overline{)130.500} \\
\underline{115} \\
15\,5 \\
\underline{11\,5} \\
4\,00 \\
\underline{3\,45} \\
550 \\
\underline{460} \\
900
\end{array}
$$

Round to 11.35

Second guess: $\frac{1}{2}(11.5 + 11.35) = \frac{1}{2}(22.85) = 11.43$

Second check: $11.43\overline{)130.50000}$ = 11.417

Third guess: $\frac{1}{2}(11.43 + 11.417) = 11.42$

Answer: $\sqrt{130.5} \cong 11.42$

Turn to **56** for a set of practice problems on calculating square roots.

Decimals

PROBLEM SET 5

56 Answers are on page 506.

Square Roots

A. Using a table of square roots find the following to two decimal digits:

1. $\sqrt{2} =$ ____ 2. $\sqrt{3} =$ ____ 3. $\sqrt{5} =$ ____ 4. $\sqrt{6} =$ ____

5. $\sqrt{10} =$ ____ 6. $\sqrt{30} =$ ____ 7. $\sqrt{150} =$ ____ 8. $\sqrt{75} =$ ____

9. $\sqrt{20} =$ ____ 10. $\sqrt{160} =$ ____ 11. $\sqrt{15} =$ ____ 12. $\sqrt{105} =$ ____

B. Calculate the following square roots to two decimal digits using the *guess n' check* method:

1. $\sqrt{68} =$ ____ 2. $\sqrt{7.5} =$ ____ 3. $\sqrt{1.75} =$ ____ 4. $\sqrt{88.5} =$ ____

5. $\sqrt{71.45} =$ ____ 6. $\sqrt{349} =$ ____ 7. $\sqrt{1245} =$ ____ 8. $\sqrt{2000} =$ ____

9. $\sqrt{450} =$ ____ 10. $\sqrt{600} =$ ____ 11. $\sqrt{2.8} =$ ____ 12. $\sqrt{20.5} =$ ____

C. Brain Boosters:

1. If a golf ball is dropped from the roof of a building and falls 88 feet to the ground, how much time is required for its fall?

   Time of fall in seconds $= \sqrt{\dfrac{\text{Distance in feet}}{16}}$

   $\left(\textit{Hint.}\ \sqrt{\dfrac{a}{b}} = \dfrac{\sqrt{a}}{\sqrt{b}}\ \text{for any numbers } a \text{ and } b.\right)$

2. Insert $+$, $-$, $\times$, $\div$, ( ), and $\sqrt{\ }$ signs in the following so as to make the equations true.

   (a) $10 = 4\ \ 4\ \ 4$ (b) $2 = 4\ \ 4\ \ 4$
   (c) $6 = 4\ \ 4\ \ 4$ (d) $8 = 4\ \ 4\ \ 4$
   (e) $12 = 4\ \ 4\ \ 4$ (f) $16 = 4\ \ 4\ \ 4$

3. The Pythagorean theorem states that the three sides of a right triangle, A, B, and C, are related this way:

   $A^2 + B^2 = C^2$ or $C = \sqrt{A^2 + B^2}$

   What is the length of side C for a right triangle where $A = 10$ inches and $B = 6$ inches?

   ____

Date

Name

Course/Section

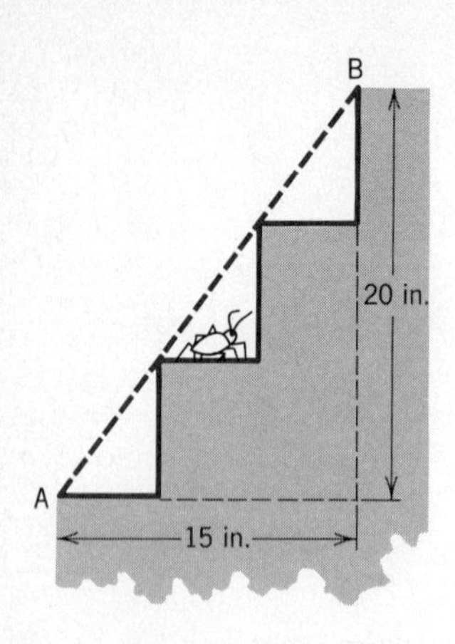

4. Friend bug is traveling to the top of the stairs.

What is the distance from A to B if he crawls up the steps?

____

What is the distance from A to B if he flies along the straight line AB?

____

(*Hint.* See problem 3.)

5. $\sqrt{12}$ is __?__.

$\sqrt{\sqrt{12}}$ is __?__.

$\sqrt{\sqrt{\sqrt{12}}}$ is __?__.

(Work them out.)

Can you guess what

$\sqrt{\sqrt{\sqrt{\cdots\sqrt{\sqrt{12}}}}}$

would equal? (The dots mean that the square root is repeated infinitely many times.)

When you have had the practice you need turn to **57** for a self-test over this unit.

Unit 3
Decimals

Self-Test

57

1. $6.2 + 13.045 =$ ________
2. $41.3 + 9.86 =$ ________
3. $16 - 7.93 =$ ________
4. $4.27 - 3.8 =$ ________
5. $8.1 \times 2.04 =$ ________
6. $5.6 \times 30 =$ ________
7. $8 \div 4.2$ (round to two decimal digits.) = ________
8. $14.2 \div 0.075$ (round to two decimal digits.) = ________
9. $0.045 \div 0.8$ (round to two decimal digits.) = ________
10. Write 0.56 as a fraction in lowest terms. ________
11. Write 3.248 as a fraction in lowest terms. ________
12. Write $\frac{7}{16}$ as a decimal. ________
13. What part of 3.8 is 4.56? ________
14. Find 0.25 of 4.8. ________
15. Find a number such that 0.35 of it is 2.45. ________
16. $-4 + 7 =$ ________
17. $-5 - (-3) =$ ________
18. $8 - 14 =$ ________
19. $17 - (-4) =$ ________
20. $(-2.1) \times (-3.1) =$ ________
21. $(-7) \times 5 =$ ________
22. $-(-16) \div (-4) =$ ________
23. $4.5 \div (-0.9) =$ ________
24. $\sqrt{6.5}$ (to two decimal digits) = ________
25. $\sqrt{425}$ (to two decimal digits) = ________

Date ________

Name ________

Course/Section ________

Answers are on page 514.

SUPPLEMENTARY PROBLEM SET 1A

Addition and Subtraction of Decimals

Decimals

A. Add or subtract as shown:

1. $0.43 + 0.59 =$ ______
2. $1.22 + 0.79 =$ ______
3. $0.81 - 0.54 =$ ______
4. $1.35 - 0.08 =$ ______
5. $2 - 0.41 =$ ______
6. $8.15 + 1.66 =$ ______
7. $1.63 + 4.78 =$ ______
8. $6.01 - 1.46 =$ ______
9. $3.75 - 1.8 =$ ______
10. $6 + .12 + 1.4 =$ ______

B. Add or subtract as shown:

1. $307.14 + 16.37 =$ ______
2. $5.5 + .55 + .055 =$ ______
3. $5.76 + .071 =$ ______
4. $6 - 3.24 =$ ______
5. $8 - 0.07 =$ ______
6. $14.25 - 6.77 =$ ______
7. $176.003 + 5.638 =$ ______
8. $0.9536 + 12.0879 =$ ______
9. $351.8 - 40.275 =$ ______
10. $2356.7 - 48.806 =$ ______

C. Word Problems:

1. Jack Jogs had $647.23 in his checking account. He wrote checks for $19.95, $107.49, and $37.68. What was his new balance?

   ______

2. At the end of the year the odometer on Dave's car reads 71,461.3 miles. At the start of the year it read 58,459.7 miles. How many miles did he drive this year

   ______

3. If your salary last year was $11,694.54 and $1927.48 was deducted for taxes and other expenses, what was your take home pay?

   ______

4. How much more is 88.08 than 80.88?

   ______

Date ______

Name ______

Course/Section ______

SUPPLEMENTARY PROBLEM SET 2A

Multiplication and Division of Decimals

Decimals

A. Multiply:

1. $4 \times 0.003 =$ ______
2. $0.2 \times 0.3 =$ ______
3. $0.004 \times 0.07 =$ ______
4. $2.3 \times .04 =$ ______
5. $3 \times 0.07 =$ ______
6. $6.5 \times 7.15 =$ ______
7. $126.4 \times .2 =$ ______
8. $120 \times 1.6 =$ ______
9. $0.214 \times 0.003 =$ ______
10. $(0.8)^2 =$ ______
11. $(0.5)^3 =$ ______
12. $(2.1)^2 =$ ______

B. Divide and round as indicated:

1. $12 \div 5 =$ ______
2. $4.6 \div 0.2 =$ ______
3. $0.008 \div 0.05 =$ ______
4. $125 \div 2.5 =$ ______

Round to two decimal digits:

5. $100 \div 17 =$ ______
6. $6 \div 7 =$ ______
7. $12 \div 0.07 =$ ______
8. $42.4 \div 6.307 =$ ______

Round to three digits:

9. $0.03 \div 2.81 =$ ______
10. $31.7 \div 5.23 =$ ______
11. $126.401 \div 0.21 =$ ______
12. $0.0007 \div 1.4126 =$ ______

C. Word Problems:

1. If you earn \$2.37 per hour, how much should you be paid for working 36.5 hours?

____

2. Which is the better buy for a TV set? (1) 23 payments at \$31.75 each or (2) 30 payments at \$21.57 each?

____

3. What is the cost of 14.6 gallons of gasoline at 67.9¢ per gallon?

____

Date ______

Name ______

Course/Section ______

(*continued on the next page*)

4. A subscription to the zoo keepers' magazine, *Gnu News,* costs \$7.85 for 13 issues. What do you save over the newstand cost of 75¢ per issue?

____

5. A case of 24 cans of vegetable soup costs \$4.72. What is the cost per can?

____

SUPPLEMENTARY PROBLEM SET 3A

Decimals

Decimal Fractions

A. Write as decimal numbers (round to two decimal digits):

1. $\frac{2}{3} =$ ______
2. $\frac{1}{7} =$ ______
3. $\frac{3}{8} =$ ______
4. $\frac{5}{8} =$ ______
5. $\frac{5}{12} =$ ______
6. $\frac{7}{8} =$ ______
7. $\frac{5}{6} =$ ______
8. $\frac{5}{9} =$ ______
9. $\frac{4}{25} =$ ______

B. Write as a fraction in lowest terms:

1. $0.46 =$ ______
2. $0.014 =$ ______
3. $2.125 =$ ______
4. $3.14 =$ ______
5. $.95 =$ ______
6. $4.0625 =$ ______

C. Solve:

1. What decimal fraction of 0.05 is 0.65? ______
2. Find 1.6 of 20. ______
3. What decimal part of 3.2 is 14.4? ______
4. If 0.06 of some number is 4.5, find the number. ______

D. Word Problems:

1. If the going rate of pay for a cashew dunker in a candy factory is $2.77 per hour, how much will he earn for $8\frac{1}{2}$ hours work?

   ______

2. One protein pill contains 2.7 grams of protein. How many pills must you take to get your daily allotment of 70 grams of protein?

   ______

3. Show that $\frac{41}{333}$ is a repeating decimal.

   ______

4. What is the cost of $8\frac{3}{4}$ feet of walnut shelving at $1.19 per foot?

   ______

5. If 19 apples sell for exactly $2, what is the cost of one?

   ______

Date ______

Name ______

Course/Section ______

SUPPLEMENTARY PROBLEM SET 4A

Signed Numbers

Decimals

A. Add or subtract as indicated:

1. $4 - 7 =$ ______
2. $-17 + 5 =$ ______
3. $-13 + (-6) =$ ______
4. $-31 - (-7) =$ ______
5. $-31 + (-7) =$ ______
6. $31 + (-7) =$ ______
7. $31 - (-7) =$ ______
8. $0.04 - 0.07 =$ ______
9. $2 - .1 =$ ______
10. $.1 - 2 =$ ______
11. $3.5 - (-1.4) =$ ______
12. $-8 + (-1.7) =$ ______
13. $7 - 12 =$ ______
14. $-2.11 - (-1.2) =$ ______
15. $-3.0 - (-2.5) =$ ______

B. Multiply or divide as indicated:

1. $2 \times (-4) =$ ______
2. $(-6) \times (-5) =$ ______
3. $(-2) \times (8) =$ ______
4. $12 \div (-4) =$ ______
5. $(-21) \div (7) =$ ______
6. $(-24) \div (-6) =$ ______
7. $1.64 \div (-0.4) =$ ______
8. $(0.05) \div (-.4) =$ ______
9. $(-124.5) \div (-.3) =$ ______
10. $(-\frac{1}{2}) \div (-1\frac{1}{3}) =$ ______
11. $(-2)^3 =$ ______
12. $(-2)^4 =$ ______

C. Calculate:

1. $(-3)(-4) - (2) =$ ______
2. $3(-2) - 2(-3) =$ ______
3. $(-2)^2(-5) + 4(-3) =$ ______
4. $4(-1) - (-2) - 3 =$ ______
5. $(-\frac{1}{2})(\frac{2}{3}) \div (-\frac{3}{4}) =$ ______
6. $(-\frac{1}{2}) + (\frac{2}{3}) - (-\frac{3}{4}) =$ ______

______
Date

______
Name

______
Course/Section

SUPPLEMENTARY PROBLEM SET 5A

Square Roots

Decimals

A. Using a table of square roots find the following to two decimal digits:

1. $\sqrt{7} =$ ______ 2. $\sqrt{8} =$ ______ 3. $\sqrt{12} =$ ______

4. $\sqrt{15} =$ ______ 5. $\sqrt{27} =$ ______ 6. $\sqrt{34} =$ ______

7. $\sqrt{64} =$ ______ 8. $\sqrt{75} =$ ______ 9. $\sqrt{94} =$ ______

10. $\sqrt{110} =$ ______ 11. $\sqrt{125} =$ ______ 12. $\sqrt{140} =$ ______

13. $\sqrt{150} =$ ______ 14. $\sqrt{162} =$ ______ 15. $\sqrt{196} =$ ______

B. Calculate the following square roots to two decimal digits using the *Guess n' Check* method:

1. $\sqrt{6} =$ ______ 2. $\sqrt{10} =$ ______ 3. $\sqrt{38} =$ ______

4. $\sqrt{55} =$ ______ 5. $\sqrt{21.75} =$ ______ 6. $\sqrt{287} =$ ______

7. $\sqrt{1000} =$ ______ 8. $\sqrt{865.4} =$ ______ 9. $\sqrt{3.754} =$ ______

10. $\sqrt{1700} =$ ______ 11. $\sqrt{0.015} =$ ______ 12. $\sqrt{12.1} =$ ______

C. Word Problems:

1. The area of a square plot of land is 4500 square yards. What is the length of a side of the square?

____

2. Use the Pythagorean theorem to find the length of the hypotenuse of a right triangle whose sides are 5 inches and 12 inches. (*Hint:* Find $\sqrt{5^2 + 12^2}$.)

____

3. Do you believe that $\sqrt{3\frac{3}{8}} = 3 \times \sqrt{\frac{3}{8}}$? Check it by changing the fractions to decimals and taking the square roots.

____

4. A roll of wall covering contains 370 square feet. If it just covers a square wall area, what is the size of the wall?

____

Date ______________

Name ______________

Course/Section ______________

PREVIEW 4

Objective	Sample Problems		Where To Go for Help Page	Frame
Upon successful completion of this program you will be able to:			Page	Frame
1. Write fractions and decimal numbers as percents.	(a) Write $1\frac{7}{8}$ as a percent.	______	261	1
	(b) Write 0.45 as a percent.	______	261	1
2. Convert percents to decimal and fraction form.	(a) Write $37\frac{1}{2}\%$ as a decimal.	______	261	1
	(b) Write 44% as a fraction.	______	261	1
3. Solve problems involving percent.	(a) Find 35% of 16.	______	271	12
	(b) Find 120% of 45.	______	271	12
	(c) What percent of 18 is 3?	______	271	12
	(d) What percent of $1\frac{2}{3}$ is $\frac{1}{2}$?	______	271	12
	(e) What percent of 0.6 is 0.25?	______	271	12
	(f) 70% of what number is 56?	______	271	12
4. Solve practical problems involving percent.	(a) How much money does a salesman earn on a $240 sale if his commission is 15%?	______	289	24
	(b) A camera normally selling for $149.50 is on sale at a discount of 25%. What is its sale price?	______	289	24
	(c) After a 10% discount a book sells for $5.40. What was the original price?	______	289	24
	(d) A coat sells for $19.75 plus 6% sales tax. What is the total cost?	______	289	24
	(e) What is the interest paid on a $1000 bank loan at $6\frac{1}{2}\%$ for 24 months?	______	289	24

______ Date

______ Name

______ Course/Section

(Answers to these sample problems are at the bottom of this sheet.)

If you are certain you can work all of these problems correctly, turn to page 305 for a self test. If you want help with any of these objectives or if you cannot work one of the sample problems, turn to the page indicated. Super-students, who are eager to learn everything in this unit, will turn to frame **1** and begin work there.

PREVIEW 4

Answers to Sample Problems

1. (a) 187.5%
   (b) 45%
2. (a) 0.375
   (b) $\frac{11}{25}$
3. (a) 5.6
   (b) 54
   (c) $16\frac{2}{3}\%$
   (d) 30%
   (e) $41\frac{2}{3}\%$
   (f) 80
4. (a) \$36.00
   (b) \$112.13
   (c) \$6.00
   (d) \$20.94
   (e) \$130

4 Percent

1 Sitting at that desk, doing math in the abstract, make-believe world of school, Sally finds it easy to lose her sense of the way it ought to be. For Sally mathematics is a guessing game played only in school, with correct guesses rewarded by a pat on the head and wrong guesses followed by another guess. When she does math for real—window shops, buys something, or counts change—she stops the wild guessing and tries to reason it out. When she gets a bit older and worries about buying a car or a house, getting a loan, paying taxes, buying on an installment plan, earning interest on savings, or shopping for bargains, she'll find that in order to reason out these new problems she must understand the concept of percent. Wild, random guesses then will mean she's going to lose a lot of money.

NUMBERS AND PERCENT

The phrase *percent* comes from the Latin words *per centum* meaning "by the hundred" or "for every hundred." A number expressed as a percent is being compared with a second number called the standard or *base* by dividing the base into 100 equal parts and writing the comparison number as so many hundredths of the base.

For example, what part of the base or standard length is length A?

Base $\qquad$ A

We could answer the question with a fraction or a decimal or a percent. First, divide the base into 100 equal parts.

(*continued on the next page*)

Then compare A with it. The length of A is 40 parts out of the 100 parts that make up the base.

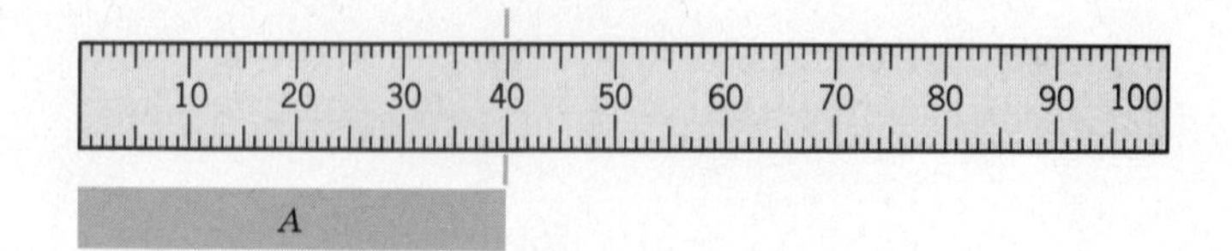

A is $\frac{40}{100}$ or 0.40 or 40% of the base.

$\frac{40}{100} = 40\%$ Thus 40% means 40 parts in 100 or $\frac{40}{100}$.

What part of this base is length B?

Base

B

Answer with a percent.

Turn to **2** to check your answer.

2

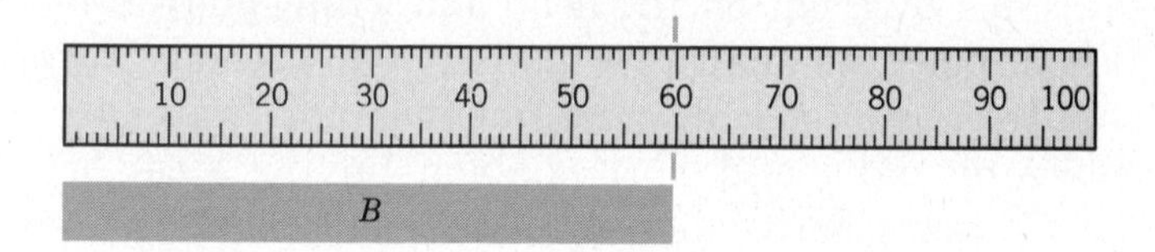

B is $\frac{60}{100}$ or 60%.

Of course the compared number may be larger than the base. For example,

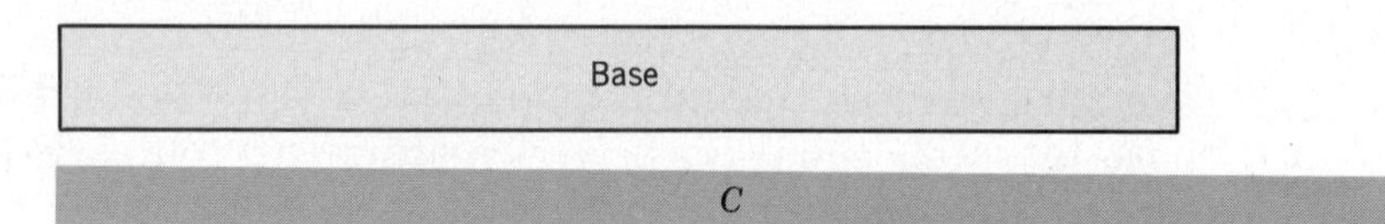

In this case, divide the base into 100 parts and extend it in length.

10 20 30 40 50 60 70 80 90 100 110 120 130 140 150

C

The length of C is 120 parts out of the 100 parts that make up the base.

C is $\frac{120}{100}$ or 120 % of the base.

Because our number system, and our money, is based on ten and multiples of ten, it is very handy to write comparisons in hundredths or percent.

What part of $1.00 is 50¢? Write your answer as a fraction, as a decimal, and as a percent.

Check in **3** for the answer.

---

3 50¢ is what part of $1.00?

$$\frac{50¢}{100¢} = \frac{50}{100} = .50 = 50\%$$

50¢ is equal to 50% of $1.00

To find the answer as a percent, write it as a fraction with denominator equal to 100.

We may also write 50¢ = $\frac{1}{2}$ of $1.00 or 50¢ = 0.50 of $1.00. Fractions, decimals, and percents are alternative ways to talk about a comparison of two numbers.

What percent part of 10 is 2? Write 2 as a fraction of 10, rename it as a fraction with denominator equal to 100, then write as a percent.

When you have completed this, go to **4**.

4

$$\frac{2}{10} = \frac{2 \times 10}{10 \times 10} = \frac{20}{100} = 20\%$$

How do you rewrite a decimal number as a percent? The procedure is simply to multiply the decimal number by 100%. For example,

$0.60 = 0.60 \times 100\% = 60\%$

$$\begin{array}{r} .60 \\ \times 100 \\ \hline 60.00 \end{array} = 60$$

$0.375 = 0.375 \times 100\% = 37.5\%$
$3.4 = 3.4 \times 100\% = 340\%$
$0.02 = 0.02 \times 100\% = 2\%$

Rewrite the following as percents.

(a) 0.70 = ______ (b) 1.25 = ______

(c) 0.001 = ______ (d) 3 = ______

Look in **5** for the answers.

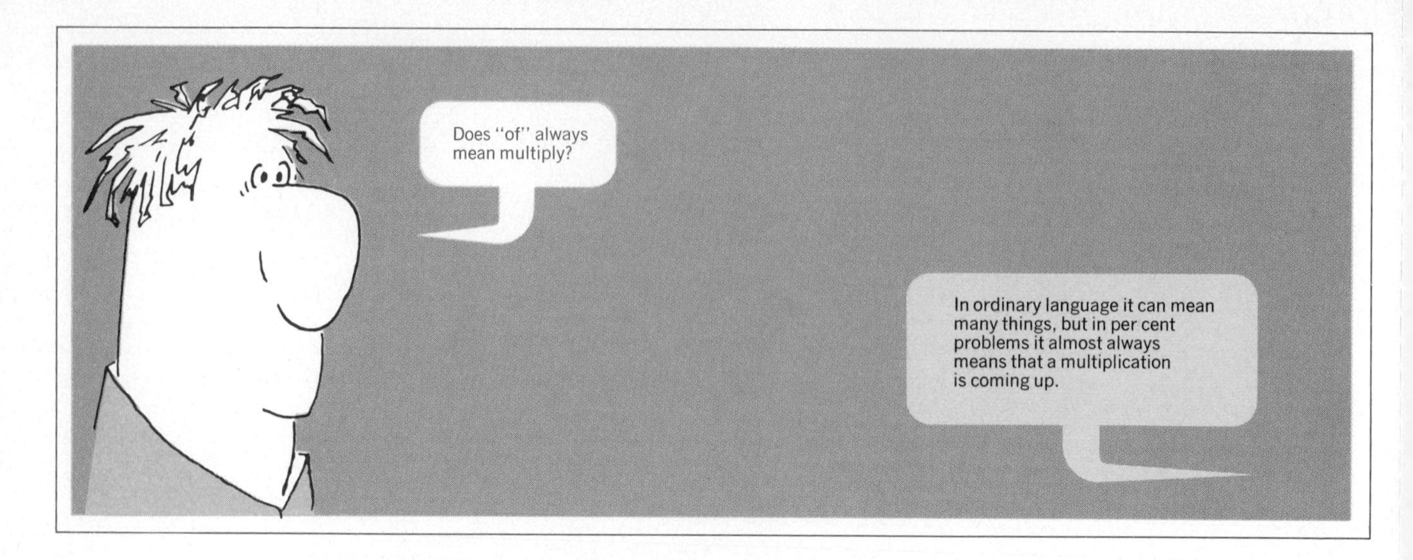

Does "of" always mean multiply?
In ordinary language it can mean many things, but in per cent problems it almost always means that a multiplication is coming up.

5

(a) $0.70 = 0.70 \times 100\% = 70\%$
(b) $1.25 = 1.25 \times 100\% = 125\%$
(c) $0.001 = 0.001 \times 100\% = 0.1\%$
(d) $3 = 3 \times 100\% = 300\%$

Notice in each of these that multiplication by 100% has the effect of moving the decimal point two digits to the right.

0.70	becomes	70.%	or 70%
1.25	becomes	125.%	or 125%
0.001	becomes	0.1%	
3 = 3.00	becomes	300.%	or 300%

To rewrite a fraction as a percent, we can always rename it as a fraction with 100 as denominator.

$$\frac{3}{20} = \frac{3 \times 5}{20 \times 5} = \frac{15}{100} = 15\%$$

However, the easiest way is to convert the fraction to decimal form by dividing and then multiply by 100%.

$$\frac{1}{2} = 0.50 = 0.50 \times 100\% = 50\% \qquad 2\overline{)1.00} = .50$$

$$\frac{3}{4} = 0.75 = 0.75 \times 100\% = 75\% \qquad 4\overline{)3.00} = .75$$

$$\frac{3}{20} = 0.15 = 0.15 \times 100\% = 15\% \qquad 20\overline{)3.00} = .15$$

$$\begin{array}{r} 2\,0 \\ \hline 1\,00 \\ 1\,00 \\ \hline \end{array}$$

$$1\frac{7}{20} = \frac{27}{20} = 1.35 = 1.35 \times 100\% = 135\%$$

$$\begin{array}{r} 1.35 \\ 20\overline{)27.00} \\ \underline{20} \\ 7\,0 \\ \underline{6\,0} \\ 1\,00 \\ \underline{1\,00} \end{array}$$

Rewrite $\frac{5}{16}$ as a percent.

Check your answer in **6**.

6

$$\frac{5}{16} = 0.3125 = 0.3125 \times 100\% = 31.25\%$$

$$\begin{array}{r} .3125 \\ 16\overline{)5.0000} \\ \underline{4\,8} \\ 20 \\ \underline{16} \\ 40 \\ \underline{32} \\ 80 \\ \underline{80} \end{array}$$

This is often written as $31\frac{1}{4}\%$.

Some fractions cannot be converted to exact decimals. For example,

$\frac{1}{3} = 0.333\ldots$ where the 3s continue to repeat endlessly.

We can round to get an approximate percent,

$\frac{1}{3} = 0.333 \times 100\% = 33.3\%$

or convert it to a fraction with 100 as denominator.

$\frac{1}{3} = \frac{?}{100}$ gives $\frac{1}{3} = \frac{1 \times 33\frac{1}{3}}{3 \times 33\frac{1}{3}} = \frac{33\frac{1}{3}}{100} = 33\frac{1}{3}\%$

Rewrite $\frac{1}{6}$ as a percent.

The answer is in **8**.

---

7 (a) $\frac{7}{5} = 1.4 \qquad 1.4 \times 100\% = 140\%$

(b) $\frac{2}{3} = \frac{?}{100} \qquad \frac{2}{3} = \frac{2 \times 33\frac{1}{3}}{3 \times 33\frac{1}{3}} = \frac{66\frac{2}{3}}{100} = 66\frac{2}{3}\%$

(c) $3\frac{1}{8} = \frac{25}{8} = 3.125 \qquad 3.125 \times 100\% = 312.5\%$

(d) $\frac{5}{12} = \frac{?}{100} = \frac{5 \times 8\frac{1}{3}}{12 \times 8\frac{1}{3}} = \frac{40\frac{5}{3}}{100} = \frac{41\frac{2}{3}}{100} = 41\frac{2}{3}\%$

In order to use percent in solving practical problems, it is often necessary to change a percent to a decimal number. The procedure is to divide by 100%. For example,

$$50\% = \frac{50\%}{100\%} = \frac{50}{100} = 0.50 \qquad 100\overline{)50.0} = .5$$

$$5\% = \frac{5\%}{100\%} = \frac{5}{100} = 0.05 \qquad 100\overline{)5.00} = .05$$

$$0.2\% = \frac{0.2\%}{100\%} = \frac{0.2}{100} = 0.002 \qquad 100\overline{)0.200} = .002$$

Fractions may be part of the percent number. If so, it is easiest to reduce to a decimal number and round if necessary.

$$6\frac{1}{2}\% = \frac{6\frac{1}{2}\%}{100\%} = \frac{6.5}{100} = 0.065 \qquad 100\overline{)6.500} = .065$$

$$33\frac{1}{3}\% = \frac{33\frac{1}{3}\%}{100\%} = \frac{33\frac{1}{3}}{100} \cong \frac{33.3}{100} = 0.333 \text{ rounded}$$

Now you try a few: Write these as decimal numbers.

(a) $4\% =$ ______ (b) $0.5\% =$ ______

(c) $16\frac{2}{3}\% =$ ______ (d) $79\frac{1}{4}\% =$ ______

Our answers are in **9**.

---

8 $\frac{1}{6} = \frac{?}{100} \qquad \frac{1}{6} = \frac{1 \times 16\frac{2}{3}}{6 \times 16\frac{2}{3}} = \frac{16\frac{2}{3}}{100} = 16\frac{2}{3}\%$

To get this ask "What number times 6 equals 100?" Answer: $100 \div 6 = 16\frac{2}{3}$

Rewrite the following fractions as percents.

(a) $\frac{7}{5} =$ ______ (b) $\frac{2}{3} =$ ______

(c) $3\frac{1}{8} =$ ______ (d) $\frac{5}{12} =$ ______

Go to **7** to check your answers.

---

9 (a) $4\% = \frac{4\%}{100\%} = \frac{4}{100} = 0.04 \qquad 100\overline{)4.00}$ with quotient $.04$

(b) $0.5\% = \frac{0.5\%}{100\%} = \frac{.5}{100} = 0.005 \qquad 100\overline{)0.500}$ with quotient $.005$

(c) $16\frac{2}{3}\% = \frac{16\frac{2}{3}}{100\%} = \frac{16\frac{2}{3}}{100} \cong \frac{16.7}{100} = 0.167$ rounded

(d) $79\frac{1}{4}\% = \frac{79\frac{1}{4}\%}{100\%} = \frac{79\frac{1}{4}}{100} = \frac{79.25}{100} = 0.7925$

To change a percent to a fraction, divide by 100% and reduce to lowest terms.

$$36\% = \frac{36\%}{100\%} = \frac{36}{100} = \frac{9 \times 4}{25 \times 4} = \frac{9}{25}$$

36% means 36 hundredths or $\frac{36}{100}$ or $\frac{9}{25}$.

$$12\frac{1}{2}\% = \frac{12\frac{1}{2}\%}{100\%} = \frac{12\frac{1}{2}}{100} = \frac{\frac{25}{2}}{100} = \frac{25}{200} = \frac{1}{8}$$

Note that $\frac{\frac{25}{2}}{100} = \frac{25}{2} \div 100 = \frac{25}{2} \times \frac{1}{100} = \frac{25}{200}$

$$125\% = \frac{125\%}{100\%} = \frac{125}{100} = \frac{5}{4} = 1\frac{1}{4}$$

Try these for practice. write as a fraction in lowest terms

(a) 72% = ______ (b) $16\frac{1}{2}\%$ = ______

(c) 240% = ______ (d) $7\frac{1}{2}\%$ = ______

You will find the answers in **10**.

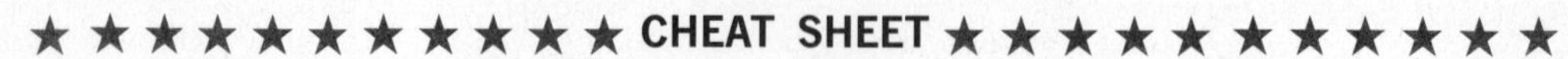

CHEAT SHEET

Percent	Decimal	Fraction
5%	.05	$\frac{1}{20}$
$6\frac{1}{4}$%	.0625	$\frac{1}{16}$
$8\frac{1}{3}$%	$.08\overline{3}$	$\frac{1}{12}$
10%	.10	$\frac{1}{10}$
$12\frac{1}{2}$%	.125	$\frac{1}{8}$
$16\frac{2}{3}$%	$.1\overline{6}$	$\frac{1}{6}$
20%	.20	$\frac{1}{5}$
25%	.25	$\frac{1}{4}$
30%	.30	$\frac{3}{10}$
$33\frac{1}{3}$%	$.3\overline{3}$	$\frac{1}{3}$
$37\frac{1}{2}$%	.375	$\frac{3}{8}$
40%	.40	$\frac{2}{5}$
50%	.50	$\frac{1}{2}$
60%	.60	$\frac{3}{5}$
$62\frac{1}{2}$%	.625	$\frac{5}{8}$
$66\frac{2}{3}$%	$.6\overline{6}$	$\frac{2}{3}$
70%	.70	$\frac{7}{10}$
75%	.75	$\frac{3}{4}$
80%	.80	$\frac{4}{5}$
$83\frac{1}{3}$%	$.8\overline{3}$	$\frac{5}{6}$
$87\frac{1}{2}$%	.875	$\frac{7}{8}$
90%	.90	$\frac{9}{10}$
100%	1.00	$\frac{10}{10}$

10

(a) $72\% = \dfrac{72\%}{100\%} = \dfrac{72}{100} = \dfrac{18 \times 4}{25 \times 4} = \dfrac{18}{25}$

(b) $16\frac{1}{2}\% = \dfrac{16\frac{1}{2}\%}{100\%} = \dfrac{16\frac{1}{2}}{100} = \dfrac{\frac{33}{2}}{100} = \dfrac{33}{200}$

(c) $240\% = \dfrac{240\%}{100\%} = \dfrac{240}{100} = \dfrac{12 \times 20}{5 \times 20} = \dfrac{12}{5} = 2\frac{2}{5}$

(d) $7\frac{1}{2}\% = \dfrac{7\frac{1}{2}\%}{100\%} = \dfrac{7\frac{1}{2}}{100} = \dfrac{\frac{15}{2}}{100} = \dfrac{15}{200} = \dfrac{3}{40}$

Now turn to **11** for a set of practice problems on what you have learned in this unit so far.

12 In all of your work with percent you will find that there are three basic types of problems. These three form the basis for all percent problems that arise in business, industry, science, or other areas. All of these problems involve three quantities:

★ 1. The *base* or *total* amount or standard used for a comparison.
★ 2. The *percentage* or part being compared with the base or total.
★ 3. The *percent* or *rate* that indicates the relationship of the percentage to the base, the part to the total.

All three basic percent problems involve finding one of these three quantities when the other two are known.

In every problem follow these seven steps:

★ **Step 1**

Translate the problem sentence into a math statement. (If you have not already read the section on Word Problems in Unit 2, do so now. It starts on page 157, frame **51**.)

For example, the question

30% of what number is 16? should be translated

$$30\% \times \square = 16$$

or $30\% \times \square = 16$

In this case, 30% is the percent or rate; $\square$, the unknown quantity, is the total or base; and 16 is the percentage or part of the total.

Notice that the words and phrases in the problem question become math symbols. The word "of" is translated *multiply*. The word "is" (and similar verbs such as "will be," and "becomes") are translated *equals*. Use a $\square$ or letter of the alphabet or ? for the unknown quantity you are asked to find.

★ **Step 2**

It will be helpful if you *label* which numbers are the base or total (T), the percent (%), and the percentage or part (P).

★ **Step 3**

Rearrange the equation so that the unknown quantity is alone on the left of the equal sign and the other quantities are on its right.

The equation $30\% \times \square = 16$

becomes $\square = 16 \div 30\% = \dfrac{16}{30\%} = \dfrac{16}{.30}$

The following *Equation Finder* may help you to do this arranging.

(*continued on the next page*)

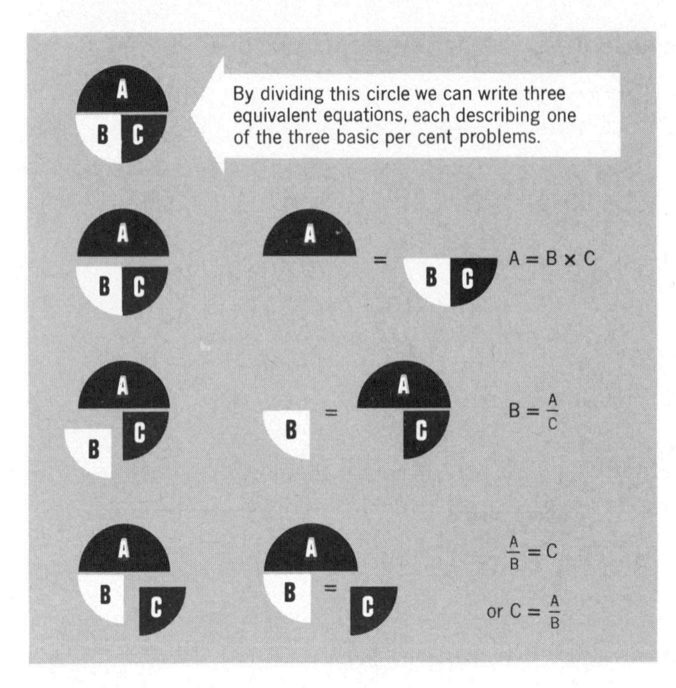

A
B
C
By dividing this circle we can write three equivalent equations, each describing one of the three basic per cent problems.
A = B × C
B = A/C
A/B = C
or C = A/B

The three equations $A = B \times C$, $B = \dfrac{A}{C}$, and $C = \dfrac{A}{B}$ are all equivalent.

★ **Step 4**

Make a reasonable estimate of the answer. *Guess,* but guess carefully. Good guessing is an art.

★ **Step 5**

Solve the problem by doing the arithmetic.

Be Careful

Never do arithmetic, never multiply or divide, with percent numbers. All percents must be rewritten as fractions or decimals before you can use them in a multiplication or division.

★ **Step 6**

Check your answer against the original guess. Are they the same, or at least close? If they do not agree, at least roughly, you have probably made a mistake and should repeat your work.

★ **Step 7**

Double-check by putting the answer number you have found back into the original problem or equation to see if it makes sense. If possible, use the answer to calculate one of the other numbers in the equation as a check.

Now let's look very carefully at each type of problem. We'll explain each, give examples, show you how to solve them, and work through a few together.

Turn to **13**.

13 *Type 1 problems* are usually stated in the form

Find 30% of 50.

or What is 30% of 50?

or 30% of 50 is what number?

Step 1

Translate: $30\% \times 50 = \square$

Step 2

Label: $\% \times T = P$

Step 3

Rearrange: $\square = 30\% \times 50$

Complete the calculation and find $\square$.

$\square = \underline{\ ?\ }$

(a) 150 Go to **14**
(b) 15 Go to **15**
(c) 1500 Go to **16**

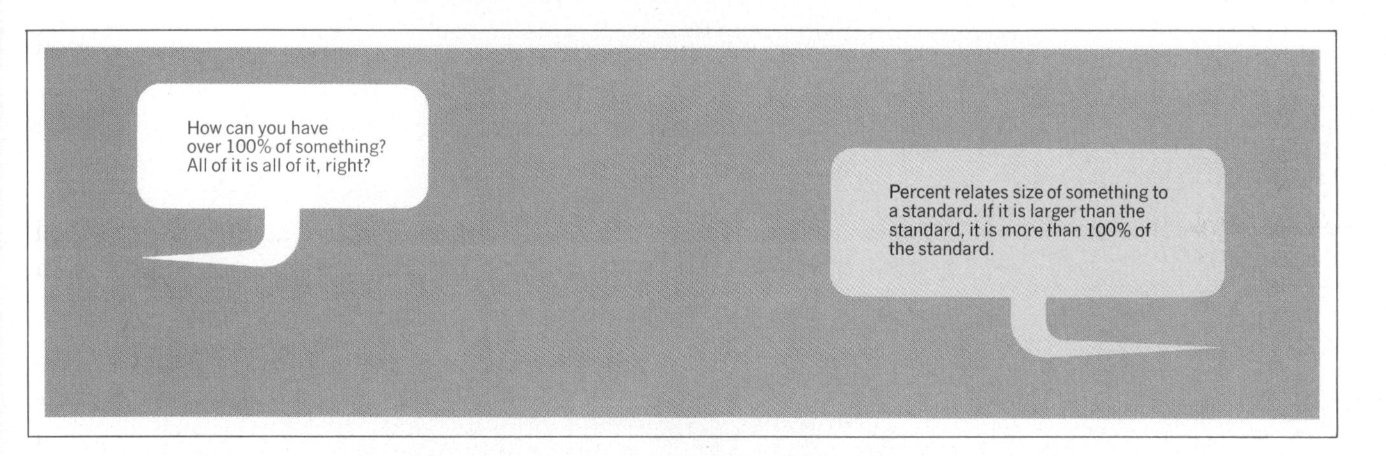

How can you have over 100% of something? All of it is all of it, right?
Percent relates size of something to a standard. If it is larger than the standard, it is more than 100% of the standard.

---

14 You answered that 30% of 50 = 150, and this is not correct. Be certain you do the following in solving this problem:

1. Guess at the answer—a careful, educated guess or estimate is an excellent check on your work. Never work a problem until you know roughly the size of the answer. 30% is roughly $\frac{1}{3}$. What is $\frac{1}{3}$ of 50, approximately?
2. *Never* multiply by a percent number. Before you multiply $30\% \times 50$, you must write 30% as a decimal number. (If you need help with this turn to **8**.)

Use these hints to solve the problem, then return to **13** and choose a better answer.

---

15 Right you are!

Step 4

Guess: The next step is to make an educated guess at the answer. For example,

30% of 50 is roughly $\frac{1}{3}$ of 50 or about 17.

Your answer will be closer to 17 than to 2 or 100.

Step 5

Solve: Never multiply by a percent number. Percent numbers are not arithmetic numbers, before you multiply write the percent number as a decimal number.

$$30\% = \frac{30\%}{100\%} = \frac{30}{100} = 0.30$$

$$30\% \times 50 = 0.30 \times 50 = 15$$

Step 6

Check: The guess (17) and the answer (15) are not exactly the same, but they are reasonably close. The answer seems acceptable.

Don't be intimidated by numbers. If the problem involves very large or very complex numbers, reduce it to a simpler problem. The problem

Find $14\frac{7}{32}\%$ of 6.4

may look difficult until you realize that it is essentially the same problem as

Find 10% of 6

which is fairly easy.

Before you begin any actual arithmetic problem involving percent you should

★ (a) Know roughly the size of the answer.
★ (b) Have a plan for solving the problem based on a simpler problem.
★ (c) Always change percents to decimals or fractions before multiplying or dividing with them. Never use percent numbers in arithmetic operations.

Now, try this problem:

Find $8\frac{1}{2}\%$ of 160.

Check your answer in **17**.

16 Your answer is incorrect.

First, it is important that you make an educated guess at the answer. Never work a problem before you know roughly the size of the answer. In this case,

Beware! 30% of 50 is roughly $\frac{1}{3}$ of 50

Second, *never* multiply by a percent number. Before you multiply $30\% \times 50$, you must write 30% as a decimal number. If you need help in writing a percent as a decimal number, turn to frame **8**, otherwise return to **13** and try again.

---

17 **Step 1.** Translate: $8\frac{1}{2}\% \times 160 = \square$

Step 2. Label: $\% \times T = P$

Step 3. Rearrange: $\square = 8\frac{1}{2}\% \times 160$

Step 4. Guess: $8\frac{1}{2}\%$ is close to 10%. 10% of 160 is $\frac{1}{10}$ of 160 or 16.

Step 5. Solve: $8\frac{1}{2}\% = \dfrac{8\frac{1}{2}\%}{100\%} = \dfrac{8.5}{100} = 0.085$

$\square = 0.085 \times 160 = 13.6$

Step 6. Check: 13.6 is approximately equal to 16, our original estimate.

Now try these for practice:

(a) Find 2% of 140. ______

(b) 35% of 20 = ______

(c) $7\frac{1}{4}\%$ of \$1000 = ______

(d) What is $5\frac{1}{3}\%$ of 3.3? ______

(e) 120% of 15 is what number? ______

The step-by-step answers are in **18**.

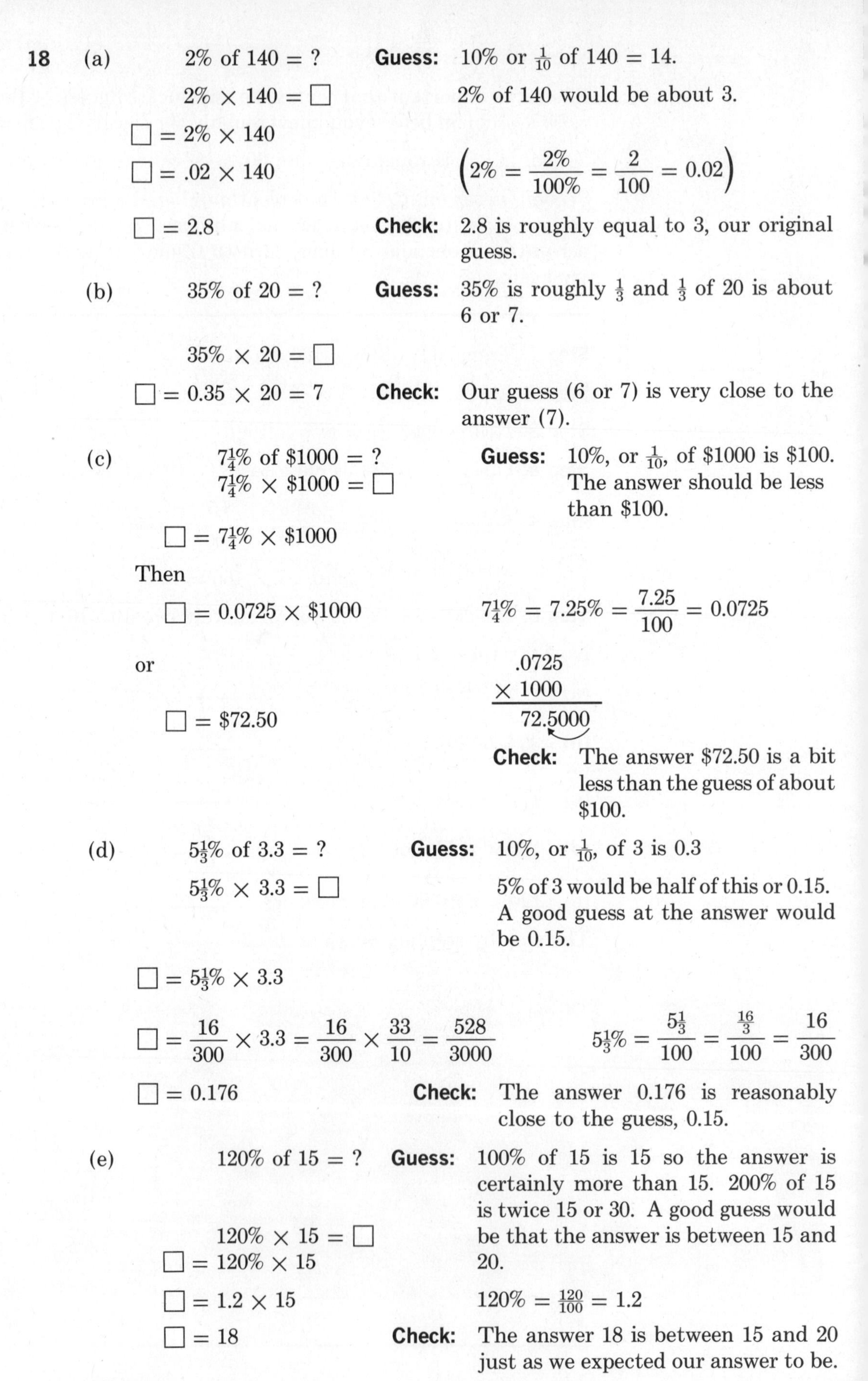

18 (a) 2% of $140 = ?$ **Guess:** 10% or $\frac{1}{10}$ of $140 = 14$.

$2\% \times 140 = \square$ — 2% of 140 would be about 3.

$\square = 2\% \times 140$

$\square = .02 \times 140$ $\left(2\% = \dfrac{2\%}{100\%} = \dfrac{2}{100} = 0.02\right)$

$\square = 2.8$ **Check:** 2.8 is roughly equal to 3, our original guess.

(b) 35% of $20 = ?$ **Guess:** 35% is roughly $\frac{1}{3}$ and $\frac{1}{3}$ of 20 is about 6 or 7.

$35\% \times 20 = \square$

$\square = 0.35 \times 20 = 7$ **Check:** Our guess (6 or 7) is very close to the answer (7).

(c) $7\frac{1}{4}\%$ of $\$1000 = ?$ **Guess:** 10%, or $\frac{1}{10}$, of \$1000 is \$100. The answer should be less than \$100.

$7\frac{1}{4}\% \times \$1000 = \square$

$\square = 7\frac{1}{4}\% \times \$1000$

Then

$\square = 0.0725 \times \$1000$ $7\frac{1}{4}\% = 7.25\% = \dfrac{7.25}{100} = 0.0725$

or

$$\begin{array}{r} .0725 \\ \times\ 1000 \\ \hline 72.5000 \end{array}$$

$\square = \$72.50$

Check: The answer \$72.50 is a bit less than the guess of about \$100.

(d) $5\frac{1}{3}\%$ of $3.3 = ?$ **Guess:** 10%, or $\frac{1}{10}$, of 3 is 0.3

$5\frac{1}{3}\% \times 3.3 = \square$ — 5% of 3 would be half of this or 0.15. A good guess at the answer would be 0.15.

$\square = 5\frac{1}{3}\% \times 3.3$

$\square = \dfrac{16}{300} \times 3.3 = \dfrac{16}{300} \times \dfrac{33}{10} = \dfrac{528}{3000}$ $5\frac{1}{3}\% = \dfrac{5\frac{1}{3}}{100} = \dfrac{\frac{16}{3}}{100} = \dfrac{16}{300}$

$\square = 0.176$ **Check:** The answer 0.176 is reasonably close to the guess, 0.15.

(e) 120% of $15 = ?$ **Guess:** 100% of 15 is 15 so the answer is certainly more than 15. 200% of 15 is twice 15 or 30. A good guess would be that the answer is between 15 and 20.

$120\% \times 15 = \square$

$\square = 120\% \times 15$

$\square = 1.2 \times 15$ $120\% = \frac{120}{100} = 1.2$

$\square = 18$ **Check:** The answer 18 is between 15 and 20 just as we expected our answer to be.

Type 2 problems require that you find the rate or percent. Problems of this kind are usually stated:

7 is what percent of 16?
or Find what percent 7 is of 16.
or What percent of 16 is 7?

Step 1

Translate: $\square\,\% \quad \times 16 = 7$

$$\square\% \times 16 = 7$$

Step 2

Label: $\% \times T = P$ All of the problem statements are equivalent to this equation.

Step 3

Rearrange: $\square\% = \dfrac{7}{16}$

To rearrange the equation and solve for $\square\%$ notice that it is in the form

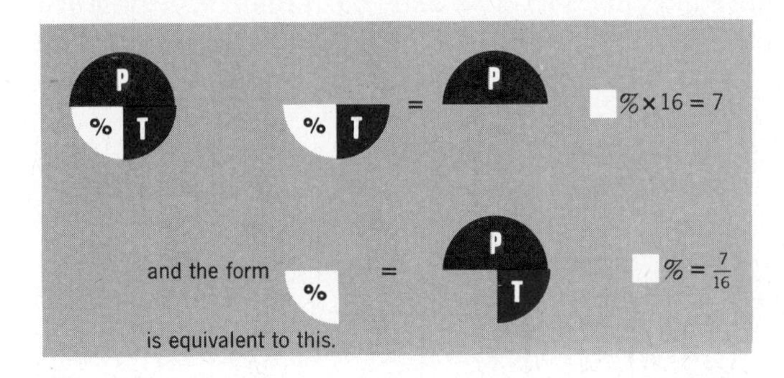

Therefore

$\square\% = \dfrac{7}{16}$ 16 is the total amount or base and 7 is the part of the base being described.

Solve this last equation.

Check your answer in **19**.

19 **Guess:** $\frac{7}{16}$ is very close to $\frac{8}{16}$ or $\frac{1}{2}$ or 50%. The answer will be a little less than 50%.

$$\square\% = \frac{7}{16} = \frac{7}{16} \times 100\% = \frac{700}{16}\%$$

$$\square\% = 43\tfrac{3}{4}\% \qquad 16\overline{)700} = 43\tfrac{12}{16} = 43\tfrac{3}{4}$$

$$\begin{array}{r} 43 \\ 16\overline{)700} \\ \underline{64} \\ 60 \\ \underline{48} \\ 12 \end{array}$$

Check: The answer $43\frac{3}{4}\%$ is reasonably close to our preliminary guess of about 50%.

If you had trouble converting $\frac{7}{16}$ to a percent, you should review this process by turning back to frame **5.**

The solution to a Type 2 or % problem will be a fraction or decimal number that must be converted to a percent.

Try these problems for practice.

(a) What percent of 40 is 16?______

(b) Find what percent 65 is of 25.______

(c) $6.50 is what percent of $18.00?______

(d) What percent of 2 is 3.5?______

(e) $10\frac{2}{5}$ is what percent of 2.6?______

Check your work in **20**.

HOW TO MISUSE PERCENT

1. In general you cannot add, subtract, multiply, or divide percent numbers. Percent helps you compare two numbers; it cannot be used in the normal arithmetic operations.

   For example, if 60% of class 1 earned A grades and 50% of class 2 earned A grades, what was the total percent of A grades for the two classes?

   The answer is that you cannot tell unless you know the number of students in each class.

2. In advertisements designed to trap the unwary, you might hear that "children had 23% fewer cavities when they used . . . ," or "50% more doctors smoke. . . ."

   Fewer than what? Fewer than the worst dental health group the advertiser could find? Fewer than the national average?

   More than what? More than a year ago? More than nurses? More than other adults? More than infants?

   There must be some reference or base given in order for the percent number to have any meaning at all.

   BEWARE of people who misuse percent!

20 (a) $\square\% \times 40 = 16$

A Type 2 or % problem $\% = \frac{P}{T}$

$$\square\% = \frac{16}{40}$$

Guess: $\frac{16}{40}$ is about $\frac{1}{3}$ or roughly 33%.

$$\square\% = \frac{16}{40} = \frac{16}{40} \times 100\% = \frac{1600}{40}\%$$

$$\square\% = 40\%$$

Check: 40% is reasonably close to 33%.

Double-check: $40\% \times 40 = ?$
$.40 \times 40 = 16$

(*continued on the next page*)

(b) $\square\% \times 25 = 65$ $\qquad \% = \dfrac{P}{T}$

$\square\% = \dfrac{65}{25}$

Guess: $\dfrac{65}{25}$ is more than 2 and 2 is 200%

The answer will be over 200%.

$\dfrac{65}{25} = \dfrac{65}{25} \times 100\% = 260\%$

$\square\% = 260\%$

Check: The answer agrees with the guess.

Double-check: $260\% \times 25 = ?$
$2.60 \times 25 = 65.00 = 65$

The most difficult part of this problem is in deciding whether the percent needed is found from $\frac{65}{25}$ or $\frac{25}{65}$. There is no magic to it. If you read the problem very carefully you will see that it speaks of 65 as a part "of 25." The base or total is 25. The percentage or part is 65.

(c) $\$6.50 = \square\% \times \$18.00$

or $\square\% \times \$18.00 = \$6.50$ $\qquad \% = \dfrac{P}{T}$

$\square\% = \dfrac{\$\ 6.50}{\$18.00}$

Guess: \$6.50 is about $\frac{1}{3}$ of \$18.00 and $\frac{1}{3}$ is roughly 30%.

$\square\% = \dfrac{6.50}{18.00} = \dfrac{6.50}{18.00} \times 100\%$

$\square\% = \dfrac{650}{18}\% = 36\frac{2}{18}\%$

$\square\% = 36\frac{1}{9}\%$

Check: $36\frac{1}{9}\%$ is reasonably close to the guess of 30%.

Double-check: $36\frac{1}{9}\% \times \$18 = ?$
$\frac{325}{900} \times 18 = 6.5$

(d) $\square\% \times 2 = 3.5$ $\qquad \% = \dfrac{P}{T}$

$\square\% = \dfrac{3.5}{2}$

Guess: $\dfrac{3}{2}$ is $1\frac{1}{2}$ or 1.5 and 1.5 is 150%.

The answer will be something more than 150%.

$\square\% = \dfrac{3.5}{2} \times 100\% = \dfrac{350}{2}\%$

$\square\% = 175\%$

Check: It will do.

Double-check: $175\% \times 2 = ?$
$1.75 \times 2 = 3.50$

(e) $10\frac{2}{5} = \square\% \times 2.6$

or $\square\% \times 2.6 = 10\frac{2}{5}$ $\qquad \% = \dfrac{P}{T}$

$\square\% = \dfrac{10\frac{2}{5}}{2.6}$ $\qquad$ **Guess:** $\frac{10}{2}$ is 5 and 5 is 500%.

$\square\% = \dfrac{10\frac{2}{5}}{2.6} = \dfrac{10.4}{2.6} = \dfrac{10.4}{2.6} \times 100\% = \dfrac{1040}{2.6}\%$

$\square\% = 400\%$ $\qquad$ **Check:** The answer and the guess are roughly the same.

Double-check: $400\% \times 2.6 = ?$
$4 \quad \times 2.6 = 10.4 = 10\frac{2}{5}$

Type 3 problems require that you find the total given the percent and the percentage or part. Typically they are stated like this:

8.7 is 30% of what number?
or Find a number such that 30% of it is 8.7.
or 8.7 is 30% of a number. Find the number.
or 30% of what number is equal to 8.7?

Step 1

Translate: $30\% \times \square = 8.7$

or $30\% \times \square = 8.7$

Step 2

Label: $\% \times T = P$

Step 3

Rearrange: $\square = \dfrac{8.7}{30\%}$

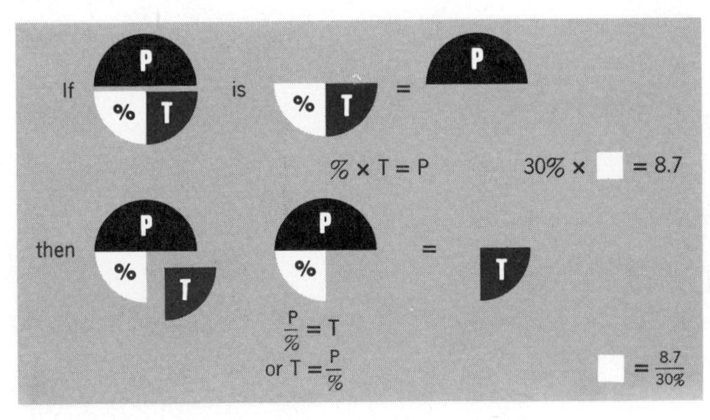

The rearranged problem is

$\square = \dfrac{8.7}{30\%}$

Solve this problem.

Check your answer in **21**.

21 $\square = \dfrac{8.7}{30\%} = \dfrac{8.7}{.30}$ (Remember, never multiply or divide by a percent number. Always convert the percent number to a decimal number.)

Guess: $\dfrac{9}{.3}$ is 30.

$$.3\overline{)9.00} = 30.$$

A reasonable guess is 30.

$\square = 29$

$$.30\overline{)8.70} = 29.$$

$$\begin{array}{r} 6\,0 \\ \hline 2\,70 \\ 2\,70 \\ \hline \end{array}$$

Check: 29 is very close to our guess.

Double-check: 30% of 29 = ?
$.30 \times 29 = 8.7$

Important

We cannot divide by 30%. We must change the percent to a decimal number before we do the division.

Here are a few practice problems to test your mental muscles:

(a) 16% of what number is equal to 5.76?
(b) 41 is 5% of what number?
(c) Find a number such that $12\frac{1}{2}\%$ of it is $26\frac{1}{4}$.
(d) 2 is 8% of a number. Find the number.
(e) 125% of what number is 35?

Check your answers against ours in **22**.

I can tell which is TOTAL and which is PART, because TOTAL is always the larger number. Right?
No, No, No! You can have a percent greater than 100%, meaning that the percentage (or part) is more than the base (or total). 125% of 40 is 50. 40 is the base or reference number and 50 is the percentage. Read the problem to tell which is which.

22 (a) $16\% \times \square = 5.76$

$$\square = \frac{5.76}{16\%}$$

$$T = \frac{P}{\%}$$

$16\% = 0.16$

Guess: $\frac{5.76}{.16} \cong \frac{500}{10}$ or about 50.

$$\square = \frac{5.76}{.16}$$

$\square = 36$

$$\begin{array}{r} 36. \\ .16\overline{)5.76} \\ 4\,8 \\ \hline 96 \\ 96 \\ \hline \end{array}$$

Check: The guess and the answer are reasonably close.

Double-check: $16\% \times 36 = ?$
$.16 \times 36 = 5.76$

(b) $41 = 5\% \times \square$

$$T = \frac{P}{\%}$$

$$\square = \frac{41}{5\%}$$

Guess: $\frac{40}{.05} \cong \frac{4000}{5}$

$$\square = \frac{41}{.05}$$

or about 800.

$\square = 820$

$$\begin{array}{r} 820. \\ 0.05\overline{)41.00} \\ 40 \\ \hline 1\,0 \\ 1\,0 \\ \hline 0 \\ 0 \\ \hline \end{array}$$

Check: The guess and answer are very close.

Double-check: $5\% \times 820 = ?$
$.05 \times 820 = 41$

(c) $26\frac{1}{4} = 12\frac{1}{2}\% \times \square$

or $\square = \frac{26\frac{1}{4}}{12\frac{1}{2}\%}$

$$T = \frac{P}{\%}$$

$12\frac{1}{2}\% = 12.5\% = 0.125$

$$\square = \frac{26.25}{0.125}$$

Guess: $\frac{26}{.1}$ is $\frac{260}{1}$ or 260.

$\square = 210$

$$\begin{array}{r} 210. \\ 0.125\overline{)26.250} \\ 25\,0 \\ \hline 1\,25 \\ 1\,25 \\ \hline 0 \\ 0 \\ \hline \end{array}$$

Check: 210 and 260 are fairly close.

Double-check: $12\frac{1}{2}\% \times 210 = ?$
$.125 \times 210 = 26.25$
$= 26\frac{1}{4}$

(continued on the next page)

(d) $2 = 8\% \times \square$

or $\square = \dfrac{2}{8\%}$ $\qquad T = \dfrac{P}{\%}$

$8\% = 0.08$

$\square = \dfrac{2}{.08}$

Guess: $\dfrac{2}{.08} \cong \dfrac{2}{.1} = 20$

$\square = 25$

$$.08\overline{)2.00} = 25.$$
$$\begin{array}{r} 16 \\ \hline 40 \\ 40 \\ \hline \end{array}$$

Check: 20 and 25 are close enough.

Double-check: $8\% \times 25 = ?$
$.08 \times 25 = 2.00$

(e) $125\% \times \square = 35$

or $\square = \dfrac{35}{125\%}$ $\qquad T = \dfrac{P}{\%}$

$125\% = 1.25$

$\square = \dfrac{35}{1.25}$

Guess: $\dfrac{35}{1.25}$ is less than 35.

Guess about 30.

$\square = 28$

$$1.25\overline{)35.00} = 28.$$
$$\begin{array}{r} 25\,0 \\ \hline 10\,00 \\ 10\,00 \\ \hline \end{array}$$

Check: The guess (30) and answer (28) agree.

Double-check: $125\% \times 28 = ?$
$1.25 \times 28 = 35$

A Review

Let's review those seven steps for solving percent problems:

★ **Step 1. Translate** the problem sentence into a math equation.

★ **Step 2. Label** the numbers as base or total (T), percentage or part (P), and percent (%).

★ **Step 3. Rearrange** the math equation so that the unknown quantity is alone on the left. Use the Equation Finder.

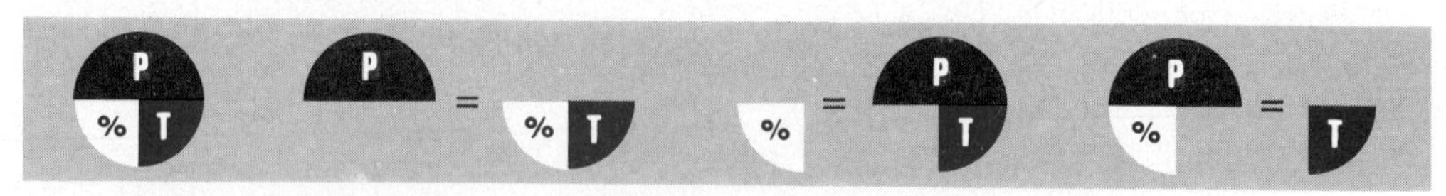

★ **Step 4. Guess.** Get a reasonable estimate of the answer.

★ **Step 5. Solve** the problem by doing the arithmetic. *Always* change percent numbers to decimal numbers first.

★ **Step 6. Check** your answer by comparing it with the guess in Step 4.

★ **Step 7. Double-check** the answer if you can by putting it back into the original problem to see if it is correct.

Are you ready for a bit of practice on the three basic kinds of percent problems? Wind your mind and turn to **23** for a problem set.

24 The simplest practical use of percent is in the calculation of a part or percentage of some total. For example, sales persons are often paid on the basis of their success at selling and receive a *commission* or share of the sales receipts. Commission is usually described as a percent of sales income.

Suppose your job as a door-to-door encyclopedia salesman pays 12% commission on all sales. How much do you earn from the sale of one $400 set of books?

12% of $400 = __?__ **Guess:** 12% is roughly $\frac{1}{10}$ and $\frac{1}{10}$ of $400 is $40.

Commission = 12% × $400
= 0.12 × $400
= $48

Try this one yourself.

At the Happy Bandit Used Car Company each salesman receives a 6% commission on his sales. What would a salesman earn if he sold a 1970 Airedale for $1299.95?

Check your answer in **25**.

---

25 6% of $1299.95 = __?__ **Guess:** 10% of $1200 is $120. The answer will be about half of this or $60.

Commission = 6% × $1299.95
= 0.06 × $1299.95
= $77.9970 or $78.00 rounded off.

The salesman earns a commission of $78 on the sale.

If this same salesman earns $255 in commissions in a given week, what was his sales total for the week?

Translate the question to a basic percent problem and solve it.

Our solution is in **27**.

---

26 Commission = percent rate × total sales cost

$$\$2000 = \square\% \times \$15{,}000$$

$$\square\% = \frac{\$2{,}000}{15{,}000} = \frac{2}{15}$$

Guess: 2 is more than 10% of 15. Somewhere between 10 and 20%.

$$= 13\tfrac{1}{3}\%$$

Ready for some practice? Try these.

(a) A real estate salesman sells a house for $34,500. His usual commission is 7%. How much does he earn on the sale?

(b) All salespeople in the Ace Junk Store receive $60 per week plus a 2% commission. If you sold $975 worth of junk in a week, what would be your income?

(c) A salesman at the Wasteland TV Company sold 5 color TV sets last week and earned $128.70 in commissions. If his commission is 6%, what does a color TV set cost?

The correct solutions are in **28**.

---

27 Commission = 6% × total

$$\$255 = 6\% \times \square$$

$$\square = \frac{\$255}{6\%} = \frac{\$255}{.06}$$

$$\square = \$4250.00$$

Guess: 10% of what equals $250? about $2500. The total sales will be almost double this or $4000 to $5000.

His week's sales total was $4250.

What rate of commission would a salesman be receiving if he sold a boat for $15,000 and received a commission of $2000?

Look in **26** for the answer.

---

28 (a) Commission $= 7\% \times \$34{,}500$
$= .07 \times \$34{,}500$
$= \$2415$

Guess: 10% of $30,000 is $3000. He'll earn almost $3000.

(b) Commission $= 2\% \times \$975$
$= .02 \times \$975$
$= \$19.50$

Income $= \$60 + \$19.50 = \$79.50$

Guess: 2% of $100 is $2. $975 is almost 10 times this or almost $20.

(c) Commission $= 6\% \times$ total

$\$128.70 = 6\% \times \square$

$$\square = \frac{\$128.70}{6\%}$$

$$\square = \frac{\$128.70}{.06}$$

$\square = \$2145$ total sales

Cost per TV set $= \$2145 \div 5$ or $\$429$.

Guess: 6% is about $\frac{1}{16}$ and $16 \times \$128$ is about $2000.

Turn to **29** for a look at another application of percent.

---

29 Another important kind of percent problem involves the idea of *discount*. In order to stimulate sales, a merchant may offer to sell some item at less than its normal price. A *discount* is the amount of money by which the normal price is reduced. It is a percentage or part of the normal price. Several new words are worth learning. *Discount* is the reduction in price. *List price* is the normal, regular, or original price before the discount is subtracted. *Sales price* is the new or discount price. The sales price is always less than the list price, of course. *Discount rate* is a percent number that enables you to calculate the discount as a part of the regular or list price.

Let's try a problem.

The list price of a lamp is $18.50. On a special sale it is offered at 20% off. What is the sale price?

Can you solve it? Try, then turn to **30** for help.

---

30 Discount $= 20\%$ of $\$18.50$
$= 0.20 \times \$18.50$
$= \$3.70$

Guess: 20% is $\frac{1}{5}$ and $\frac{1}{5}$ of $18 is about $3. The sale price will be about $15.

Sale price $=$ list price $-$ discount
$= \$18.50 - \$3.70$
$= \$14.80$

(continued on the next page)

Think of it this way:

Ready for another problem?

After a 25% discount, the sale price of a camera is $144. What was its original or list price?

Check your answer in **31**.

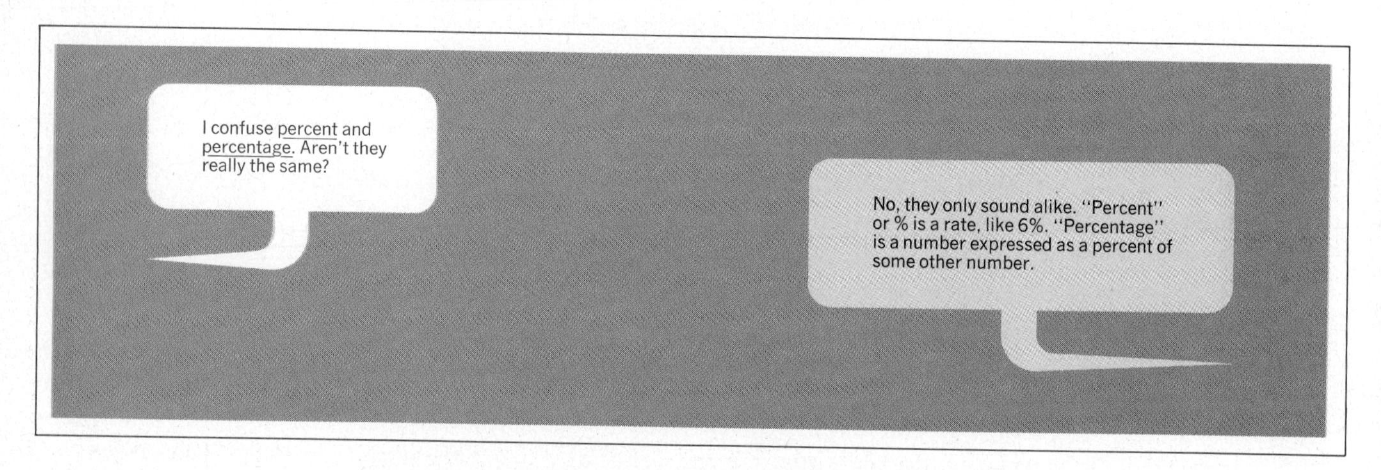

31

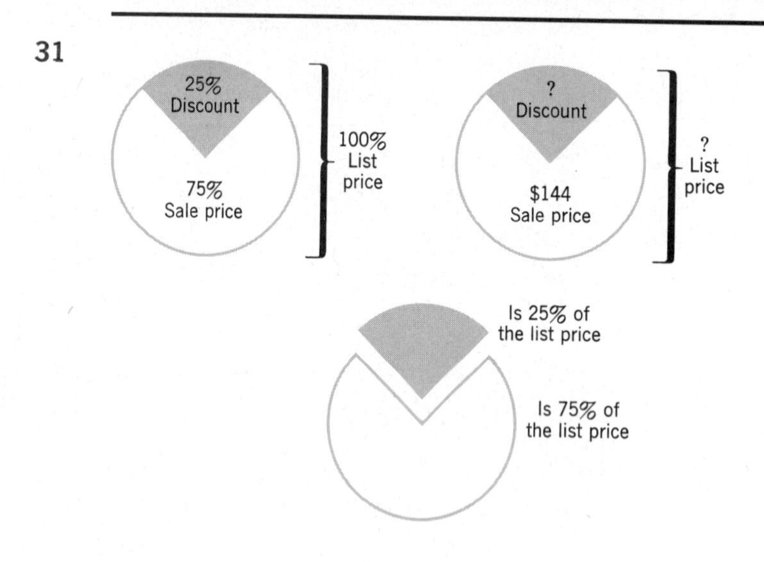

75% of list price = $144

$75\% \times \square = \$144$

$\square = \dfrac{\$144}{75\%} = \dfrac{\$144}{.75}$

$\square = \$192$

Guess: If it cost $200 list price, a 25% or $\frac{1}{4}$ discount would give a sale price of $150.
The list price must be about $200.

Here are a few problems to test your understanding of the idea of discount.

(a) An after-Christmas sale advertises all toys 70% off. What would be the sale price of a model spaceship that cost $19.95 before the sale?
(b) A refrigerator is on sale for $376 and is advertised as "12% off regular price." What was its regular price?
(c) A set of four 740-15 automobile tires is on sale for 15% off list price. What would be the sale price if their list price is $20.80 each?

Work hard at these. Knowing how to do them may save you a lot of money one day.

Check your answers in **32**.

SMALL LOANS

Sooner or later everyone finds it necessary to borrow money. When you do you will want to know beforehand how it works. Suppose you borrow $200 and the loan company specifies that you repay it at $25 per month plus interest at 3% per month on the unpaid balance. What interest do you actually pay?

Month 1	$200 × 0.03 = $ 6.00	you pay	$25 + $6.00 = $ 31.00
Month 2	$175 × 0.03 = $ 5.25	you pay	$25 + $5.25 = $ 30.25
Month 3	$150 × 0.03 = $ 4.50	you pay	$25 + $4.50 = $ 29.50
Month 4	$125 × 0.03 = $ 3.75	you pay	$25 + $3.75 = $ 28.75
Month 5	$100 × 0.03 = $ 3.00	you pay	$25 + $3.00 = $ 28.00
Month 6	$ 75 × 0.03 = $ 2.25	you pay	$25 + $2.25 = $ 27.25
Month 7	$ 50 × 0.03 = $ 1.50	you pay	$25 + $1.50 = $ 26.50
Month 8	$ 25 × 0.03 = $ .75	you pay	$25 + $0.75 = $ 25.75
	$27.00		$227.00

Total interest is $27.00 (↑ $27.00)

Total of 8 loan payments (↑ $227.00)

They might also set it up as 8 equal payments of $227.00 ÷ 8 = $28.375 or $28.38 per month.

(continued on the next page)

The 3% interest rate seems small, but it amounts to about 20% per year.

A bank loan for \$200 at 7% for 8 months would cost you

$$\$200 \times 7\% \times \tfrac{8}{12}$$

or $\$200 \times .07 \times \tfrac{8}{12}$

or \$9.34 Quite a difference.

The loan company demands that you pay more, and in return they are less worried about your ability to meet the payments. For a bigger risk, they want a higher interest.

Calculate the total interest for \$200 borrowed at

a. 2% per month and 8 months
b. 3% per month and 12 months

32 (a) Discount = 70% of \$19.95
= .70 × \$19.95
= \$13.985
= \$13.99 rounded

Sale price = list price − discount
= \$19.95 − \$13.99
= \$5.96

(b) Discount = 12% of list price

Sale price = 88% of list price

$$\$376 = 88\% \times \square$$

$$\square = \frac{\$376}{88\%} = \frac{\$376}{.88}$$

$$\square = \$427.28 \quad \text{rounded}$$

(c) Discount = 15% of list price
= 15% × (4 × \$20.80)
= .15 × \$83.20
= \$12.48

Sale price = list price − discount
= \$83.20 − \$12.48
= \$70.72

Taxes are almost always calculated as a percent of some total amount. *Property taxes* are written as some fraction of the value of the property involved. *Income taxes* are most often calculated from complex formulas that depend on many factors. We cannot consider either income or property taxes here.

A *sales tax* is an amount calculated from the actual price of a purchase and added to the buyer's cost. Retail sales tax rates are set by the individual states in the United States and vary from 0 to 7% of the sales price. A sales tax of 6% is often stated as "6¢ on the dollar," since 6% of $1.00 equals 6¢.

If the retail sales tax rate is 6% in California, how much would you pay in Los Angeles for a pair of shoes costing $16.50?

Try it, then check your work in **33**.

---

33 Sales tax = 6% of $16.50
= 0.06 × $16.50
= $0.99

Guess: 6¢ on each dollar; $16; then the sales tax should be about 6 × 16 or 96¢.

Actual cost = list price + sales tax
= $16.50 + $0.99
= $17.49

Most stores and sales clerks use tables to look up the sales tax and, therefore, do not need to do the arithmetic shown above except on large purchases beyond the range of the tables. However, it is your best interest to be able to check their work.

Here are a few problems to test your ability to calculate sales tax.

If sales tax is 5%, find the sales tax on each of the following.

(a) A pen priced at .39¢ ______

(b) A chair priced at $27.50 ______

(c) A toy priced at $2.95 ______

(d) A new car priced at $2785 ______

(e) A bicycle priced at $126.50 ______

(f) A tube of toothpaste priced at .79¢ ______

Check your answers in **34**.

LIVING AND DYING ON THE INSTALLMENT PLAN

The biggest financial deal most people ever undertake is the purchase of a car. Suppose you buy a new car for \$3600 and pay $\frac{1}{6}$ down and obtain a loan from the finance company for the remainder at $7\frac{1}{2}\%$ to be paid in 24 monthly payments.

The down payment is $\$3600 \div 6 = \$600$
The loan is $\$3600 - \$600 = \$3000$
Interest = $\$3000 \times 7\frac{1}{2}\% \times 2$

(\$3000: Amount of the loan; $7\frac{1}{2}\%$: Interest rate per year; 2: Time in years)

Then

$$\text{Interest} = \$3000 \times 0.075 \times 2 = \$450$$

$$\text{Total to be paid} = \$3000 + \$450 = \$3450$$

$$\text{Payments} = \$3450 \div 24 = \$143.75$$

You pay $\$450 \div 24 = \$18.75$ each month for 24 months to use the "easy payment plan," and you may find it worthwhile to have the car now rather than wait until you can save up the money to buy it with cash.

Repeat the calculation above with a down payment of only \$100 and see what a large difference in total cost that makes. Repeat the calculation using an interest rate of 6% or 9%.

34

	Tax	Total Cost	Calculation
(a)	2¢	41¢	$.05 \times 39¢ = 1.95¢ \cong 2¢$
(b)	\$ 1.38	\$ 28.88	$.05 \times \$27.50 = \$1.375 \cong \$1.38$
(c)	\$ 0.15	\$ 3.10	$.05 \times \$2.95 = \$0.1475 \cong \$0.15$
(d)	\$139.25	\$2924.25	$.05 \times \$2785 = \$139.25$
(e)	\$ 6.33	\$ 132.83	$.05 \times \$126.50 = \$6.325 \cong \$6.33$
(f)	4¢	83¢	$.05 \times 79¢ = 3.95¢ \cong 4¢$

An important characteristic of modern society is that we have set up complex ways to enable you to use someone else's money. A *lender,* with money beyond his needs, supplies cash to a *borrower* whose needs exceed his money. The money is called a loan. *Interest* is the amount the lender is paid for the use of his money. Interest is the money you pay to use someone else's money. The more you use and the longer you use it, the more interest you must pay. When you purchase a house with a bank loan, a car or refrigerator on an installment loan, or gasoline on a credit card, you are using someone else's money and you pay interest for that use. If you are on the other end of the money game, you may earn interest for money you invest in a savings account or in shares of a business.

Suppose you have \$500 in savings earning $5\frac{1}{4}\%$ annually in your local bank, how much interest do you receive in a year?

Do it this way:

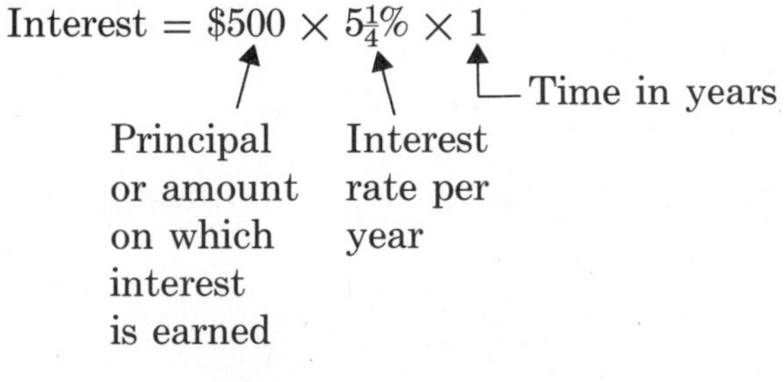

Interest = \$500 × $5\frac{1}{4}\%$ × 1

- \$500: Principal or amount on which interest is earned
- $5\frac{1}{4}\%$: Interest rate per year
- 1: Time in years

Interest = \$500 × 0.0525
= \$26.25

By placing your \$500 in a bank savings account, you allow the bank to use it in various profitable ways, and you are paid \$26.25 per year for its use.

Most of us play the money game from the other side of the counter. Suppose you find yourself in need of cash and arrange to obtain a loan from a bank. You borrow \$600 at 8% per year for 3 months. How much interest must you pay?

Try to set up and solve this problem exactly as we did in the problem above. Our worked solution is in **35**.

HOW DOES A CREDIT CARD WORK?

Many forms of small loans, such as credit card loans, charge interest by the month. These are known as "revolving credit" plans. (If you buy very much money this way, *you* do the revolving and may run in circles for years trying to pay it back!) Essentially, you buy now and pay later. If you repay the full amount borrowed within 25 or 30 days, there is no charge for the loan.

After the first pay period of 25 or 30 days, you pay a percent of the unpaid balance each month, usually between $\frac{1}{2}$ and 2%. In addition you must pay some minimum amount each month, usually \$10 or 10% of the unpaid balance, whichever is larger. Generally you are also charged a small monthly amount for insurance premiums. (The credit card company insures themselves against your defaulting on the loan or disappearing, and you pay for their insurance.)

Let's see how it works. Suppose you go on a short vacation and pay for gasoline, lodging, and meals with your Handy Dandy credit card.

(*continued on the next page*)

A few weeks later you receive a bill for $100. You can pay it within 30 days and owe no interest or you can pay over several months as follows:

Month 1

$100 × 1\frac{1}{2}\% = $100 × 0.015 = $1.50 owe $101.50,

pay $10, and carry $91.50 over to next month

Month 2

$91.50 × 0.015 = $1.38 owe $92.88,

pay $10, and carry $82.88 over to next month

. . . and so on.

A year later you will have repaid the $100 loan and all interest.

The $1\frac{1}{2}\%$ per month interest rate seems small, but it is equivalent to between 15% and 18%. You pay at a high rate for the convenience of using the credit card and the no-questions-asked ease of getting the loan.

35

$$\text{Interest} = \$600 \times 8\% \times \frac{3}{12}$$

Principal or amount loaned — Interest rate per year — Time in years $\frac{3 \text{ months}}{12 \text{ months}} = \frac{3}{12}$

$$\text{Interest} = \$600 \times 0.08 \times \frac{3}{12} = \$12$$

Depending on how you and the bank decide to arrange it, you may be required to pay the total principal ($600) plus interest ($12) all at once at the end of three months, or you may use some sort of regular payment plan—for example, pay $204 each month for three months.

What interest do you pay if you borrow $1200 for two years at $8\frac{1}{2}\%$ per year?

The answer is in **36**.

36

$$\begin{aligned}\text{Interest} &= \$1200 \times 8\tfrac{1}{2}\% \times 2 \text{ years}\\ &= \$1200 \times 0.085 \times 2\\ &= \$204\end{aligned}$$

You repay $\$1200 + \$204 = \$1404$ over two years, perhaps in 24 monthly payments of $\$1404 \div 24$, or \$58.50 each.

Now that we have finished our very brief excursion into the mysteries of high finance, turn to **37** for a set of practice problems on these important concepts.

Unit 4
Percent

38

Self-Test

1. Write $3\frac{1}{6}$ as a percent. ________
2. Write $\dfrac{5}{12}$ as a percent. ________
3. Write 0.08 as a percent. ________
4. Write 6.43 as a percent. ________
5. Write 2% as a decimal. ________
6. Write $112\frac{1}{2}\%$ as a decimal. ________
7. Write 68% as a fraction. ________
8. Find 48% of 250. ________
9. Find 63% of 12. ________
10. Find 165% of 70. ________
11. Find $6\frac{1}{2}\%$ of 134. ________
12. Find 46.5% of 13.4. ________
13. What percent of 26 is 9.1? ________
14. What percent of 75 is 112.5? ________
15. What percent of 0.72 is 1.62? ________
16. What percent of 1.45 is 0.609? ________
17. What percent of $\dfrac{3}{8}$ is $\dfrac{1}{9}$? ________
18. 15% of what number is 12? ________
19. If 120% of a number is 0.85 find the number. (Round to two decimal digits.) ________
20. A gallon of gas costs 42.9¢. If the price is increased 20% find the new cost. ________
21. After a 30% discount an article costs $43.40. Find the original price. ________

April 17, 1978
Date

Stephanie Catanzaro
Name

050/003
Course/Section

22. What commission does a salesman earn on a $370 sale if his commission rate is 16%? ________

23. What is the sale price of a book marked 40% off if its list price is $8.95? ________

24. What is the interest on a $400 loan at $7\frac{1}{4}$% over 30 months? ________

25. If the retail sales tax is 6%, what would be the total cost of a $1.85 toy? ________

Answers are on page 514.

Measurement

PREVIEW 5

Objective	Sample Problems	Where To Go for Help: Page	Frame
Upon successful completion of this program you will be able to:			
1. Explain the nature of measurement.	Which is the more accurate measurement: 8.6 lb or 8.60 lb?	315	1
2. Use information on units to convert a measurement from the units in which it is given to any other appropriate units.	(a) 3 ft 8 in. = ____ ft = ____ yd	319	4
	(b) 1 week = ____ minutes	323	8
	(c) 14 pints = ____ gallons	328	16
	(d) 8 lb = ____ oz	327	15
3. Add, subtract, multiply, and divide measurement numbers.	(a) $2\frac{1}{2}$ lb + 3 lb 4 oz + 25 oz = ____	339	22
	(b) $3\frac{1}{2}$ gal − 1 qt = ____	339	22
	(c) Area of a floor 12 ft by 9 ft 6 in. = ____	341	26
	(d) What is the rate at which you use gasoline if you travel 150 miles on 12 gal? ____	341	26
4. Work with the common metric units and convert from one metric unit to another.	(a) Estimate your height in meters ____	347	30
	(b) 40 kilometers = ____ m = ____ cm = ____ miles	349	32
	(c) 75 kilograms = ____ grams, weighs ____ lb	359	39
	(d) 3 qt = ____ liters	367	43
	(e) A speed of 50 mph is equal to ____ km/hr (to the nearest 10 km/hr)	354	37
	(f) 5 ft 6 in. = ____ m	347	30
	(g) 100° F ≅ ____ ° C.	372	48

Date ____________

Name ____________

Course/Section ____________

(continued on the next page)

Objective	Sample problem	Page	Frame
5. Calculate the area and volume of simple shapes and objects.	(a) The area of a piece of fabric 50 cm wide and $3\frac{1}{2}$ meters long is ____ sq m.	385	53
	(b) The volume of a box 2 ft wide by 1 ft 6 in. high by $3\frac{1}{2}$ ft long is ____ ft^3.	391	60
	(c) The area of a lot 150 ft by 318 ft is ____ yd^2.	385	53
	(d) The volume of a cylinder container of water 20 cm in diameter and 40 cm long is ____ cm^3.	391	60
	(e) The perimeter of the lot in (c) is ____ ft.	390	59

(Answers to these problems are at the bottom of this sheet.)

If you are certain you can work all of these problems correctly, turn to page 399 for a self-test. If you want help with any of these objectives or if you cannot work one of the sample problems, turn to the page indicated. Super-students, who want to be certain they learn everything in this unit, will turn to frame 1 and begin work there.

PREVIEW 5

Answers to Sample Problems

1. 8.60 lb.
2. (a) $3\frac{2}{3}$ ft, $1\frac{2}{9}$ yd
   (b) 10,080 minutes
   (c) $1\frac{3}{4}$ gallons
   (d) 128 oz
3. (a) 7 lb 5 oz
   (b) 13 qt or $3\frac{1}{4}$ gal
   (c) 114 sq ft
   (d) 12.5 miles/gal
4. (a) 5 ft 4 in. = 1.62 m
   6 ft = 1.83 m
   (b) 40,000 m, 4,000,000 cm, about 24.9 mi
   (c) 75,000 g, 165 lb
   (d) about 2.8 liters
   (e) 80 km/hr
   (f) about 1.68 m
   (g) 38° C (rounded)
5. (a) 1.75 sq m
   (b) 10.5 cu ft
   (c) 5300 yd^2
   (d) 12.6 liters (rounded)
   (e) 936 ft

Measurement

1 Charlie Brown and his friends will spend most of their lives in the twenty-first century, living and working in a very complex society based on technology and the numbers that go with it. They will use measurement numbers where they work, find them attached to things they buy, and worry about them in the games they play. Like most people, they will find that the most important numbers in their everyday lives are those that answer the questions "How big?" "How small?" "How much?" "How far?"—6 feet tall, 3 seconds, $20, 26 miles. Sally will soon learn that for most of the world a gram is not a nice little old lady married to a grampa, but part of the language used to describe measurements.

Unit Conversion

A *measurement* is a description of some quantity. It tells you the amount or size or number or strength of something. We cannot measure everything—no one has measured love or joy, the beauty of spring or the promise of a summer's day—but things that can be measured are often important. It may be important to know that you weigh 140 pounds, that you drive 8 miles to work, or that the pills your doctor prescribed contain 5 milligrams of the needed drug and not 5 grams.

(*continued on the next page*)

What is the length of this line?

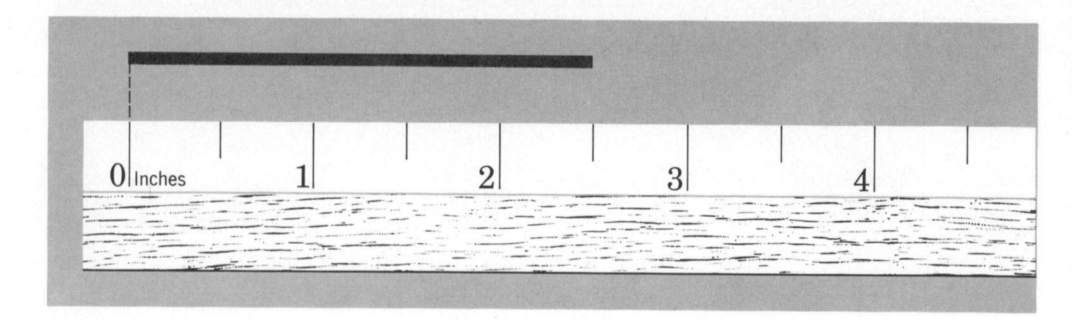

Write your answer here.

Length = ______

and then turn to 2.

2

Length = 2.5 inches

Every measurement description must answer three questions.

1. What is the size or *amount* of the measurement?

   In this case the number part of the measurement is 2.5. Of course, the number portion of the description may be a whole number, fraction, or decimal number.

2. What is the comparison standard or *unit* used in the measurement?

   In this case the units are inches. The length being measured is 2.5 times as long as a 1-inch standard unit.

   $$\text{Length} = \underbrace{2.5}_{\text{Amount}} \times \underbrace{1\ \text{inch}}_{\text{Units}}$$

3. What is the *accuracy* of the measurement?

   The number part of the description has a built-in measure of accuracy.

When a length is written as 7 inches it implies that the measurement scale used is only accurate enough to tell you that the actual length is somewhere between $6\frac{1}{2}$ and $7\frac{1}{2}$ inches. The actual length is closer to 7 inches than to 6 or 8 inches.

Any of the lengths shown on page 317 would be measured as 7 inches by the scale shown.

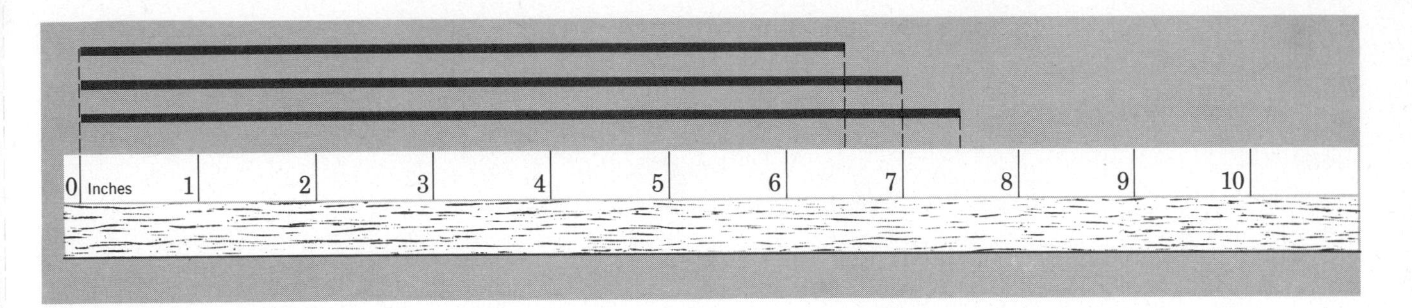

We know the length to 0.5 inch either way.

If a length is written as 7.0 inches, it implies that the measurement scale used is marked to tenths of an inch. Such a measurement assures us that the actual length is between 6.95 inches and 7.05 inches. The length is closer to 7.0 inches than to 6.9 inches or 7.1 inches.

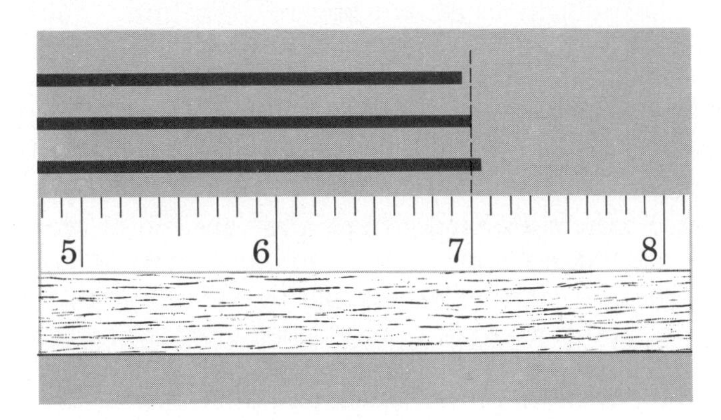

We know the length to 0.05 inch either way.

A football coach claims that his star halfback runs the 100-yard dash in "ten flat." How accurate is a time of 10 seconds? Compare it with the track coach's measurement of a time of 10.0 for his star sprinter.

Write your answer here . . .

. . . then turn to 3.

---

3 When a time is given as 10 seconds, we should understand this to mean that the time has been measured to the nearest second (perhaps with a wristwatch) and the actual time is somewhere between 9.5 and 10.5 seconds. A time given as 10.0 seconds has been measured to the nearest tenth of a second (with a stopwatch), and the actual time is between 9.95 and 10.05 seconds.

This kind of information about how the measurement was made is a natural part of every measurement description.

In addition to a number telling its size and accuracy, every measurement has a word or symbol giving the units used. A measurement *unit* compares the size of the quantity being measured to some standard. For example, when we write that the length of a car is 14 feet, we are saying that the basic length unit, 1 foot, must be repeated 14 times to equal the length of the car.

(*continued on the next page*)

Length = 14×1 foot

The size of the measurement is 14 and the units are feet.

Money units are the simplest, most familiar units in everyday use. The cost of a book might be \$8, but never 8. The unit "dollars," abbreviated \$, must be given as part of the number—the cost is not 8 cents or 8 pesos or 8 francs.

Cost = 8 dollars = 8×1 dollar

We have many different measurement units for each of the quantities length, weight, time, area, volume, and so on. Most units have no scientific basis; none of them are God-given. They are simply chosen for convenience. We can use *natural units* such as a *pinch* of salt, a *handful* of sand, an *armload* of groceries, or the *blink* of an eye and everyone knows roughly what we mean. But if we want to buy or sell a *truckload* of coal or a *tank* of gasoline, we must be certain everyone agrees on the size of our *truckload* and *tank* units. To be useful to everyone, units must be standardized.

Most units started as natural, body-related measures. A length of one *digit* in early England was the width of a man's finger. Four digits was equal to one *hand,* and even today the height of horses is measured in hands. And so the units piled up:

2 hands = 1 *span,* an open hand from tip of thumb to tip of little finger.
2 spans = 1 *cubit,* the distance from fingertips to elbow (*cubitum* is Latin for elbow).
2 cubits = 1 arm or 1 *yard,* the distance from chin to outstretched fingertips.

and so on.

The *foot* is probably the oldest length unit, dating back to the ancient Egyptians, Greeks, and Romans. The problem with such units is that their size depends on whose finger, hand, arm, or foot you use. The foot was a different unit in every town. By the eighteenth century, business, science, and travel were being done on a worldwide basis, and standard units were needed. The *yard* was defined as the distance between marks on a brass bar kept in a vault in London. The *foot* was defined as exactly one-third of this standard yard. Today in almost every nation of the world, including the United States, units of length such as the inch, foot, yard, and mile are defined in terms of the *meter,* a unit we shall discuss later in this chapter.

Distance

The basic English units of length or distance and their usual abbreviations are related as shown below.

1 foot (ft) = 12 inches (in.)
1 yard (yd) = 3 feet
1 mile (mi) = 5280 feet

These three equations give all the information you need to know to work with English length units. But you should develop the ability to translate from one unit to another. For example, write the length 5 feet in inches.

5 feet = ______ inches

Work it out, then turn to 4.

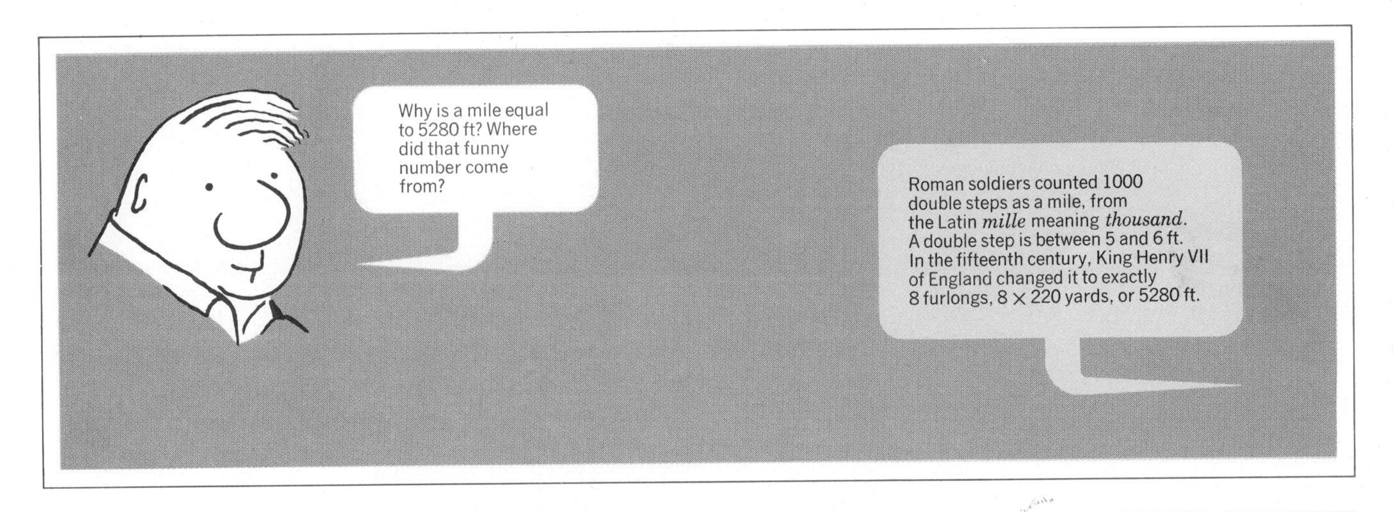

---

4 5 feet = $\underline{60}$ inches

You probably worked this problem by having a little conversation with yourself that went something like this: "Twelve inches in a foot, so 5 feet would be 5 times 12 or . . . 60 inches." The method is simple and quite correct, but it may not help you when the problem is more difficult.

Try it this way:

First, write 5 feet as 5 feet = 5×1 ft,

then, multiply to change the units:

(*continued on the next page*)

$$5 \text{ feet} = (5 \times \cancel{1 \text{ ft}}) \times \left(\frac{12 \text{ in.}}{\cancel{1 \text{ ft}}}\right) = 5 \times 12 \text{ in.} = 60 \text{ in.}$$

Because 1 ft = 12 in. we know that the fraction $\frac{12 \text{ in.}}{1 \text{ ft}}$ is equal to 1. Any fraction in which numerator and denominator are equal, is equal to 1.

$$\frac{2}{2} = 1 \qquad \frac{3}{3} = 1 \qquad \frac{4760}{4760} = 1$$

In fact $\frac{\square}{\square} = 1$ where you can put any number you like in place of $\square$ except zero.

The product of any number and one is the original number.

$$2 \times 1 = 2$$
$$73 \times 1 = 73$$

or $\square \times 1 = \square$ for any number in place of $\square$

Multiplying by the fraction $\left(\frac{12 \text{ in.}}{1 \text{ ft}}\right)$ enables us to cancel out the 1-ft units we don't want and end up with the inch units we do want.

A fraction such as $\left(\frac{12 \text{ in.}}{1 \text{ ft}}\right)$, with a value of one, is called a *unity fraction.* Unity fractions are very helpful when you work with measurement numbers.

Try this problem:

3.5 yards = ______ ft

Work it out using unit fractions, then check your answer in **5**.

I write 6 ft as 6′ and 4 in. as 4″. Isn't that ok?
That's an old abbreviation and many people still use it, but it can get confusing. In this book we'll use the abbreviations in. and ft for inches and feet.

5 $3.5 \text{ yards} = (3.5 \times 1\,\cancel{\text{yd}}) \times \left(\frac{3 \text{ ft}}{1\,\cancel{\text{yd}}}\right) = 3.5 \times 3 \text{ ft} = 10.5 \text{ ft}$

Of course, the equation 1 yd = 3 ft can be used to create two unity fractions, $\left(\frac{1 \text{ yd}}{3 \text{ ft}}\right)$ and $\left(\frac{3 \text{ ft}}{1 \text{ yd}}\right)$.

The trick is to choose the unity fraction that allows you to cancel the unwanted units on top and bottom of the fraction, leaving you with only the units you need.

This may seem like unnecessary work to solve such a simple problem, but without unity fractions some problems are very difficult. For example, try this one:

In Martian length measurements there are 3 gronks in a smersh, and 2 pflugs equal one smersh. Translate a length of 6 gronks to pflugs.

6 gronks = ______ pflugs

Try it with unity fractions.

(If you want to tie your brain in a knot, try it without unity fractions.)

The answer is in **7**.

---

6 $4 \text{ yards} = 4 \times 1\,\cancel{\text{yd}} \times \left(\frac{3\,\cancel{\text{ft}}}{1\,\cancel{\text{yd}}}\right) \times \left(\frac{12 \text{ in.}}{1\,\cancel{\text{ft}}}\right)$

$= 4 \times 3 \times 12 \text{ in.} = 144 \text{ in.}$

Here are a few practice problems to help you get the use of unity fractions into your muscles.

(a) 6.25 ft = ______ in.

(b) $5\frac{1}{4}$ yd = ______ ft

(c) 2 mi = ______ yd

(d) 45 in. = ______ yd

The solutions are in **8**.

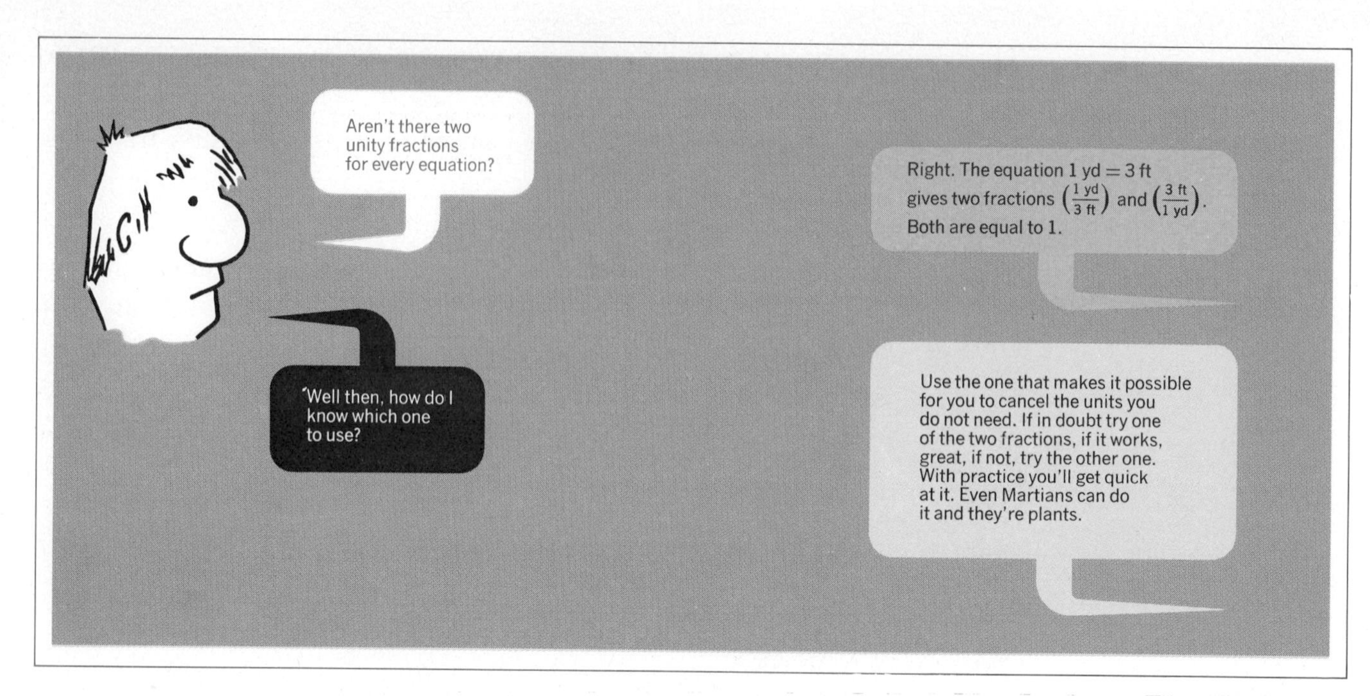

7 The smart Martian does it this way:

$$6 \text{ gronks} = 6 \times (1 \cancel{\text{gronk}}) \times \left(\frac{1 \cancel{\text{smersh}}}{3 \cancel{\text{gronks}}}\right) \times \left(\frac{2 \text{ pflugs}}{1 \cancel{\text{smersh}}}\right)$$

$$= \frac{6 \times 2}{3} \text{ pflugs} = 4 \text{ pflugs}$$

Write a length of 4 yards in inches.

4 yards = ______ inches.

Check your answer back in **6**.

---

8 (a) $6.25 \text{ ft} = 6.25 \times 1 \cancel{\text{ft}} \times \left(\frac{12 \text{ in.}}{1 \cancel{\text{ft}}}\right) = 6.25 \times 12 \text{ in.} = 75 \text{ in.}$

(b) $5\frac{1}{4} \text{ yd} = 5\frac{1}{4} \times 1 \cancel{\text{yd}} \times \left(\frac{3 \text{ ft}}{1 \cancel{\text{yd}}}\right) = 5\frac{1}{4} \times 3 \text{ ft} = 15\frac{3}{4} \text{ ft}$

(c) $2 \text{ mi} = 2 \times 1 \cancel{\text{mi}} \times \left(\frac{5280 \cancel{\text{ft}}}{1 \cancel{\text{mi}}}\right) \times \left(\frac{1 \text{ yd}}{3 \cancel{\text{ft}}}\right)$

$$= \frac{2 \times 5280}{3} \text{ yd} = 3520 \text{ yd}$$

(d) $45 \text{ in.} = 45 \times 1 \cancel{\text{in.}} \times \left(\frac{1 \cancel{\text{ft}}}{12 \cancel{\text{in.}}}\right) \times \left(\frac{1 \text{ yd}}{3 \cancel{\text{ft}}}\right)$

$$= \frac{45 \times 1 \text{ yd}}{12 \times 3} = \frac{45}{36} \text{ yd} = 1.25 \text{ yd}$$

Time The basic *time* units and their usual abbreviations are related as shown below:

1 minute (min)	= 60 seconds (sec)
1 hour (hr)	= 60 minutes
1 day (d)	= 24 hours
1 week (wk)	= 7 days
1 calendar year (yr)	= 365 days

The mean solar day, the average time needed for the earth to turn once on its axis, is the basic practical unit of time. It is a convenient sized time interval and available to everyone, so we define all other time units in terms of it.

Use unity fractions to convert a time of 3 weeks into minutes.

3 weeks = ______ minutes

Check your work in **9**.

---

9 $3 \text{ weeks} = 3 \times 1 \cancel{\text{wk}} \times \left(\frac{7 \cancel{\text{days}}}{1 \cancel{\text{wk}}}\right) \times \left(\frac{24 \cancel{\text{hr}}}{1 \cancel{\text{day}}}\right) \times \left(\frac{60 \text{ min}}{1 \cancel{\text{hr}}}\right)$

$= 3 \times 7 \times 24 \times 60 \text{ min} = 30{,}240 \text{ min}$

Three weeks is roughly equal to 30,000 minutes. Exactly 3 weeks is exactly equal to 30,240 minutes.

A certain machine in a factory has been operating for a total of 408 hours. How many weeks would this be? Solve the problem and round your answer to the nearest hundredth of a week.

Turn to **11** when your work is done.

---

10 Right you are!

The area measurement number 274 sq ft is accurate to the nearest 1 sq ft or 1 part in 274 or roughly 1 part in 200.

An easy way to determine the accuracy of a measurement number is to count significant digits. A *significant digit* is one arrived at by actual measurement.

35 sec has 2 significant digits
256 ft has 3 significant digits
4052 miles has 4 significant digits

To find the number of significant digits start counting digits on the left of the number, and

(1) Ignore any zeros on the left, as in 0.4 or 0.025.
(2) Count as significant all digits on the right of a decimal number.
(3) Ignore any zeros on the right of a whole number.

(continued on the next page)

For example,

402 ft	has 3 significant digits (zeros in the body of a number always count).
0.0025 sec	has 2 significant digits (zeros on the left are never significant).
1.20 lb	has 3 significant digits (zeros on the right of the decimal point are always significant).
250 miles	has 2 significant digits (zeros on the right of a whole number do not count as significant unless you know more about the measurement).

Use these rules to find the number of significant digits in each of the following measurement numbers:

(a) 24 weeks ______ (b) 2.4 yd ______

(c) 2.40 in. ______ (d) 204 days ______

(e) 0.24 yr ______ (f) 0.024 sec ______

(g) 0.0240 sec ______ (h) 24.0 inches ______

(i) 240.0 miles ______ (j) 240 ft ______

(k) $24,000,000 ______ (l) 0.0004 sec ______

Check your answers in **13**.

11

$$408 \text{ hours} = 408 \times 1 \cancel{\text{hr}} \times \left(\frac{1 \cancel{\text{day}}}{24 \cancel{\text{hr}}}\right) \times \left(\frac{1 \text{ week}}{7 \cancel{\text{days}}}\right)$$

$$= \frac{408}{24 \times 7} \text{ weeks} = 2.42857\ldots \text{ weeks.}$$

$$\cong 2.43 \text{ weeks.}$$

The answer should be rounded to the nearest hundredth of a week. The answer 2.43 weeks agrees in accuracy with the original measurement 408 hours. (If you need to review rounding, return to page 204, frame **20** for a quick refresher.

Rounding measurement numbers is almost always necessary and is often quite tricky. Let's look at it in detail.

When you make a calculation with measurement numbers, you must always round your answer so that it agrees in accuracy with the least accurate measurement number used in the calculation. In the problem above, the numbers 24 and 7 come from the definition of hour and week and are exact of course, but the number 408 is a measurement number. Someone measured the time to the nearest hour, an accuracy of 1 hour in 408, or about 1 part in 400. If we write the answer as 2.4285 weeks, we would be stating the accuracy as 1 part in 24285, or about 1 part in 24,000. That is much more accurate than the actual measurement. We would be exaggerating and misleading. In order to be honest about it and not deceive anyone, we round the answer to about the same accuracy as the original number. We round the answer to 2.43 weeks because 243 is accurate to 1 part in 243 and that is roughly the same as 1 part in 400.

Suppose you calculated the area of a room to be 273.62 sq ft. How would you write this area if the numbers used to calculate it were accurate to only 1 part in 200?

Choose an answer.

(a) 274 sq ft   Go to **10**.
(b) 270 sq ft   Go to **12**.

---

12 You are incorrect.

An area of 270 sq ft is accurate to within 10 sq ft or 10 parts in 270 or 1 part in 27. Because the numbers used to calculate this area were accurate to 1 part in 200, you have rounded too much. Your answer should be as accurate as the numbers used to get it, no more, no less.

Return to **11** and try again.

13

Measurement Number	Number of Significant Digits	
(a) 24 weeks	2	Count all non-zero digits.
(b) 2.4 yd	2	
(c) 2.40 in.	3	Count all zeros on the right of a decimal number.
(d) 204 days	3	Count all zeros in the body of the number.
(e) 0.24 yr	2	Zeros on the left never count. This zero is designed to keep you from overlooking the decimal point.
(f) 0.024 sec	2	
(g) 0.0240 sec	3	Count all zeros on the right of a decimal number.
(h) 24.0 inches	3	
(i) 240.0 miles	4	
(j) 240 ft	2	Do not count as significant any digit on the right of a whole number unless you know more about the number. This number tells us that the length was measured to the nearest 10 ft.
(k) $24,000,000	2	The measurement is "about 24 million" or to the nearest million.
(l) 0.0004 sec	1	

Try this:

28.60857 ft = 28.6086 ft rounded to 6 significant digits

= ______ ft rounded to 5 significant digits

= ______ ft rounded to 4 significant digits

= ______ ft rounded to 3 significant digits

= ______ ft rounded to 2 significant digits

= ______ ft rounded to 1 significant digit

Work carefully, then check your answers in **14**.

---

14 28.60857 ft = 28.6086 ft rounded to 6 significant digits

= 28.609 ft rounded to 5 significant digits

= 28.61 ft rounded to 4 significant digits

= 28.6 ft rounded to 3 significant digits

= 29 ft rounded to 2 significant digits

= 30 ft rounded to 1 significant digit

Use unity fractions made from the equations in frame **8** to convert each of these time measurements. Round your answer to agree with the original measurement.

(a) 4.0 days = ______ minutes

(b) 4 days = ______ minutes

(c) 2500 minutes = ______ days

(d) 1 million seconds = ______ weeks

Look in **15** for the solutions when you have finished.

---

15 (a) $4.0 \text{ days} = 4.0 \times (1 \cancel{\text{day}}) \times \left(\frac{24 \cancel{\text{hr}}}{1 \cancel{\text{day}}}\right) \times \left(\frac{60 \text{ min}}{1 \cancel{\text{hr}}}\right)$

$= 4.0 \times 24 \times 60 \text{ min} = 5760 \text{ min}$

$\cong 5800$ min rounded to two significant digits to agree in accuracy with 4.0 days.

(The symbol $\cong$ means that 4.0 days is "approximately equal" to 5800 min. We have rounded the answer and they are not exactly equal. They are measurement numbers with the same accuracy.)

(b) 4 days ≅ 6000 min   rounded to one significant digit to agree in accuracy with 4 days.

(c) $2500 \text{ min} = 2500 \times (1 \cancel{\text{min}}) \times \left(\frac{1 \cancel{\text{hr}}}{60 \cancel{\text{min}}}\right) \times \left(\frac{1 \text{ day}}{24 \cancel{\text{hr}}}\right)$

$= \frac{2500}{60 \times 24} \text{ days} = 1.736\ldots \text{ days}$

$\cong 1.7$ days   rounded to two significant digits to agree in accuracy with 2500 minutes.

(d) $1 \text{ million sec} = 1{,}000{,}000 \times (1 \cancel{\text{sec}}) \times \left(\frac{1 \cancel{\text{min}}}{60 \cancel{\text{sec}}}\right) \times \left(\frac{1 \cancel{\text{hr}}}{60 \cancel{\text{min}}}\right) \times \left(\frac{1 \cancel{\text{day}}}{24 \cancel{\text{hr}}}\right) \times \left(\frac{1 \text{ wk}}{7 \cancel{\text{days}}}\right)$

$= \frac{1{,}000{,}000}{60 \times 60 \times 24 \times 7} \text{ weeks} = 1.653\ldots \text{ weeks}$

$\cong 2$ weeks   rounded to one significant digit to agree in accuracy with 1 million seconds.

Weight   The basic *weight* units and their usual abbreviations are related as shown below:

1 ounce (oz) = 437.5 grains (gr)
16 ounces = 1 pound (lb)
2000 pounds = 1 ton (t)

These are the common units used for most ordinary weight measurements by grocery stores, factories, housewives, and engineers.

Unfortunately, and for no very good reason, there is a second set of weight units used to measure drugs and precious metals such as gold, silver, and platinum. This *druggist system* is used by doctors prescribing drugs and by pharmacists or nurses preparing or weighing drugs and medicine. This is the *Troy* or *Apothecaries* system of weight units.

60 grains = 1 dram (*t*)
8 drams (*t*) = 1 ounce (*t*)
12 ounces (*t*) = 1 pound (*t*)

Notice that the letter *t*, for Troy, is added to distinguish the druggist's ounce or pound from the ordinary ounce or pound.

Both of these sets of units start with the *grain* as the smallest unit. Centuries ago the grain was defined as the weight of a single barley seed. The phrases "grain of truth" and "take it with a grain of salt" come from this very old unit. Both the common English weight units and the druggist's weight units are defined in terms of the grain. An aspirin tablet weighs 5 grains and a drop of water weighs about 1 grain.

Use unity fractions to solve this problem involving weight:

In order to decide the postage needed for mailing, find the weight in ounces of a $3\frac{1}{2}$ pound package.

Check your answer in **16**.

Why 2000 lb in
a ton? Why
not a nice round
number like 1000?

The word *ton* comes from
the word *tub* and a
winemakers *tub* or *tun*
traditionally fills 250
one gallon bottles of
wine each weighing about 8 lb.

---

16 $3\frac{1}{2}\text{ lb} = 3\frac{1}{2} \times (1\ \cancel{\text{lb}}) \times \left(\frac{16\text{ oz}}{1\ \cancel{\text{lb}}}\right) = 3\frac{1}{2} \times 16\text{ oz}$

$= 56\text{ oz}$

Two basic kinds of volume units are commonly used in the United States: measures of dry volume and measures of liquid volume. The common units for *dry measure* are

2 pints = 1 quart
8 quarts = 1 peck
4 pecks = 1 bushel

A peck of material would fill a cubic box about 8 inches on a side. A bushel is the volume of a cubic box about 13 inches on a side.

These volume units are used to measure the volume of any sort of dry material, such as grain, fruits and vegetables. When the United States was a farming nation these units were very much used on farms and in grocery stores. Today such materials are measured and sold by weight—you buy strawberries or tomatoes by the pound rather than the peck.

The basic volume units for liquid measure are

16 fluid ounces (fl oz) = 1 pint (pt)
2 pints = 1 quart (qt)
4 quarts = 1 gallon (gal)
$31\frac{1}{2}$ gallons = 1 barrel (bl)

One gallon of liquid fills a cubic container about $6\frac{1}{8}$ inches on each edge. One *quart* is one-*quarter* of a gallon.

Even though the words are the same, a fluid *ounce* is not the same as an *ounce* of weight. A gallon of water does not weigh the same as a gallon of gasoline or a gallon of paint. A volume measure tells you the space the material occupies; a weight measure tells you how heavy it is.

Complete this statement: $3\frac{1}{2}$ gallons of water contains ______ fluid ounces.

Check your answer in **17**.

---

17 $3\frac{1}{2}$ gallons $= 3\frac{1}{2} \times (1 \text{ gal}) \times \left(\frac{4 \text{ qt}}{1 \text{ gal}}\right) \times \left(\frac{2 \text{ pt}}{1 \text{ qt}}\right) \times \left(\frac{16 \text{ fl oz}}{1 \text{ pt}}\right)$

$= 3\frac{1}{2} \times 4 \times 2 \times 16$ fl oz

$= 448$ fl oz

It may help you to translate these units into common kitchen units:

3 teaspoonfuls (tsp) = 1 tablespoonful (tbsp)
2 tablespoonfulls = 1 fluid ounce
8 fluid ounces = 1 cup
2 cups = 1 pint

(*continued on the next page*)

The cup unit is an 8-oz drinking glass sized cup and not a little tea cup.

Doctors, druggists, and nurses use two smaller units of liquid volume when working with liquid medicines:

60 minims (m) = 1 fluid dram (fl dr)
8 fluid drams = 1 fluid ounce (fl oz)

In kitchen units one minim is about a drop and one dram is roughly equal to a teaspoonful. The doctor gives instructions to the druggist in drams and minims, and the druggist writes the label on the prescription bottle in teaspoonfuls and drops.

We have listed only a few of the hundreds of volume units that have been invented and used since the Egyptians started it all off 3000 years ago when they measured grain by the handful and beer by the mouthful. If you are confused by the number of units, discouraged by the number of equations relating units, lost in the 2s, 4s, 8s, 16s, and 60s, and upset over the use of the same name to represent different units, you are beginning to understand why most of the world is either shifting to or already using metric units. We shall look at the metric system in detail later in this unit. First let's get a little more practice in working with units.

Complete these:

(a) 400 fluid ounces = ______ quarts

(b) 2.5 cups = ______ teaspoonfuls

(c) 1 quart = ______ tablespoonfuls

(d) 1 gallon = ______ cups

Check your answers in **19**.

18

$$12 \text{ oz} = 12 \times (1 \not{\text{oz}}) \times \left(\frac{1 \not{\text{lb}}}{16 \not{\text{oz}}}\right) \times \left(\frac{\$1.12}{1 \not{\text{lb}}}\right)$$

$$= \frac{12 \times \$1.12}{16} = \$0.84 \text{ or } 84¢$$

The phrase "\$1.12 per pound" can be written as the unity fraction $\frac{\$1.12}{1 \text{ lb}}$.

Numerator and denominator of the fraction are equivalent quantities.

Try these for practice:

(a) A certain kind of candy is on sale at 6 lb for \$10. What should a 20-ounce box cost? ______

(b) If you sell lemonade for 5¢ per 8-ounce glass at the office picnic, how much profit will you earn from a 5-gallon bottle of lemonade costing $1.98?______

(c) At the supermarket an 18-oz. package of berries sells for 63¢, but they are selling for only 45¢ per pound at a roadside market. How much do I save buying 2 lb at the roadside stand instead of at the supermarket?______

(d) If health-food cookies sell for $1.25 per lb and they weigh 0.8 oz each, how much should you pay for one?______

(e) A certain cookie recipe calls for 2 tsp of vanilla for each batch of cookies. How many batches of cookies can you make from a bottle containing 4 fl oz of vanilla?______

When you have finished these turn to **20** for the correct solutions.

19

(a) $$400 \text{ fl oz} = 400 \times (1 \text{ fl } \cancel{\text{oz}}) \times \left(\frac{1\ \cancel{\text{pint}}}{16 \text{ fl } \cancel{\text{oz}}}\right) \times \left(\frac{1 \text{ qt}}{2\ \cancel{\text{pints}}}\right)$$

$$= \frac{400}{16 \times 2} \text{ qt} = 12.5 \text{ qt}$$

(b) $$2.5 \text{ cups} = 2.5 \times (1\ \cancel{\text{cup}}) \times \left(\frac{8 \text{ fl } \cancel{\text{oz}}}{1\ \cancel{\text{cup}}}\right) \times \left(\frac{2\ \cancel{\text{tbsp}}}{1 \text{ fl } \cancel{\text{oz}}}\right) \times \left(\frac{3 \text{ tsp}}{1\ \cancel{\text{tbsp}}}\right)$$

$$= 2.5 \times 8 \times 2 \times 3 \text{ tsp}$$

$$= 120 \text{ tsp}$$

(c) $$1 \text{ quart} = 1\ \cancel{\text{qt}} \times \left(\frac{2\ \cancel{\text{pints}}}{1\ \cancel{\text{qt}}}\right) \times \left(\frac{8 \text{ fl } \cancel{\text{oz}}}{1\ \cancel{\text{pt}}}\right) \times \left(\frac{2 \text{ tbsp}}{1 \text{ fl } \cancel{\text{oz}}}\right)$$

$$= 2 \times 8 \times 2 \text{ tbsp}$$

$$= 32 \text{ tbsp}$$

(d) $$1 \text{ gallon} = 1\ \cancel{\text{gal}} \times \left(\frac{4\ \cancel{\text{qt}}}{1\ \cancel{\text{gal}}}\right) \times \left(\frac{2\ \cancel{\text{pint}}}{1\ \cancel{\text{qt}}}\right) \times \left(\frac{2 \text{ cups}}{1\ \cancel{\text{pint}}}\right)$$

$$= 4 \times 2 \times 2 \text{ cups}$$

$$= 16 \text{ cups}$$

Some problems involve cost or other factors in addition to standard units. For example,

What is the cost of a 12-ounce package of cheese that sells for $1.12 per pound?

Work it out using unity fractions, then turn to **18**.

Is it ok to abbreviate 6 pounds as 6#?

You will confuse less people if you write it out or use the standard abbreviation lb.

20

(a) $\text{Cost} = 20 \times (1\ \cancel{\text{oz}}) \times \left(\dfrac{1\ \cancel{\text{lb}}}{16\ \cancel{\text{oz}}}\right) \times \left(\dfrac{\$10}{6\ \cancel{\text{lb}}}\right)$

$= \dfrac{20 \times \$10}{16 \times 6} = \$2.083\ldots$

$= \$2.09$ (REMEMBER, with money we always round *up*. If in doubt, the buyer loses.)

(b) $5 \text{ gallons} = 5 \times (1\ \cancel{\text{gal}}) \times \left(\dfrac{4\ \cancel{\text{qt}}}{1\ \cancel{\text{gal}}}\right) \times \left(\dfrac{2\ \cancel{\text{pints}}}{1\ \cancel{\text{qt}}}\right) \times \left(\dfrac{16\ \cancel{\text{oz}}}{1\ \cancel{\text{pint}}}\right) \times \left(\dfrac{5¢}{8\ \cancel{\text{oz}}}\right)$

$= \dfrac{5 \times 4 \times 2 \times 16 \times 5¢}{8} = 400¢ = \$4.00$

Profit $= \$4.00 - 1.98 = \$2.01$

(c) Cost of 2 lb at supermarket $= 2 \times (1\ \cancel{\text{lb}}) \times \left(\dfrac{45¢}{1\ \cancel{\text{lb}}}\right)$

$= 90¢$

Cost of 2 lb at roadside stand $= 2 \times (1\ \cancel{\text{lb}}) \times \left(\dfrac{16\ \cancel{\text{oz}}}{1\ \cancel{\text{lb}}}\right) \times \left(\dfrac{63¢}{18\ \cancel{\text{oz}}}\right)$

$= \dfrac{2 \times 16 \times 63¢}{18}$

$= 112¢ = \$1.12$

Savings $= \$1.12 - \$.90 = \$.22$ or $22¢$

(d) Cost of one cookie $= 0.8 \times (1 \not{oz}) \times \left(\frac{1 \not{lb}}{16 \not{oz}}\right) \times \left(\frac{\$1.25}{1 \not{lb}}\right)$

$$= \frac{0.8 \times \$1.25}{16} = \$0.0625$$

$$= \$0.07 \text{ or } 7¢ \qquad \text{(With money we round up.)}$$

(e) $4 \text{ oz} = 4 \times (1 \not{oz}) \times \left(\frac{2 \not{tbsp}}{1 \not{oz}}\right) \times \left(\frac{3 \not{tsp}}{1 \not{tbsp}}\right) \times \left(\frac{1 \text{ batch of cookies}}{2 \not{tsp}}\right)$

$$= \frac{4 \times 2 \times 3}{2} \text{ batches} = 12 \text{ batches of cookies}$$

Now for a set of practice problems on converting units turn to 21.

ARITHMETIC WITH MEASUREMENT NUMBERS

22 Measurement numbers are very like the ordinary numbers of arithmetic in that they can be added, subtracted, multiplied, and divided. For example, add

3 ft + 2 ft = ?

Easy enough. We are adding two lengths and our answer is the total length of the two parts

3 ft + 2 ft = 5 ft

We add the number part of the measurements

$3 + 2 = 5$

and then attach the common units. Three 1-ft units added to two 1-ft units is equivalent to five 1-ft units.

Try this one:

3 ft + 5 in. = ? (Change feet to inches and then add.)

Check your work in **23**.

---

23 3 ft + 5 in. = 36 in. + 5 in. = 41 in.

When adding or subtracting measurement numbers follow these rules:

1. Convert all measurement numbers to be added or subtracted into the same units. (We cannot add 3 ft and 5 in. to get 8 of anything. Convert so that both measurements are either in foot units or inch units.)
2. Add or subtract the number part of the measurements as indicated. ($36 + 5 = 41$)
3. Attach the common units. (41 in.)
4. Round the answer to agree with the least accurate number in the sum or difference.

Add:

3 yd + 4 ft 2 in. + 27.4 in. = ?

Follow the rules above in adding these measurements, then check your answer in **24**.

---

24 3 yd + 4 ft 2 in. + 27.4 in.

First, convert all measurements to inches

36 in. + 50 in. + 27.4 in.

Second, add the number parts

$36 + 50 + 27.4 = 113.4$

Third, attach the common units

113.4 in.

Fourth, round to agree with the least accurate number in the sum.

If the first measurement is 3 yards to the nearest inch, we must round 113.4 in. to the nearest inch, giving *113 in.* or *9 ft 5 in.*

Here are a few practice problems:

(a) 56 min + 1 hr 4 min + 70 sec = ______

(b) 4 gal − 7 qt = ______

(c) 4 lb 6 oz + 2 lb 15 oz + 40 oz = ______

(d) 3 ft 8 in. − 21 in. = ______

(e) 3 hr 45 min − 1 hr 56 min = ______

(f) 3 qt + $1\frac{1}{2}$ pints + 20 fl oz = ______

Our step-by-step answers are in **26**.

---

25 Area = 34 ft × 63 ft

$= (34 \times 63)\ \text{ft}^2 = 2142\ \text{ft}^2$ or 2142 sq ft

$\cong 2100\ \text{ft}^2$ or 2100 sq ft

Notice that the product number 2142 must be rounded to two significant digits to agree with the two significant digits in 34 or 63.

The units for length, weight, or time are named with a single word—ft, lb, or sec. The unit for area is named with several words. In the problem above the area unit is ft × ft or ft^2 or sq ft. The unit for speed is miles per hour or mph and it involves both distance (miles) and time (hour) units. Units such as these, named using more than one of the simple, one word units, are called *compound units.*

For example, if a car travels 75 miles at a constant rate in 1.5 hours, it is moving at a speed of

$$\frac{75 \text{ miles}}{1.5 \text{ hours}} = \frac{75 \times 1 \text{ mile}}{1.5 \times 1 \text{ hour}} = \frac{75}{1.5} \times \frac{1 \text{ mile}}{1 \text{ hour}}$$

$$= 50 \frac{\text{mile}}{\text{hour}}$$

We write this as 50 mi/hr or 50 mph and read it as "50 miles per hour." Ratios or rates are always written with compound units.

In this case we have divided units, miles by hours, to arrive at a new unit, miles per hour or mph.

Use the rules above to find the distance a car travels if it moves for 2 hours at a speed of 35 miles/hr.

Hop to **27** to check your answer.

---

26 (a) 56 min + 1 hr 4 min + 70 sec
= 56 min + 64 min + 1.2 min
= 121.2 min = 121 min, when rounded to the nearest minute.

$$\left(70 \cancel{\text{sec}} \times \frac{1 \text{ min}}{60 \cancel{\text{sec}}} = \frac{70}{60} \text{ min} = 1.17\ldots \text{min} \cong 1.2 \text{ min}\right)$$

(b) 4 gal − 7 qt = 16 qt − 7 qt = 9 qt

(c) 4 lb 6 oz + 2 lb 15 oz + 40 oz = 70 oz + 47 oz + 40 oz
= 157 oz ≅ 10 lb

(d) 3 ft 8 in. − 21 in. = 44 in. − 21 in. = 23 in.

(e) 3 hr 45 min − 1 hr 56 min = 225 min − 116 min
= 109 min = 1 hr 49 min

(f) 3 qt + $1\frac{1}{2}$ pints + 20 fl oz = 96 fl oz + 24 fl oz + 20 fl oz
= 140 fl oz or 4 qt 12 fl oz

Multiplying and dividing

Multiplication or division of a measurement number by an ordinary number is easy.

8 × 1 inch = 8 inches

8 × □ = □□□□□□□□

or 6 inches ÷ 2 = 3 inches

□□□□□□ ÷ 2 = □□□

The multiplication of one measurement number by another is a little more difficult. The area of a rectangular rug 9 ft wide and 12 ft long is

Area = 9 ft × 12 ft

To multiply two measurement numbers, follow these rules:

1. Find the product of the number parts

   9 × 12 = 108

(continued on the next page)

2. Multiply the units

   ft × ft = ft^2 or   square feet or sq ft

3. Attach the new units to the product from step 1

   Area = 108 ft^2 or   108 sq ft

4. Round the product so that it has the same number of significant digits as the least accurate measurement number.

   9 ft is a measurement number with one significant digit; therefore, we round the answer to

   Area ≅ 100 sq ft

What is the area of a garden 34 ft wide by 63 ft long?

Follow the rules above and check your work in **25**.

---

27 Distance = 2 hours × 35 miles/hr

$$= 2 \times (1\ \cancel{hr}) \times 35 \times \left(\frac{1\ \text{mi}}{1\ \cancel{hr}}\right) = 70\ \text{mi}$$

The phrase "35 miles per hour" is translated $35\ \dfrac{\text{mile}}{\text{hour}}$.

The word *per* tells you to *divide*.

Solve these problems using unity fractions:

(a) What is the cost of 32 ft of rope if it sells for 6¢ per foot?
(b) If you are paid $18.40 for an 8-hour workday, what is your rate of pay?
(c) A 5-lb block of iron has a measured volume of 20 cubic inches (cu in.). What is its density? (*Hint.* Divide weight by volume.)
(d) What is the cost of 15 lb of sugar if it sells for 34¢ per pound?

Check your solutions in **28**.

---

28 (a) $32\ \text{ft} \times 6¢\ \text{per foot} = 32 \times (1\ \cancel{ft}) \times 6\left(\frac{\text{cents}}{1\ \cancel{ft}}\right)$

$$= 192\ \text{cents} = \$1.92$$

(b) $\$18.40 \div 8\ \text{hr} = \dfrac{18.40}{8}\ \dfrac{\$}{\text{hr}}$

$$= 2.30\ \text{dollars per hr} \quad \text{or} \quad \$2.30\ \text{per hr}$$

(c) $\dfrac{5\ \text{lb}}{20\ \text{cu in.}} = \dfrac{5 \times (1\ \text{lb})}{20 \times (1\ \text{cu in.})} = 0.25\ \dfrac{\text{lb}}{\text{cu in.}}$

(d) $15\ \text{lb} \times 34¢\ \text{per pound} = 15 \times (1\ \cancel{lb}) \times 34 \times \left(\frac{1¢}{1\ \cancel{lb}}\right)$

$$= 510¢ = \$5.10$$

For a set of practice problems on arithmetic with measurement numbers turn to **29**.

---

30 Why worry about the metric system? Why use it? First, about 95% of the world's people already measure, think, buy, and sell in metric units. The few who don't—Canada, Great Britain, and the United States—are switching. They have learned that you sell more tractors in Poland, computers in Brazil, and aspirin in Turkey if you use the language of those countries, and when it comes to numbers they speak metric. Second, the United States already uses metric units in many ways: most electrical units such as volts, amperes, or watts are metric units; cameras, film (8 mm, 16 mm, 35 mm), lenses (50 mm, 260 mm), and other optical parts are measured in metric units; ball bearings, spark plugs, even skis and filter cigarettes come in metric sizes. Most important of all, the metric system is easier to use and simpler to learn. Working with metric units means there is much less use of fractions, fewer units to remember, less to memorize.

In a few years traffic signs, road maps, U.S. National Park information, grade school textbooks, automobiles, and machine parts will all be in metric units. A pinch of salt, a teaspoonful of sugar, or a cup of flour will probably stay the same in the kitchen and inch worms won't become centimeter worms, but commercial, industrial, and government measurements will be metric. It is important that you start learning to read, use, and think metric.

The most important common units in the metric system are those for length or distance, speed, weight or mass, volume, and temperature. Other metric units are used in science but are not generally needed in day-to-day activities. Time units—year, day, hour, minute, second—are the same in the metric as in the English system.

The basic unit of length in the metric system is the *meter* (pronounced *meet-ur* and abbreviated *m*).* It is the people-sized metric unit of length. One meter is roughly equal to one yard. Originally, in the eighteenth century, the meter was defined as one ten-millionth of the distance from the North Pole to the equator on a line through Paris. Later it was defined as the distance between two lines ruled on a metal bar kept at the International Bureau of Weights and Measures in Paris.

In 1960 an international agreement defined the meter in terms of the wavelength of a particular color of light emitted from excited krypton-86 gas. This gave scientists a very accurate length unit available to everyone and not dependent on scratches on a metal bar hidden in a vault.

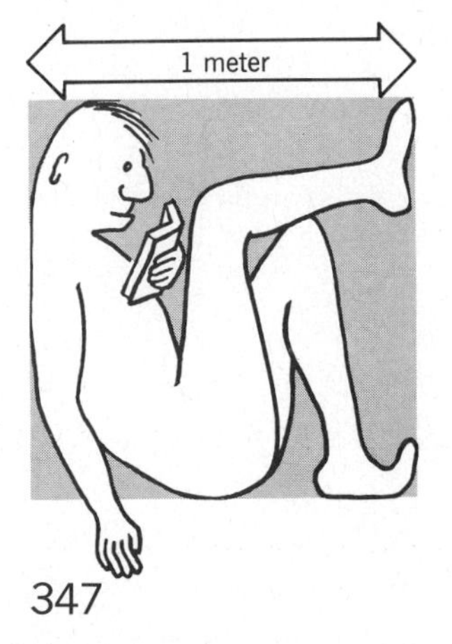

* This metric unit is sometimes spelled *metre*—either spelling is correct and both are pronounced *meet-ur*.

(continued on the next page)

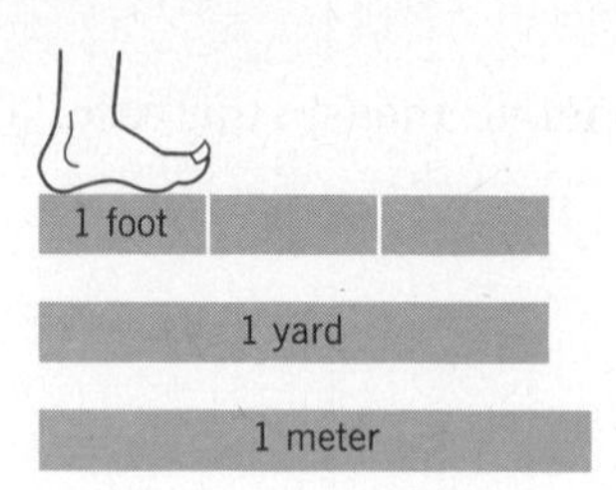

The meter is the appropriate unit for measuring your height, the width of a room, the height of a flagpole, or the length of a football field. Use it in any situation where you would normally use feet or yards.

Estimate the following lengths in meters:

(a) The length of the room in which you are now sitting = ______ m

(b) Your height = ______ m

(c) The length of a ping-pong table = ______ m

(d) Length of a football field = ______ m

Guess as closely as you can, then turn to **31** and continue.

---

31 (a) A small room might be 3 or 4 meters long, and a large one might be 6 or 8 meters long.
(b) Your height is probably between $1\frac{1}{2}$ and 2 meters. Two meters is roughly 6 ft 7 in. A meter and a half is about 4 ft. 11 in.
(c) A ping-pong table is a little less than 3 meters in length.
(d) A football field is about 100 meters long.

Several other length units are defined from the meter. The *centimeter* (pronounced *cent-uh-meter* and abbreviated *cm*) is a finger-sized unit.

1 cm = $\frac{1}{100}$ meter

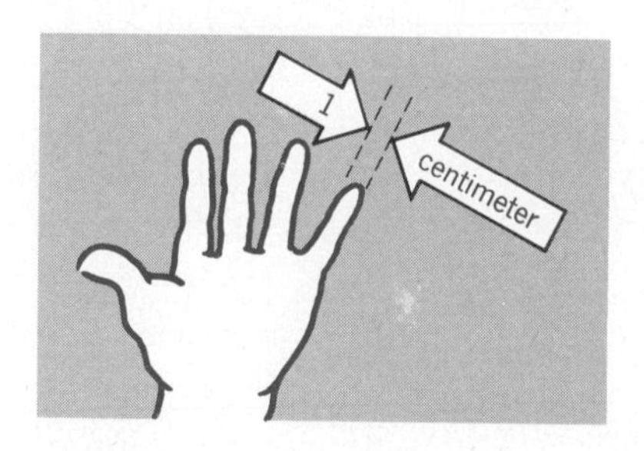

It is the appropriate unit to use for measuring the size of a paper clip, pencil, coin, or book, the length of your shoe or the distance around your waist. One inch is roughly equal to $2\frac{1}{2}$ cm.

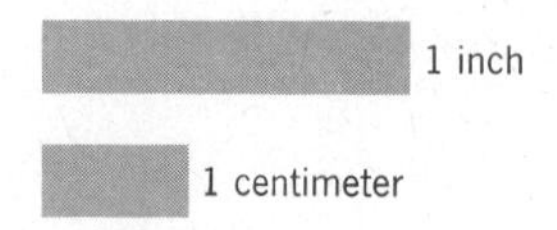

On page 526 you will find a metric measuring rule. Cut it out as shown and use it to measure the following:

(a) Length of this page = ______ cm

(b) Length of a dollar bill = ______ cm

(c) Length of a pencil = ______ cm

(d) Length of your shoe = ______ cm

(e) Your height = ______ cm

(f) Diameter of a 25¢ piece = ______ cm

(g) Width of a postage stamp = ______ cm

(h) Distance from seat of your chair to the floor = ______ cm

Do it carefully, record your measurements here, then turn to **32**.

---

32 (a) This page is 27.94 centimeters long.
(b) The length of a dollar bill is about 15.6 cm.
(c) A new pencil is about 21 cm long.
(d) Your shoe is probably 20 to 30 cm long.
(e) You are probably between 150 and 200 cm tall.
(f) A 25¢ coin is about 2.5 cm in diameter.
(g) A non-commemorative United States stamp is about 2.2 cm wide.
(h) The seat of your chair is probably about 42 cm from the floor.

The *kilometer* (pronounced *kill-o-meter* and abbreviated *km*) is a jogger's unit.

1 km = 1000 meters

One kilometer is roughly equal to 0.6 mile. It is the appropriate unit for traffic signs, road maps, geography books, Olympic bicycle races, and the distance to the moon.

One very nice feature of the metric system is that there are no 2s, 3s, 4s, 12s, 16s, or 5280s involved when we change from one unit to another. All metric units are multiples of ten times the basic unit. Just as with money, metric units increase in steps of ten.

(continued on the next page)

Money	Metric Length Unit	Prefix	Multiplier
$1000	kilometer	kilo-	1000×1 meter
$ 100	hectometer	hecto-	100×1 meter
$ 10	decameter	deca-	10×1 meter
$ 1	meter		1 meter
10¢	decimeter	deci-	0.1×1 meter
1¢	centimeter	centi-	0.01×1 meter
$\frac{1}{10}$¢	millimeter	milli-	0.001×1 meter

It is much easier to remember that 100 cm equals 1 meter than that 63,360 in. equals 1 mile. The most often used metric length units are

1 kilometer (km)	=	1000 meters (m)
1 meter	=	100 centimeters (cm)
1 centimeter	=	10 millimeters (mm)

The millimeter, $\frac{1}{10}$th of a centimeter, is a useful unit for very small lengths less than a centimeter. One millimeter is roughly the thickness of a dime.

The following *Think Metric* quiz will help you get the feel of metric units. Choose an answer using the information you have been given in the past few frames. Work quickly. If you are thinking metric, you will be able to choose the correct answer immediately with no need for pencil-and-paper calculations.

Wind your mind, here we go:

1. Width of a door = ______ (a) 2 m (b) 80 cm (c) 10 cm
2. Height of the Statue of Liberty = ______ (a) 4.6 km (b) 460 m (c) 46 m
3. One-hour drive on a super-highway = ______ (a) 100 km (b) 10 km (c) 1000 m
4. Diameter of a penny = ______ (a) 1.9 mm (b) 19 cm (c) 1.9 cm
5. Hundred mile bike race = ______ (a) 600 km (b) 60 km (c) 160 km

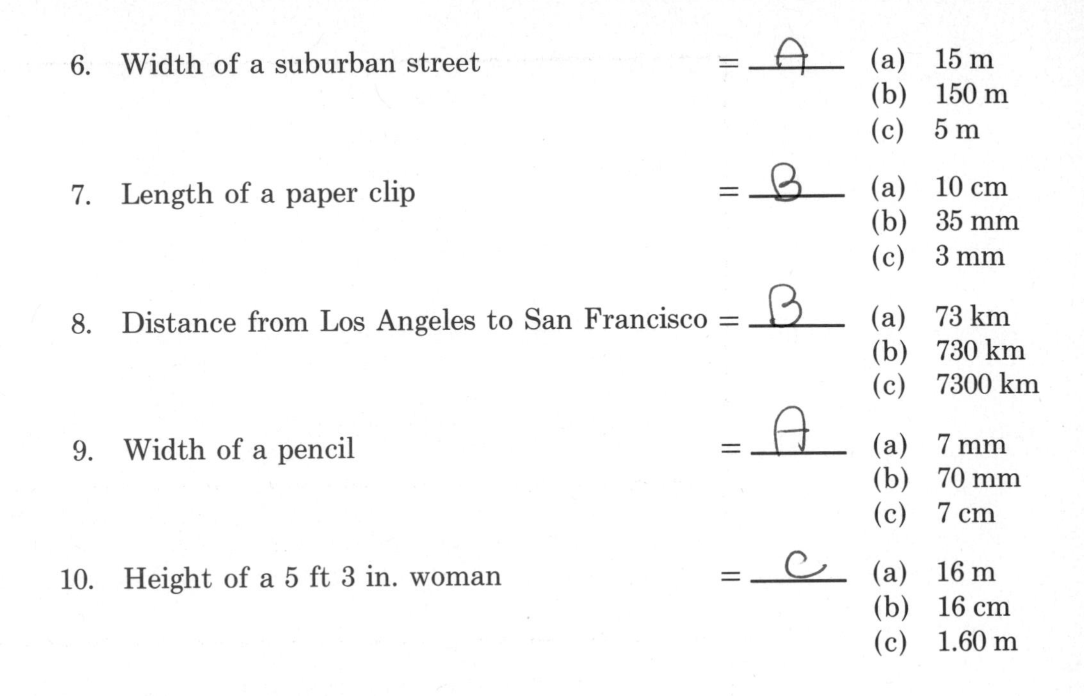

6. Width of a suburban street = A
(a) 15 m
(b) 150 m
(c) 5 m

7. Length of a paper clip = B
(a) 10 cm
(b) 35 mm
(c) 3 mm

8. Distance from Los Angeles to San Francisco = B
(a) 73 km
(b) 730 km
(c) 7300 km

9. Width of a pencil = A
(a) 7 mm
(b) 70 mm
(c) 7 cm

10. Height of a 5 ft 3 in. woman = C
(a) 16 m
(b) 16 cm
(c) 1.60 m

Check your answers in **33**

33

1. b	2. c	3. a	4. c	5. c
6. a	7. b	8. b	9. a	10. c

If you are thinking metric you would immediately recognize in question 10 that 16 cm is the size of a paperback book and 16 m is the size of a swimming pool. A woman's height would be about 1.6 m.

In order to convert from English to metric units or metric to English units, you need to remember only one relationship:

1 inch = 2.54 centimeter

This is the exact, legal definition of the inch in the United States.

To save time when you are working with larger distances you may want to remember that roughly

$$1 \text{ meter} \simeq 3.3 \text{ feet}$$

and

$$1 \text{ kilometer} \simeq 0.62 \text{ miles}$$

Converting units involves multiplying by unity fractions and cancelling units. (If you need to review unity fractions, return to frame **4** on page 319.) For example, the height of the Eiffel Tower in Paris is

1049 ft = _______ m

Work it out, then turn to **34**.

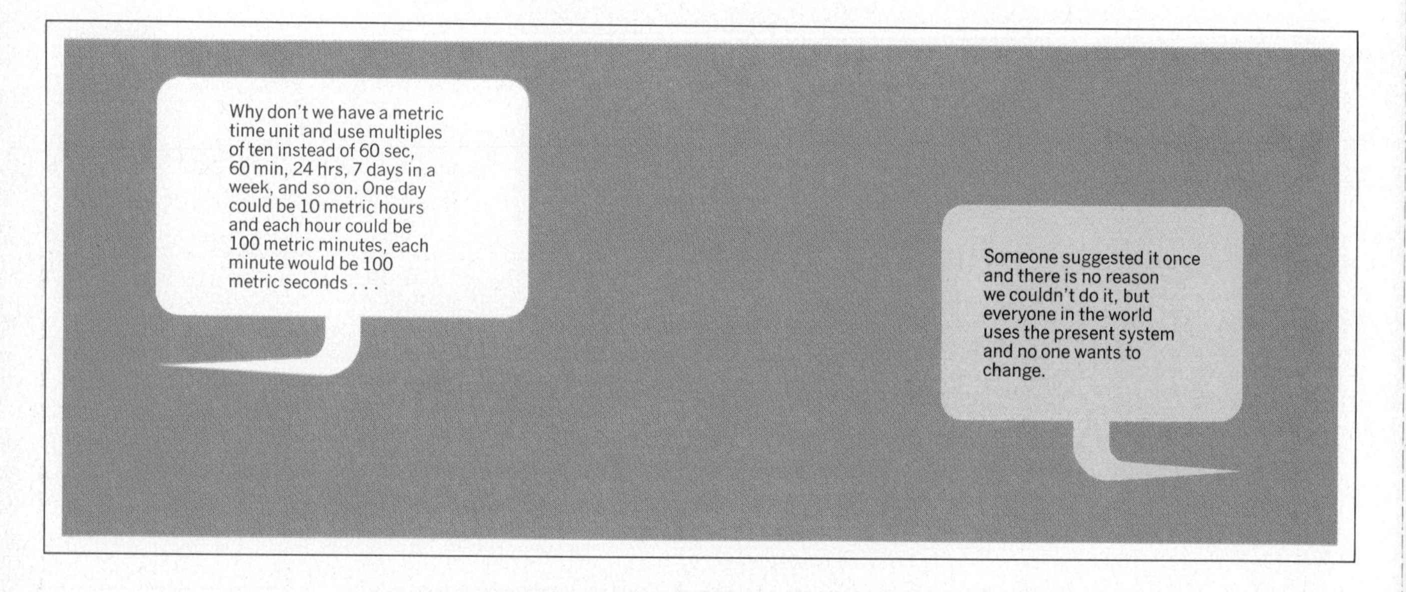

34 $1049 \text{ ft} = 1049 \times (1 \cancel{\text{ft}}) \times \left(\frac{1 \text{ m}}{3.3 \cancel{\text{ft}}}\right) = \frac{1049}{3.3} \text{ m}$

$= 320 \text{ m (rounded)}$

Convert the following measurements as shown.

(a) A 30-km footrace = ______ miles.

(b) A 6-ft tall man is ______ cm tall = ______ m.

(c) The Empire State building is 1260 ft = ______ m tall.

(d) A 35-mm roll of color film is ______ in. wide.

Check your work in **36**.

---

35 $80 \frac{\text{km}}{\text{hr}} = 80 \times \left(1 \frac{\cancel{\text{km}}}{\cancel{\text{hr}}}\right) \times \left(\frac{62 \text{ mph}}{100 \cancel{\text{km/hr}}}\right)$

$= \frac{80 \times 62}{100} \text{ mph} = 0.8 \times 62 \text{ mph}$

$\cong 50 \text{ mph}$ If you are driving 55 mph in an 80 km/hr zone you are speeding.

This table of equivalent speeds someday may save you much more than the price of this book in traffic tickets. Think metric when you drive.

EQUIVALENT SPEEDS
(approximate)

km/hr	*mph*	
100	62	
90	56	super-highway
80	50	main roads
70	43	
60	38	suburbs
50	31	
40	25	city driving
30	19	
25	15	school zone
20	12	run
10	6	jog

Now turn to **38** for a set of practice problems on metric length and speed units.

---

36 (a) $$30 \text{ km} = 30 \times (1 \cancel{\text{km}}) \times \left(\frac{0.62 \text{ mile}}{1 \cancel{\text{km}}}\right) = 30 \times 0.62 \text{ mile}$$
$$= 18.6 \text{ miles}$$

(b) $$6 \text{ feet} = 6 \times (1 \cancel{\text{ft}}) \times \left(\frac{12 \cancel{\text{in.}}}{1 \cancel{\text{ft}}}\right) \times \left(\frac{2.54 \text{ cm}}{1 \cancel{\text{in.}}}\right)$$
$$= 6 \times 12 \times 2.54 \text{ cm} = 182.88 \text{ cm} \cong 183 \text{ cm}$$

$$183 \text{ cm} = 183 \times (1 \cancel{\text{cm}}) \times \left(\frac{1 \text{ m}}{100 \cancel{\text{cm}}}\right)$$
$$= \frac{183}{100} \text{ m} = 1.83 \text{ m}$$

(c) $$1260 \text{ ft} = 1260 \times (1 \cancel{\text{ft}}) \times \left(\frac{1 \text{ meter}}{3.3 \cancel{\text{ft}}}\right) = \frac{1260}{3.3} \text{ meter}$$
$$= 382 \text{ m (rounded)}$$

(d) $$35 \text{ mm} = 35 \times (1 \cancel{\text{mm}}) \times \left(\frac{1 \cancel{\text{cm}}}{10 \cancel{\text{mm}}}\right) \times \left(\frac{1 \text{ in}}{2.54 \cancel{\text{cm}}}\right)$$
$$= \frac{35}{10 \times 2.54} \text{ in.} = 1.4 \text{ in. (Always round it.)}$$

To help you convert from English to metric units and from metric to English units quickly and easily, with no arithmetic, we have devised the METRIC MINDER. This handy little gadget and instructions for using it are inside. The METRIC MINDER is the best thing since the invention of peanut butter. If you want to learn how to use the METRIC MINDER turn to the front inside cover now. When you are ready to continue here turn to **37**.

(Mark this page so you know where to return, we don't want you wandering around lost in this book.)

---

37 The basic unit of *speed* in the metric system is the km/hr. A good approximation to remember is that

100 km/hr ≅ 62 mph

From this relationship you can quickly translate speed units from metric to English or English to metric. For example, if you are driving through Paris and see a traffic sign warning you that the speed limit is 50 km/hr, how fast should you drive according to the speedometer on your American-made car?

If you think metric, you would have a little conversation with yourself like this:

> "Well I know that 100 km/hr is about 62 mph. 50 km/hr is half of 100 km/hr, so 50 km/hr must be about half of 62 mph or 31 mph."

Solve this one using unity fractions: if the speed limit is marked as 80 km/hr and you are driving 55 mph, are you likely to be stopped by a policeman?

Go to **35** to check your answer.

---

39 In the everyday world of people and things the question "How much?" usually means "How heavy?" We usually measure how much of something we have by finding its weight. The *weight* of an object is the gravitational pull exerted by the earth on that object when it is sitting at rest near the surface of the earth. In the English system of measurement the basic unit of weight is the pound.

In the metric system we answer the question "How much?" by telling the *mass* of the object in kilograms or grams. The mass of an object is a measure of how it behaves when it is pushed or pulled, and for scientists the difference between mass and weight is important. For everyday use mass and weight are equivalent properties: both mass and weight units are used to tell us how much of something we have or how heavy it is. We can measure both on the same scale and easily translate units back and forth. Unless you are enrolled in a physics class or plan to do your shopping on Mars, you need not worry about the difference.

The basic unit of mass in the metric system is the kilogram (pronounced *kill-o-gram* and abbreviated *kg*). The kilogram is defined as the mass of a standard metal cylinder kept at the International Bureau of Weights and Measures in Paris.

A mass of 1 kg weighs roughly 2.2 lb.

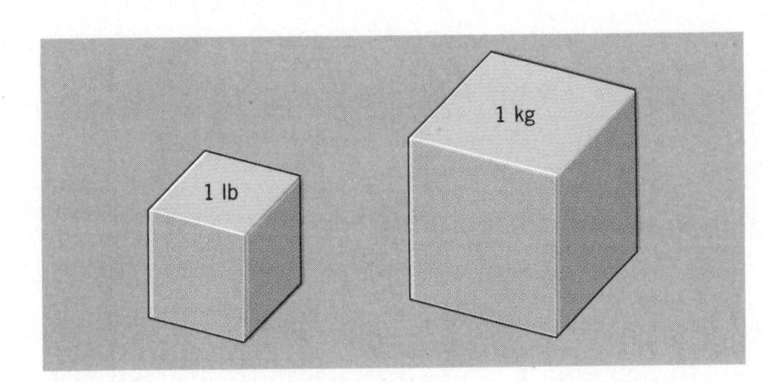

The kilogram is the appropriate unit for measuring the weight of a person, a sack of flour, a bag of potatoes, or luggage at the airline ticket counter. Use it in any situation where you would normally use pounds. It is the appropriate weight or mass unit for grocers, weightlifters, construction workers, factory workers, or dieters.

Estimate the following weights in kilograms:

(a) Your weight = ______ kg

(b) Loaf of bread = ______ kg

(c) Small car = ______ kg

Guess as closely as you can, then turn to **40** and continue.

40 (a) You probably have a mass between roughly 50 kg (110 lb) and 100 kg (220 lb).

(b) A loaf of bread weighs a bit less than 1 kg (2.2 lb).

(c) A small car weighs between 1000 kg (2200 lb) and 1500 kg (3300 lb).

An important unit of mass closely related to the kilogram is the *gram* (abbreviated *g*) defined as

$$1 \text{ gram} = \tfrac{1}{1000} \text{ kilogram}$$

and the milligram (pronounced *mill-e-gram* and abbreviated *mg*)

$$1 \text{ milligram} = \tfrac{1}{1000} \text{ gram}$$

The *gram* is a very small unit of mass or weight, about $\frac{1}{500}$ of a pound, and is much too small to be useful for measuring groceries or football players. It is the appropriate unit for measuring the weight or mass of an aspirin tablet, a can of soup, a package of breakfast cereal, a tube of toothpaste or a bar of soap. Druggists already measure out prescriptions in grams, and North American shoppers will soon find small packaged foods labeled in grams.

The *milligram* is used by druggists and nurses to measure drugs and other medications. An aspirin tablet has a mass of 300 mg. The average daily adult need for vitamin C is about 30 mg. The milligram is much too small a unit for most ordinary uses.

Thinking metric will be easier if you remember that a paper clip has a mass of about one gram, and a 5¢ piece has a mass of 5 grams.

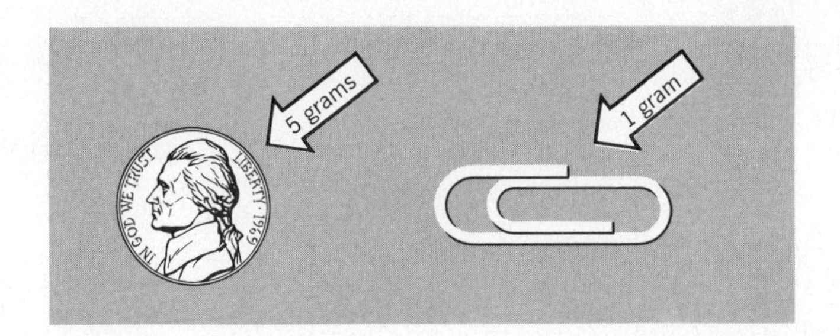

Choose the best answer to each question in the following *Think Metric* quiz without paper and pencil calculations.

What is the mass of each of the following?

1. Football player = ______ (a) 120 kg
   (b) 12 kg
   (c) 1200 g

2. Can of soup = ______ (a) 3 g
   (b) 30 g
   (c) 300 g

3. Bar of soap = ______ (a) 15 g
   (b) 150 g
   (c) 1.5 kg

4. Sugar cube = ______ (a) 1 kg
   (b) 2 grams
   (c) 20 grams

5. Quart of milk = ______ (a) 1 kg
   (b) 10 kg
   (c) 10 g

6. Teaspoonful of water = ______ (a) .5 kg
   (b) 50 g
   (c) 5 g

7. Flashlight battery = ______ (a) 1 g
   (b) 1 kg
   (c) 100 g

8. Pair of shoes = ______ (a) 10 kg
   (b) 1 kg
   (c) 10 g

9. Five-pound sack of sugar = ______ (a) 5 kg
   (b) 1 kg
   (c) 2 kg

10. Hundred-pound barbell = ______ (a) 50 kg
    (b) 5 kg
    (c) 500 kg

Check your answers in **41**.

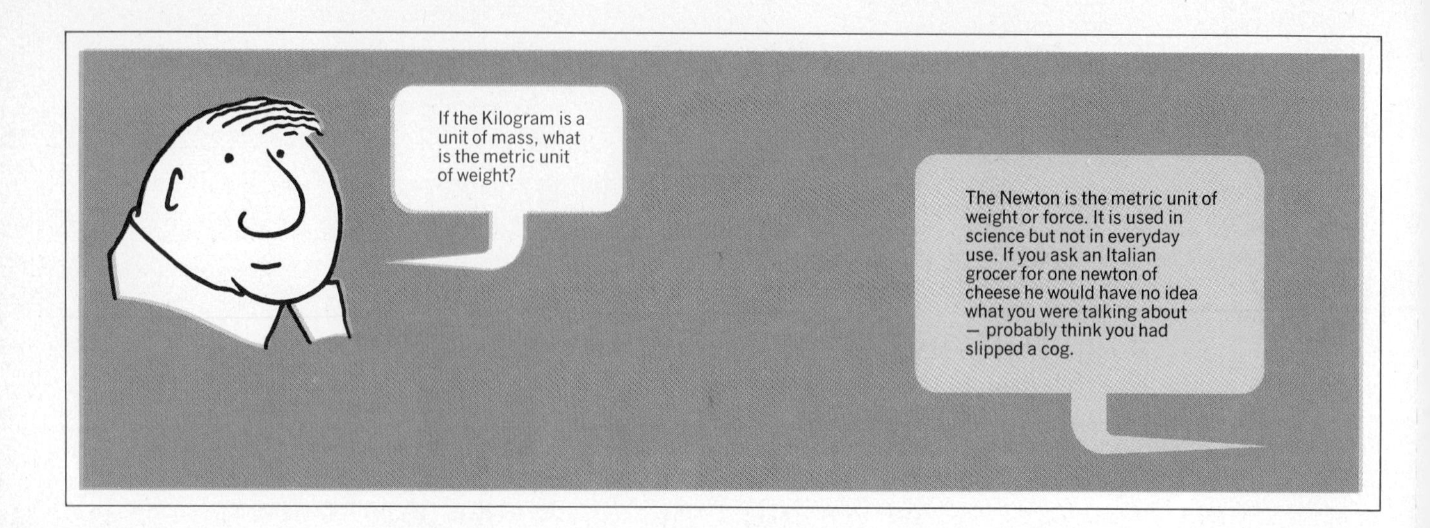

41

1. (a)	2. (c)	3. (b)	4. (b)	5. (a)
6. (c)	7. (c)	8. (b)	9. (c)	10. (a)

In order to convert from English to metric or metric to English weight units you need to remember only one relationship:

Remember it

> 1 kilogram (kg) weighs about 2.20 pounds (lb)

To save time when you are working with smaller units, you may want to remember that

> 1 ounce (oz) is roughly the weight of 28 grams (g)

Converting units involves either multiplying by unity fractions and cancelling or using the METRIC MINDER. (If you need to review unity fractions, return to frame **4** on page 319. The METRIC MINDER is a quick and easy device for converting units. You will find it inside the front cover.)

Convert the following measurements as shown using unity fractions and check your answers with the METRIC MINDER.

(a) 10 lb = ______ kg (b) 1 lb = ______ g

(c) 70 kg = ______ g (d) 1 ton = ______ kg

(e) 100 g = ______ oz (f) 2 lb 10 oz = ______ kg

(g) 6 kg = ______ lb (h) 350 g = ______ lb

(i) 20 oz = ______ g (j) 2.5 kg = ______ oz

The correct answers are in **44**.

(h) $350 \text{ g} = 350 \times (1\ \cancel{\text{g}}) \times \left(\dfrac{1\ \cancel{\text{kg}}}{1000\ \cancel{\text{g}}}\right) \times \left(\dfrac{2.2 \text{ lb}}{1\ \cancel{\text{kg}}}\right) = \dfrac{350 \times 2.2}{1000} \text{ lb} \cong 0.77 \text{ lb}$

(i) $20 \text{ oz} = 20 \times (1\ \cancel{\text{oz}}) \times \left(\dfrac{1\ \cancel{\text{lb}}}{16\ \cancel{\text{oz}}}\right) \times \left(\dfrac{1\ \cancel{\text{kg}}}{2.2\ \cancel{\text{lb}}}\right) \times \left(\dfrac{1000 \text{ g}}{1\ \cancel{\text{kg}}}\right)$

$= \dfrac{20 \times 1000}{16 \times 2.2} \text{ g} \cong 570 \text{ g}$

(j) $2.5 \text{ kg} = 2.5 \times (1\ \cancel{\text{kg}}) \times \left(\dfrac{2.2\ \cancel{\text{lb}}}{1\ \cancel{\text{kg}}}\right) \times \left(\dfrac{16 \text{ oz}}{1\ \cancel{\text{lb}}}\right)$

$= 2.5 \times 2.2 \times 16 \text{ oz} \cong 88 \text{ oz}$

All answers have been rounded.

Now turn to **42** for a set of practice problems on metric weight-mass units.

---

45 (a) $\frac{1}{2}$ gallon of milk $\cong$ 2 liters

(b) 8-ounce glass of water $= \frac{1}{4}$ quart $\cong \frac{1}{4}$ liter

(c) 12-gallon tank of gasoline $= 12 \times 4 \text{ qt} = 48 \text{ qt} \cong 48$ liters

If gasoline costs 20¢ per liter in Spain, what is the cost per English gallon?

Use unity fractions to work it out, then turn to **46** to check your answer.

---

46 Cost = 20¢ per liter $= 20 \times \left(\dfrac{¢}{\cancel{\text{liter}}}\right) \times \left(\dfrac{1\ \cancel{\text{liter}}}{1.06\ \cancel{\text{qt}}}\right) \times \left(\dfrac{4\ \cancel{\text{qt}}}{1 \text{ gal}}\right)$

$= \left(\dfrac{20 \times 4}{1.06}\right) \dfrac{¢}{\text{gal}}$

$= 75.5¢/\text{gal}$

The appropriate metric volume unit to use in measuring small, teaspoon-size volumes is the *milliliter* (pronounced *mill-e-leter* and abbreviated *ml*). The milliliter is exactly one-thousandth of a liter.

1 liter = 1000 milliliter (ml)

The milliliter is exactly equal to the volume of a cube one centimeter on each edge. One milliliter is exactly equal to one *cubic centimeter* (abbreviated *cc* or cm^3) and the two units *milliliter* (ml) and *cubic centimeter* (cc) are used interchangeably.

1 cm
= 1 ml = 1 cc
1 cm
1 cm

(*continued on the next page*)

Doctors, druggists, and nurses now use milliliters or cubic centimeters, and most liquid medicines are packaged, labeled, and sold in ml or cc. It will help you to think metric if you remember that

1 teaspoonful $\cong$ 5 ml
1 fluid ounce $\cong$ 30 ml

The third metric volume unit, the *cubic meter,* is used less often than the others, but is useful for very large volumes. A cubic meter (abbreviated m^3) is exactly equal to the volume of a cube one meter on each edge.

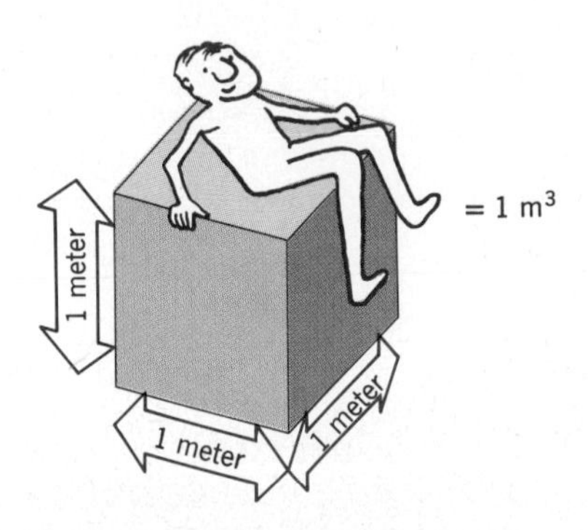

A refrigerator might have a volume of roughly 0.5 m^3. This unit might be used by construction workers to specify the volume of a swimming pool, a water tank, a railroad car, or a quantity of sand, gravel, or concrete.

Think metric: quickly estimate the volume of each of the following.

(a) 2-oz bottle of ink = ______ ml

(b) 1 tablespoonful of cough syrup = ______ ml

(c) Closet = ______ m^3

(d) Cup of tea = ______ ml

Check your answers in **47**.

47 (a) 2-oz bottle of ink $\cong$ 60 ml (One fluid ounce is about 30 ml.)

(b) 1 tablespoonful of cough syrup $\cong$ 15 ml (1 tbsp = 3 tsp)

(c) Closet $\cong$ 2 or 3 m^3

(d) Cup of tea $\cong$ 240 ml

Convert the following volume measurements using unity fractions.

(a) 5 gal $\cong$ ______ liter (b) 2 qt $\cong$ ______ liters

(c) 3 cups $\cong$ ______ ml (d) 4 fl oz = ______ ml

(e) 3 liters $\cong$ ______ ml (f) $\frac{1}{2}$ qt $\cong$ ______ liters

(g) 2.5 liters $\cong$ ______ qt (h) 100 ml $\cong$ ______ fl oz

The correct solutions are in **48**.

---

48 (a) 5 gal $\cong 5 \times (1 \text{ gal}) \times \left(\frac{4 \text{ qt}}{1 \text{ gal}}\right) \times \left(\frac{1 \text{ liter}}{1.06 \text{ qt}}\right)$

$\cong \left(\frac{5 \times 4}{1.06}\right)$ liter = 19 liters

(b) 2 qt $\cong 2 \times (1 \text{ qt}) \times \left(\frac{1 \text{ liter}}{1.06 \text{ qt}}\right) = \left(\frac{2}{1.06}\right)$ liter = 1.9 liters

(c) 3 cups $\cong 3 \times (1 \text{ cup}) \times \left(\frac{8 \text{ fl oz}}{1 \text{ cup}}\right) \times \left(\frac{30 \text{ ml}}{1 \text{ fl oz}}\right) = 3 \times 8 \times 30$ ml

$\cong$ 720 ml

(d) 4 fl oz $\cong 4 \times (1 \text{ fl oz}) \times \left(\frac{30 \text{ ml}}{1 \text{ fl oz}}\right) = 4 \times 30 \text{ ml} = 120$ ml

(continued on the next page)

(e) 3 liters $= 3 \times (1 \text{ liter}) \times \left(\frac{1000 \text{ ml}}{1 \text{ liter}}\right) = 3 \times 1000 \text{ ml} = 3000 \text{ ml}$

(f) $\frac{1}{2}$ qt $\cong \frac{1}{2} \times (1 \text{ qt}) \times \left(\frac{1 \text{ liter}}{1.06 \text{ qt}}\right) = \frac{1}{2} \times \frac{1}{1.06} \text{ liter} = 0.47 \text{ liter}$

(g) 2.5 liter $\cong 2.5 \times (1 \text{ liter}) \times \left(\frac{1.06 \text{ qt}}{1 \text{ liter}}\right) = 2.5 \times 1.06 \text{ qt} = 2.65 \text{ qt}$

(h) 100 ml $\cong 100 \times (1 \text{ ml}) \times \left(\frac{1 \text{ fl oz}}{30 \text{ ml}}\right) = \frac{100}{30} \text{ fl oz} = 3.3 \text{ fl oz}$

Temperature

When we read in the newspapers that it is 125 in the shade in Death Valley or learn from the doctor that we have a fever of 102 degrees or set the oven at 375 to bake a cake, we are using the concept of temperature. Each of us has thousands of built-in temperature sensors in our skin, miniature thermometers telegraphing to our brain the temperature of the surface of our skin. These skin thermometers enable us to feel hot and cold but do not provide measurement numbers or units for the sensation. Standard thermometers are useful because they are marked with numerical temperature scales agreed upon across the world.

The first modern temperature scale was devised in 1714 by Gabriel Robert Fahrenheit, a German scientist. This is the Fahrenheit temperature scale in common use today in the United States. It is the temperature of weather reports, medical thermometers, and cookbooks—the temperature numbers listed above. This scale was improved by the Swedish astronomer Anders Celsius and his new temperature scale, the Celsius scale (pronounced *sell-see-us*) was quickly adopted in most countries of the world. The United States and most English-speaking countries continued to use the Fahrenheit scale and referred to the Celsius scale as the Centigrade, or 100-unit, temperature scale. In 1948 the name Centigrade was changed to Celsius across the world, and now the Celsius scale of temperature is part of the set of internationally accepted units of which the metric system is a part.

Because most of the world has been using Celsius temperatures for many years, and because the United States is planning to shift to it in the not-too-distant future, it is important for you to learn to think metric with temperatures.

The Celsius temperature units were set up using two reference temperatures: the temperature at which water normally freezes to ice was set as 0° C (read it "zero degrees Celsius") and the temperature at which water normally boils was set as 100° C. The drawing compares the Celsius and Fahrenheit temperature scales.

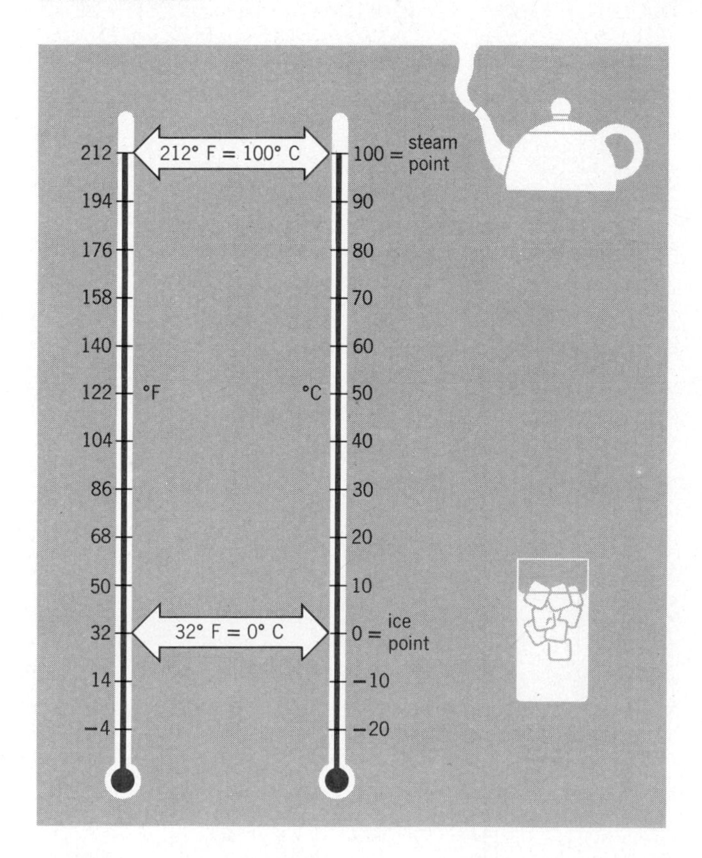

Notice that 100 units on the Celsius scale corresponds to 180 units on the Fahrenheit scale. A temperature change of 5° C is equal to a change of 9° F. The zero does not mean "no temperature"—it is just another point on the scale. Temperatures lower than zero are possible, and they are labeled with negative numbers.

Most people have a body temperature of 98.6° F measured with an oral thermometer. Using the drawing above, estimate this in Celsius degrees.

Turn to **49** to check your estimate.

49 Average normal body temperature is 98.6° F or 37° C.

Thinking Celsius is important. This diagram should help.

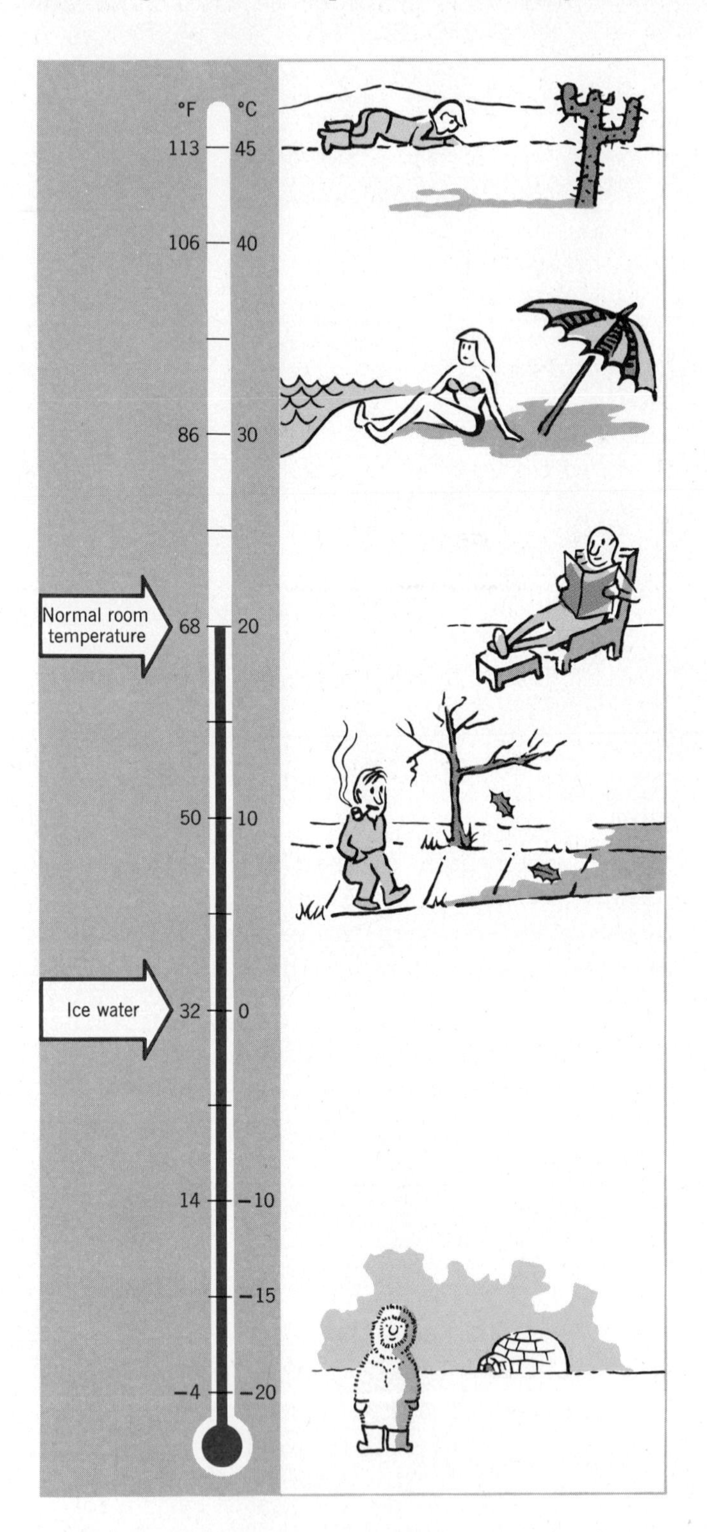

An outdoors temperature of 32° F is icy cold, but 32° C is a hot summer day.

Converting temperatures from Fahrenheit to Celsius or from Celsius to Fahrenheit exactly requires a bit of tricky arithmetic (see box on page 377) and you probably do not need to be able to do it except in a science class. It is much more important that you be able to "think Celsius." Look carefully again at that drawing on page 374, then try to estimate the following temperatures in Celsius degrees.

(a) Cold drink $\cong$ _____ °C

(b) Cup of hot coffee $\cong$ _____ °C

(c) Hot July day in Phoenix, Arizona $\cong$ _____ °C

(d) Cold winter day in New York $\cong$ _____ °C

(e) Normal body temperature $\cong$ _____ °C

(f) Comfortable room temperature $\cong$ _____ °C

(g) Dishwater $\cong$ _____ °C

(h) A nice day at the beach in Miami $\cong$ _____ °C

(i) Ice freezing $\cong$ _____ °C

(j) Fall day in Chicago $\cong$ _____ °C

Check your answers in 50.

---

50 (a) About 3 to 5° C
(b) 50° C or higher
(c) 40 to 45° C
(d) 0° C to −15° C
(e) 37° C
(f) 20° C
(g) 50° C or so
(h) 30 to 35° C
(i) 0° C
(j) About 10° C

For your convenience here is a temperature conversion chart showing corresponding values of Celsius and Fahrenheit temperatures.

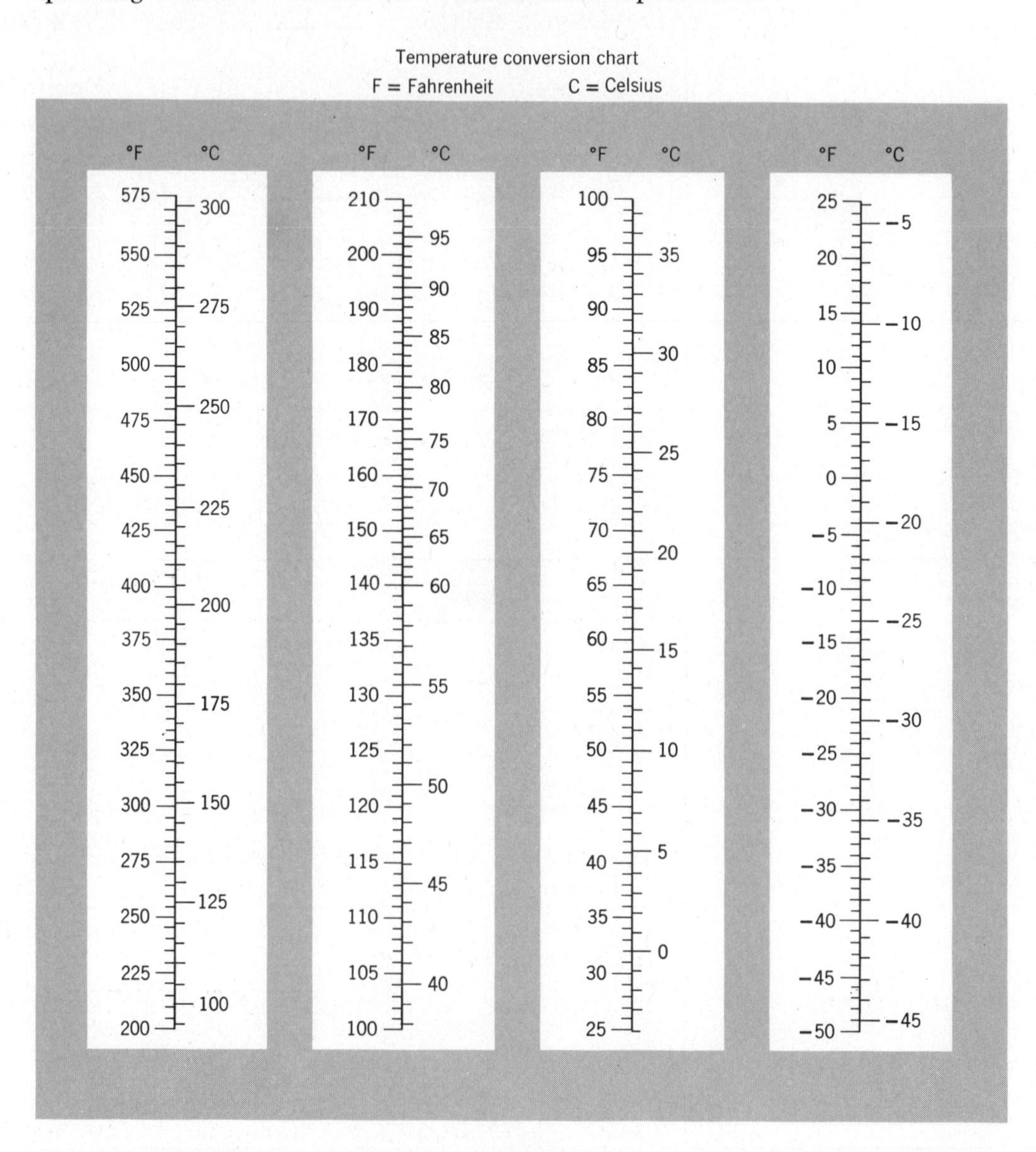

On this chart equal temperatures are placed side by side. To use the chart,

First, find the temperature scale given. **Example:** 50° F = ______° C

°F is in the left column

Second, locate the given temperature number.

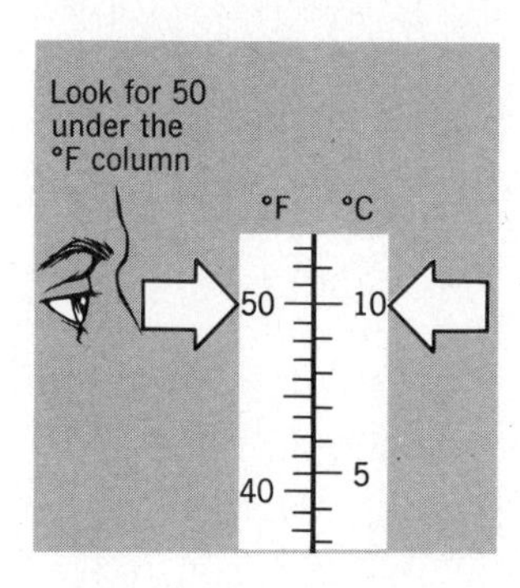

Third, read off the equivalent temperature next to the given temperature.

Use the Temperature Conversion Chart to find these temperatures.

(a) Oven temperature for baking bread, 375° F = ______ ° C

(b) New Year's Day in Athens, Greece, 14° C = ______ ° F

(c) A Finnish sauna bath, 80° C = ______ ° F

(d) A cold day in St. Paul, Minn. −20° F = ______ ° C

Check your answers in **51**.

CONVERTING TEMPERATURES

In science courses it is often necessary to convert with accuracy from one temperature scale to the other, from Celsius to Fahrenheit or from Fahrenheit to Celsius. The two temperature scales are connected by a simple algebraic equation.

$$\frac{C + 40}{F + 40} = \frac{5}{9}$$

where C stands for the Celsius temperature and F is the equivalent Fahrenheit temperature.

If you are given a temperature in Fahrenheit degrees and want to convert it to its value in Celsius degrees, the above equation can be rearranged as

$$C = \frac{5(F - 32)}{9}$$

(continued on the next page)

For example,

$50° F =$ ______ $° C$

$$C = \frac{5(50 - 32)}{9} = \frac{5(18)}{9} = 10$$

Therefore

$50° F = 10° C$

If you are given a temperature in Celsius degrees and want to convert it to its value in Fahrenheit degrees, the first equation above can be rearranged as

$$F = \frac{9C}{5} + 32$$

For example,

$60° C =$ ______ $° F$

$$F = \frac{9(60)}{5} + 32$$

$$= 108 + 32 = 140$$

Therefore

$60° C = 140° F$

Here are a few practice problems in the use of these equations for converting temperatures.

1. $120° C =$ ______ $° F$
2. $200° F =$ ______ $° C$
3. $40° C =$ ______ $° F$
4. $5° F =$ ______ $° C$
5. $-10° C =$ ______ $° F$

The answers are on page 509.

51 (a) $375° \text{ F} \cong 191° \text{ C}$
(b) $14° \text{ C} \cong 57° \text{ F}$
(c) $80° \text{ C} \cong 176° \text{ C}$
(d) $-20° \text{ F} \cong -29° \text{ C}$

It is worth knowing body temperatures in Celsius degrees. A mistake here in thinking Celsius could be very serious. The following diagram should help.

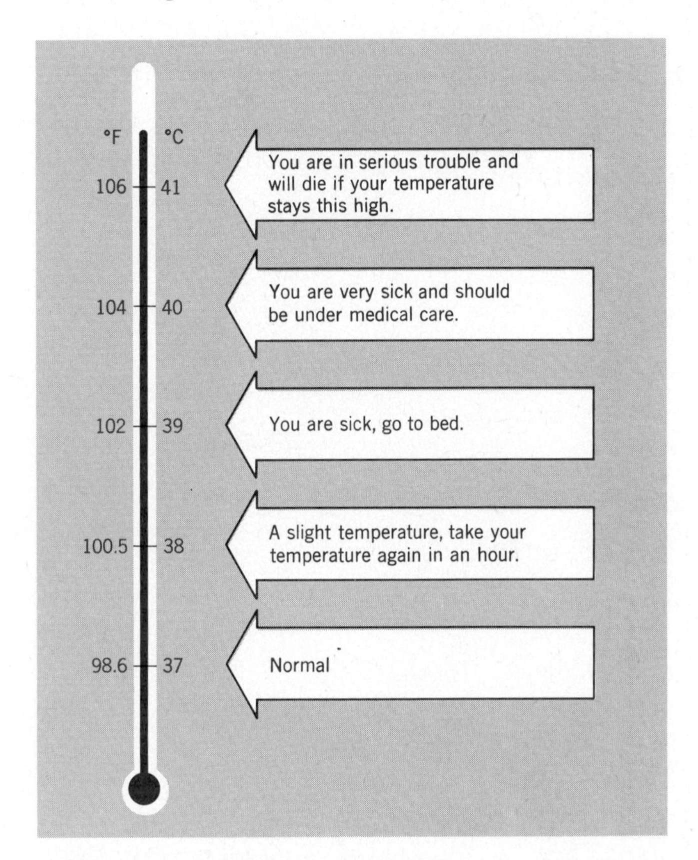

The average healthy adult will have a body temperature of 37° C. Every degree rise in Celsius temperature is equivalent to a rise of about 2° on the Fahrenheit scale.

Turn to **52** for a set of practice problems on metric volume and Celsius temperature units.

Measurement

PROBLEM SET 5

52

Metric Units: Volume and Temperature

Answers are on page 509.

A. Think Metric. Circle the measurement closest to the first one given:

		(a)	(b)	(c)
1.	60° C	20° F	140° F	100° F
2.	3 liters	3000 ml	1 qt	300 ml
3.	1 gallon	0.5 liters	2 liters	4 liters
4.	1 pint	0.5 liters	1 liter	2 liters
5.	2 tsp	10 ml	30 ml	60 ml
6.	1 cup	10 ml	100 ml	240 ml
7.	85° F	3° C	30° C	300° C
8.	6 fl oz	180 ml	30 ml	1 liter
9.	10 gal	40 liters	20 liters	80 liters
10.	6 qts	2 liters	3 liters	6 liters
11.	1 liter	10 cups	4 cups	3 pints
12.	50° F	0° C	10° C	30° C
13.	100 ml	$\frac{1}{2}$ cup	1 cup	2 cups
14.	212° F	100° C	400° C	50° C
15.	0° C	100° F	32° F	−30° F

B. Convert to the units shown using either unity fractions or the METRIC MINDER:

1. 4 fl oz = ______ ml
2. 100 ml = ______ tbsp
3. 3.5 qt = ______ liters
4. 2000 ml = ______ qt
5. 20 ml = ______ cup
6. 10 gallons = ______ liter
7. 5 liters = ______ fl oz
8. 2 liters = ______ tsp
9. 6 qt = ______ liter
10. 10 liters = ______ qt

Date

Stephanie Catanzaro

Name

050/003

Course/Section

C. Convert the temperatures given using the temperature conversion chart in frame 49:

1. 20° C = ______ ° F
2. 200° F = ______ ° C
3. −10° C = ______ ° F
4. 400° F = ______ ° C
5. 40° C = ______ ° F
6. 136° F = ______ ° C
7. 15° C = ______ ° F
8. 50° F = ______ ° C
9. 140° C = ______ ° F
10. −10° F = ______ ° C

D. Brain Boosters:

1. If your body temperature was 40° C, should you stay in bed? Ignore it? Call a doctor?

   ____

2. George's little Fiat sports car has a temperature gauge that usually reads about 85° C. What is the normal operating temperature of his car in Fahrenheit degrees?

   ____

3. Is 10° C suitable for bath water? How about 288° F?

   ____________, ____________

4. The setting for broiling steaks in my oven is 250° C. What is this on the Fahrenheit scale?

   ____

5. Estimate the volume of each of the following in metric units (ml or liters).

   (a) 16 fl oz of root beer ≅ ______ ml = ______ liters

   (b) 8 gallons of gasoline ≅ ______ liters

   (c) One-half gallon of milk ≅ ______ liters

   (d) 2 tablespoons of syrup ≅ ______ ml

   (e) 1 cup of sugar ≅ ______ liter

   (f) 400 ml of milk ≅ ______ qt

   (g) 4 cubic yards of dirt = ______ m^3

6. Translate these well-known phrases into the metric system.

(a) Peter Piper picked ______ liters of pickled peppers. (a peck = ______ liters)

(b) A cowboy in a ______ liter hat (10 gallons = ______ liters)

(c) He's in ______ ml of trouble. (a peck = ______ ml)

7. Which is performing more efficiently, a car getting 12 km per liter of gas or one getting 25 miles per gallon of gas?

____

8. If the mass of one liter of water is 1 kg, what would be the mass in grams of 800 ml?

____

9. A certain paint sells for $5.00 per gallon. What would 1 liter cost?

____

10. The highest temperature ever recorded in Death Valley is 134° F. Translate this to °C.

____

11. Write the following temperatures in both °C and °F.

(a) Normal body temperature ______

(b) Boiling water ______

(c) Freezing water ______

12. What part of a quart is 200 ml?

____

Date ______________

Name ______________

Course/Section ______________

When you have had the practice you need, either return to the preview test on page 313 or continue in 53 with the calculation of area, perimeter, and volume.

53 The area of a square 1 cm on each edge is defined as 1 cm $\times$ 1 cm or 1 cm^2 (pronounced *one square centimeter*) of area. One cm^2 is the basic unit of area in the metric system. In exactly the same way we can define 1 $in.^2$, 1 ft^2, and 1 yd^2 as area units in the English system and 1 m^2 as a larger area unit in the metric system.

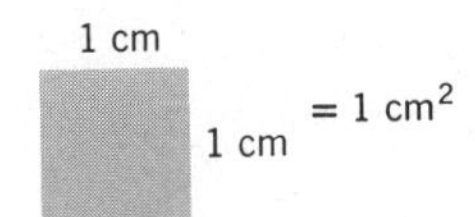

The measure of the area of any plane geometric figure is equal to the number of square units it contains. For example, the area of a rectangle* 4 cm long and 3 cm wide is 12 cm^2. The rectangle contains 12 one cm^2 unit areas—count them.

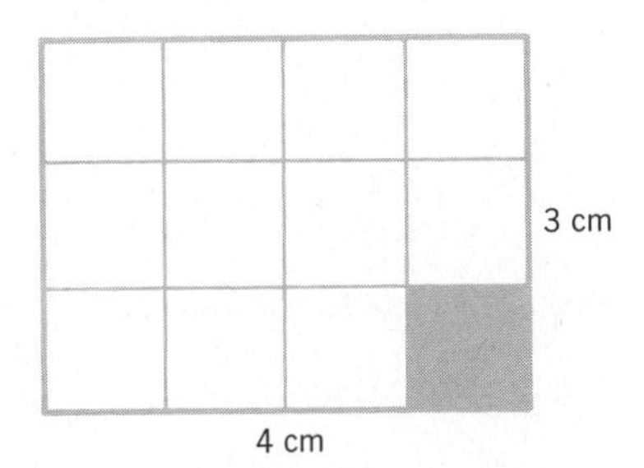

Think of the rectangle as 3 rows with 4 unit square areas in each row. Then the area is equal to

Area = number of rows $\times$ number of unit squares per row

In general, the area of a rectangle is

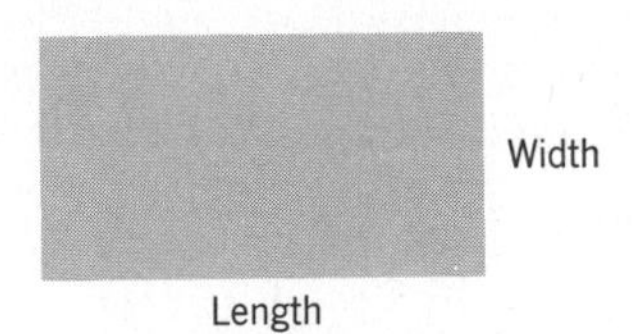

Area of rectangle = length $\times$ width

where both length and width are given in the same units (cm, in., ft, m) and the area is given in square unit areas (cm^2, $in.^2$, ft^2, m^2). For example, for the rectangle above

$$\begin{aligned}\text{Area} &= 4 \text{ cm} \times 3 \text{ cm} = (4 \times 1 \text{ cm}) \times (3 \times 1 \text{ cm}) \\ &= 4 \times 3 \times (1 \text{ cm} \times 1 \text{ cm}) \\ &= 12 \times 1 \text{ cm}^2 \\ &= 12 \text{ cm}^2\end{aligned}$$

What is the area of the floor of a rectangular room exactly 5 meters wide by 8 meters long?

Work it out, then turn to **54** to check your answer.

---

* A rectangle is a geometric figure with four straight sides, opposite sides parallel, and each corner angle a right or 90° angle.

54 Area = 5 m × 8 m = 40 m²

It may be helpful to know the relationships between various area units. Here are a few:

$1\ \text{m}^2 = 10{,}000\ \text{cm}^2$
$1\ \text{ft}^2 = 144\ \text{in}^2$
$1\ \text{yd}^2 = 9\ \text{ft}^2$

Calculate the area of a plot of ground 250 ft wide by 420 ft long. Write your answer in square yards and round it to the nearest square yard.

Check your work in 55.

---

55 Area $= 250\ \text{ft} \times 420\ \text{ft} = (250 \times 420)\ \text{ft}^2 = 105{,}000\ \text{ft}^2$

$$105{,}000\ \cancel{\text{ft}^2} \times \left(\frac{1\ \text{yd}^2}{9\ \cancel{\text{ft}^2}}\right) = 11{,}667\ \text{yd}^2$$

In English units large areas are often measured in *acres.*

$1\ \text{acre} = 4840\ \text{yd}^2$

The area calculated above is

$$\text{Area} = 11{,}667\ \cancel{\text{yd}^2} \times \left(\frac{1\ \text{acre}}{4840\ \cancel{\text{yd}^2}}\right) = 2.45\ \text{acres}$$

In metric units large areas are measured in *hectares,* the area of a square 100 meters on each side.

$1\ \text{hectare} = 10{,}000\ \text{m}^2$
$1\ \text{hectare} = 2.47\ \text{acres}$

The area of a *triangle* is equal to one-half of the area of a rectangle whose length is the base of the triangle and whose width is the height of the triangle.

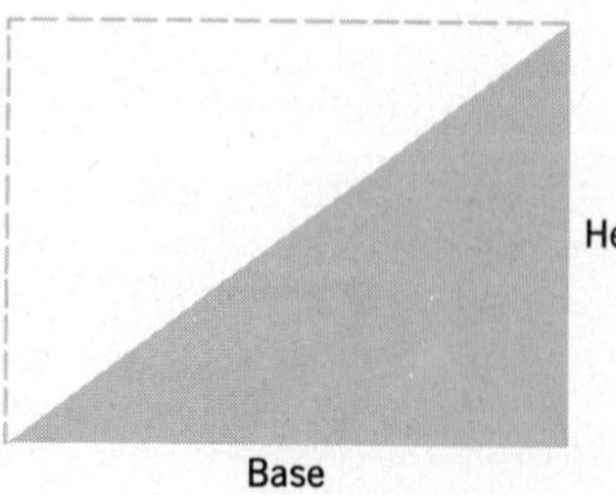

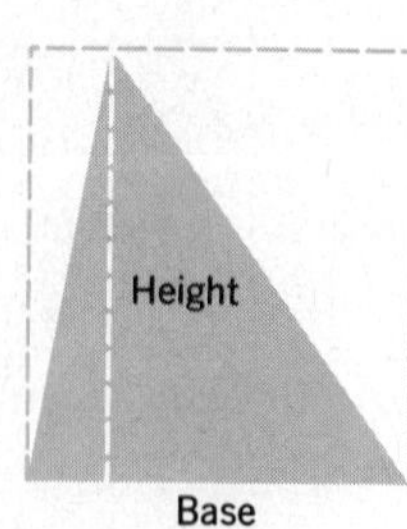

The area of any triangle is

$$\text{Area of triangle} = \tfrac{1}{2} \times \text{base} \times \text{height}$$

This is true no matter what the form of the triangle.

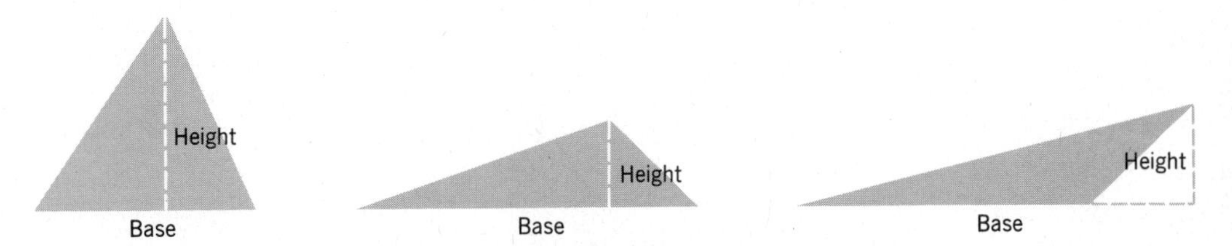

Calculate the area of a triangular piece of metal 15 cm high and 23 cm along the base.

Turn to **56** when you have finished your arithmetic.

---

56 Area of triangle $= \frac{1}{2} \times 15 \text{ cm} \times 23 \text{ cm} = \frac{1}{2} \times 345 \text{ cm}^2 = 172 \text{ cm}^2$

or 170 cm^2 when we round the answer to agree in accuracy with the numbers we used to calculate it.

The area of other simple straight edge geometric figures, or polygons, can be found by dividing them into rectangles and triangles and adding the areas. For example,

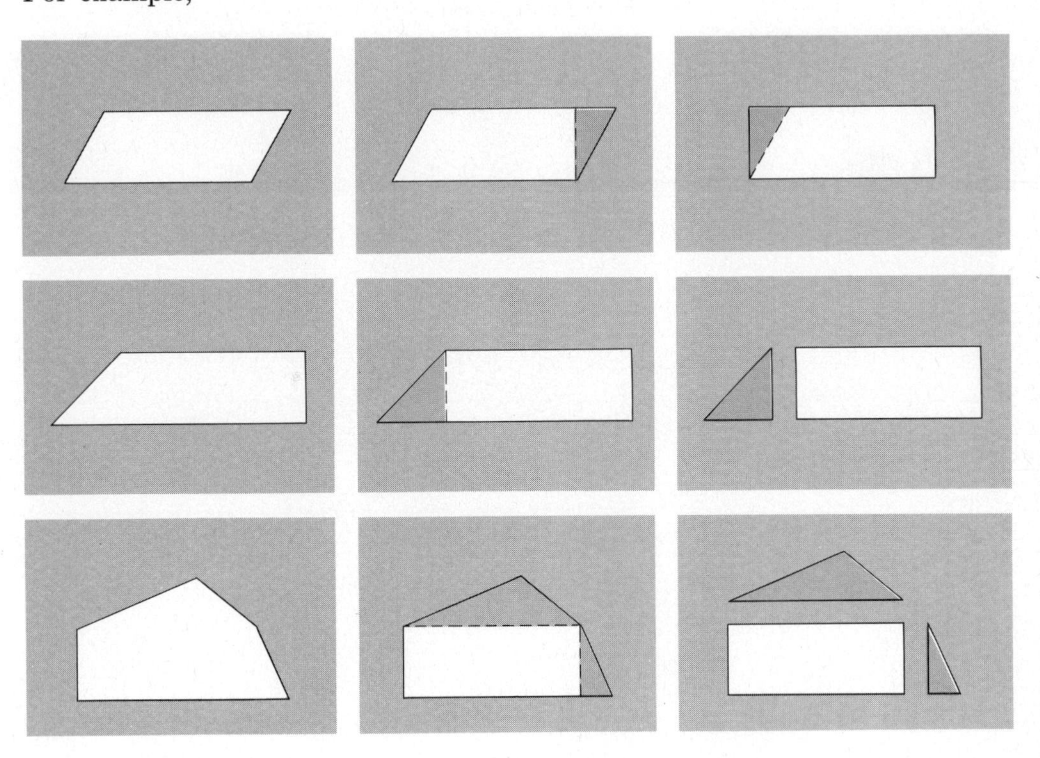

(continued on the next page)

The area of a *circle* is approximately equal to

$$\text{Area of circle} \cong 3.14 \times (\text{radius})^2$$

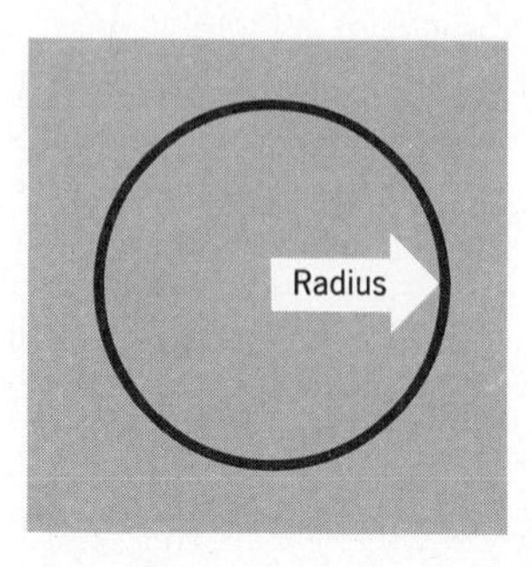

The *radius* of a circle is the distance from its center to its edge. Notice that the equation is only approximate. The number 3.14 is the approximate value of the ratio of the circumference (distance around) a circle to its diameter (distance across). This number is usually named with the Greek letter π (pronounced *pi*). It cannot be written exactly as a decimal or a fraction.

What is the area of a circular disk of radius 22 cm?

Use the formula above to calculate the area, round it properly, then turn to **57** to check your answer.

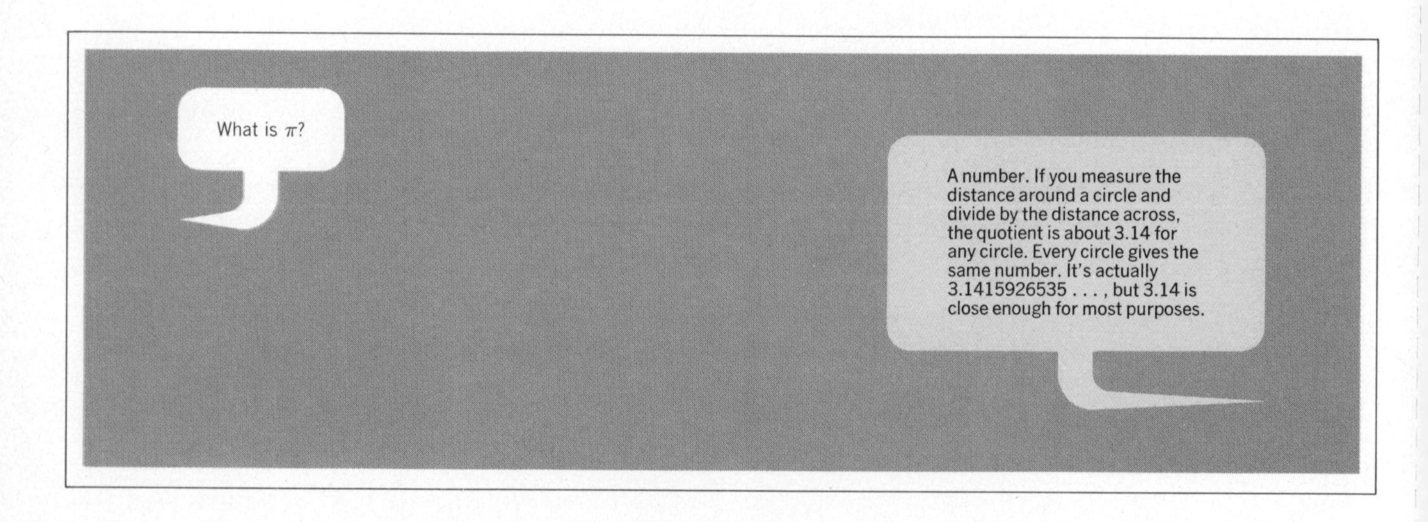

What is π?
A number. If you measure the distance around a circle and divide by the distance across, the quotient is about 3.14 for any circle. Every circle gives the same number. It's actually 3.1415926535 . . . , but 3.14 is close enough for most purposes.

57 Area of circle $= 3.14 \times 22 \text{ cm} \times 22 \text{ cm}$
$= 3.14 \times (22 \times 22) \text{ cm}^2 = 3.14 \times 484 \text{ cm}^2$
$\cong 1500 \text{ cm}^2$ rounded to two significant digits

Find the following areas:

1. Area of a square table top 65 cm on each edge = ______ cm^2

2. Area of a rectangular room 4.2 m wide and 5.1 m long = ______ m^2

3. Area of a triangular piece of cloth 18 in. high and 2.0 ft along the base = ______ ft^2.

4. Area of a circular opening with radius 8 in. = ______ in.^2

The correct answers are in **59**.

---

58 The perimeter is $47 \text{ m} + 75 \text{ m} + 68 \text{ m} + 85 \text{ m} + 83 \text{ m} = 358 \text{ m}$

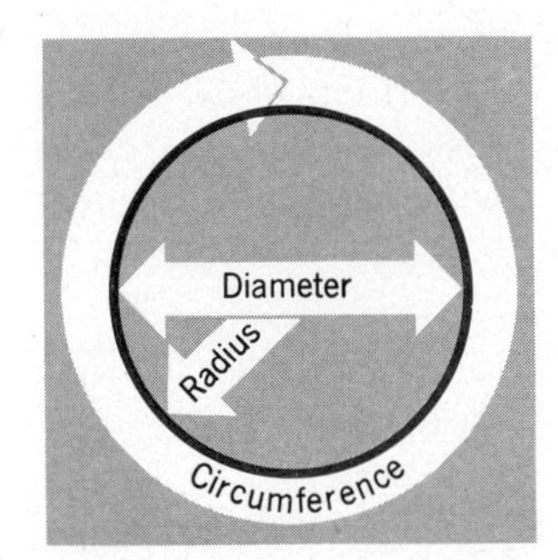

Each of the lengths has been measured to the nearest meter; therefore when we add them our answer is also accurate to the nearest meter and we do not need to round it. (For a quick review of how to add measurement numbers look in **23** on page 339.)

The distance around a circle is called its *circumference*. It is difficult to measure the actual distance along the circle, but we can calculate the circumference if we know the diameter of the circle.

Circumference of a circle $\cong 3.14 \times$ diameter

The diameter of the circle is the distance across it on a line through the center. The diameter is twice the radius.

What is the circumference of a circular lake 240 ft in diameter?

Check your answer in **60**.

---

59 1. Area $= 65 \text{ cm} \times 65 \text{ cm} = 4225 \text{ cm}^2$
$\cong 4200 \text{ cm}^2$ rounded to two significant digits

2. Area $= 4.2 \text{ m} \times 5.1 \text{ m} = 21.42 \text{ m}^2$
$\cong 21 \text{ m}^2$ rounded

3. Area $= \frac{1}{2} \times 18 \text{ in.} \times 2 \text{ ft} = \frac{1}{2} \times 1.5 \text{ ft} \times 2 \text{ ft}$
$= 1.5 \text{ ft}^2$ (Notice that both numbers must be in the same units before we can multiply.)

4. Area $= 3.14 \times 8 \text{ in.} \times 8 \text{ in.} = 200.96 \text{ in.}^2$
$\cong 200 \text{ in.}^2$ rounded to one significant digit

The distance around a polygon* is called its *perimeter*. The value of the perimeter is found by simply adding the lengths of the sides. For example, the perimeter of this square is

$5 \text{ cm} + 5 \text{ cm} + 5 \text{ cm} + 5 \text{ cm} = 20 \text{ cm}$

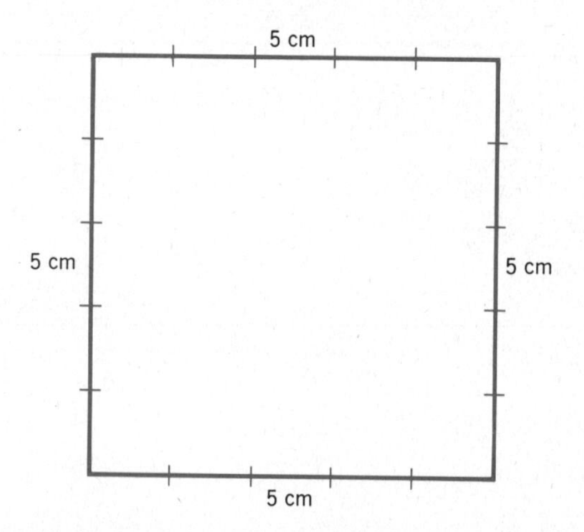

Find the perimeter of the plot of land shown below.

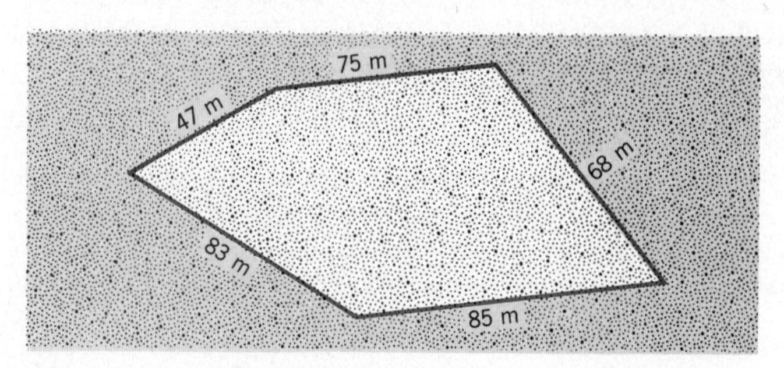

Check your answer in **58**.

---

* A *Polygon* is a flat, closed figure whose sides are straight lines. A triangle is a polygon; so is a square or a rectangle. A circle is not a polygon because it has a curved boundary. A cube is not a polygon because it is a solid figure rather than a plane or flat figure.

60 Circumference = 3.14×240 ft. = 753.6 ft

$\cong$ 750 ft. rounded (We must round the multiplication.)

Volume Any real physical object occupies space, and the measure of the space it occupies is called its *volume.* The volume of a cube 1 cm on each edge is defined as 1 cm $\times$ 1 cm $\times$ 1 cm or 1 cm^3 (pronounced *one cubic centimeter*). One cm^3 is a basic metric unit of volume. Other useful volume units are the cubic inch (in^3 or cu in.), cubic foot (ft^3 or cu ft), cubic yard (yd^3 or cu yd), and cubic meter (m^3). (The unit cm^3 is sometimes written cc. It is the same size volume as the milliliter discussed earlier in this chapter.)

The measure of the volume of any object is equal to the number of basic units of volume it contains. For example, a block can be thought of as built of unit volumes, as shown. The volume of a block 4 cm long, 3 cm high, and 2 cm wide is 24 cm^3. It contains 24 cubic cm volume units—count them.

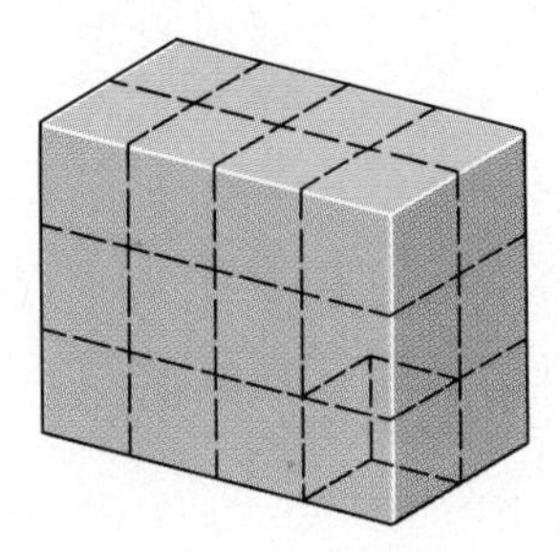

Think of the block as three layers, each with two rows of four unit volumes.

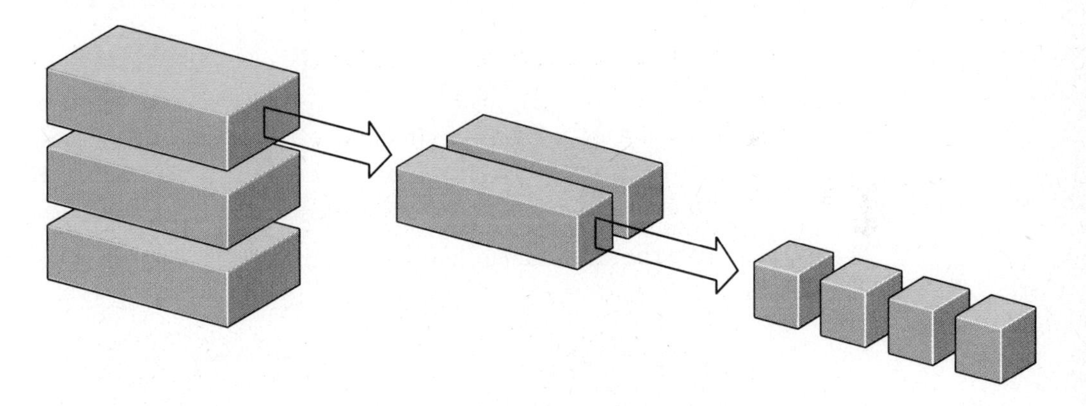

The volume is equal to

number of layers $\times$ number of rows $\times$ number of unit volumes per row

(*continued on the next page*)

In general, the volume of a rectangular solid is

$$\text{Volume of a rectangular solid} = \text{height} \times \text{length} \times \text{width}$$

where height, length, and width are given in the same units (cm, in., ft, m) and the volume is given in cubic units (cm^3, $in.^3$, ft^3, m^3).

For example, the block on page 391 has volume equal to

$$\begin{aligned}\text{Volume} &= 4\text{ cm} \times 3\text{ cm} \times 2\text{ cm} = (4 \times 1\text{ cm}) \times (3 \times 1\text{ cm}) \times (2 \times 1\text{ cm})\\ &= (4 \times 3 \times 2) \times (1\text{ cm} \times 1\text{ cm} \times 1\text{ cm})\\ &= 24\text{ cm}^3\end{aligned}$$

Calculate the volume of an aquarium exactly 9 in. wide, 1 ft high, and 2 ft long.

Check your work in **61**

---

61 $\text{Volume} = \frac{9}{12}\text{ ft} \times 1\text{ ft} \times 2\text{ ft}$

$= 1.5\text{ ft}^3$

It is often useful to know that

$$\begin{aligned}1\text{ m}^3 &= 1{,}000{,}000\text{ cm}^3\\ 1\text{ ft}^3 &= 1728\text{ in.}^3\\ 1\text{ yd}^3 &= 27\text{ ft}^3\end{aligned}$$

How many cubic yards of concrete are needed to fill a wooden box form 18 in. thick, 3.5 ft wide, and 10 ft long?

Do the arithmetic, convert your answer to cubic yards, and round it.

Our work is in **62**.

---

62 $\text{Volume} = \frac{18}{12}\text{ ft} \times 3.5\text{ ft} \times 10\text{ ft} = 52.5\text{ ft}^3$

$$52.5\text{ ft}^3 = 52.5 \times (1\,\cancel{\text{ft}^3}) \times \left(\frac{1\text{ yd}^3}{27\,\cancel{\text{ft}^3}}\right)$$

$$= \frac{52.5}{27}\text{ yd}^3 \cong 1.9\text{ yd}^3 \qquad \text{or about 2 cubic yards.}$$

The volume of a rectangular box is

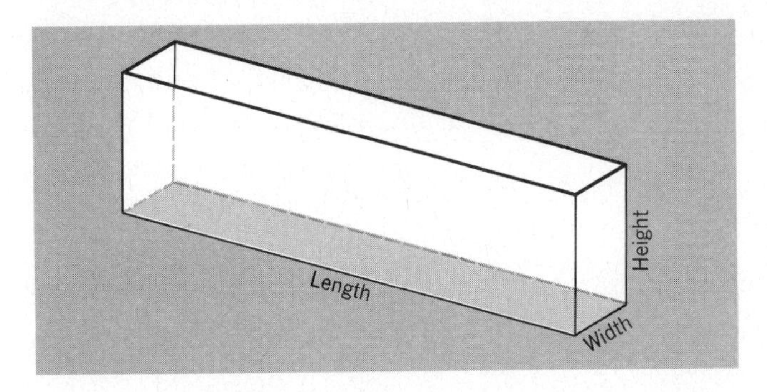

Volume = height × length × width

But (length × width) is the area of the base of the box, therefore we can calculate the volume of the box as

Volume = height × area of base

This is true for any solid object with parallel sides and parallel ends.

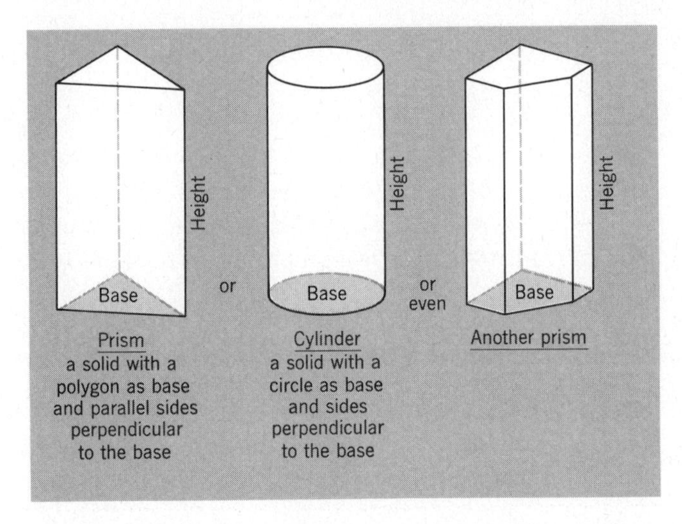

For example, the volume of a cylindrical container 4 ft long with a 6 in. radius base is

$$\begin{aligned}\text{Volume} &= \text{height} \times \text{area of base}\\ &= 4\text{ ft} \times (3.14 \times \tfrac{1}{2}\text{ ft} \times \tfrac{1}{2}\text{ ft})\\ &= 4 \times 3.14 \times \tfrac{1}{2} \times \tfrac{1}{2}\text{ ft}^3 = 3.14\text{ ft}^3\\ &\cong 3\text{ ft}^3 \text{ rounded}\end{aligned}$$

Find the volume of a cylindrical wastebasket 1 ft high and with a bottom 7 in. in radius.

Check your work in **63**.

BOARD FEET

Lumber is often sold according to a special measure of volume called the *board foot.*

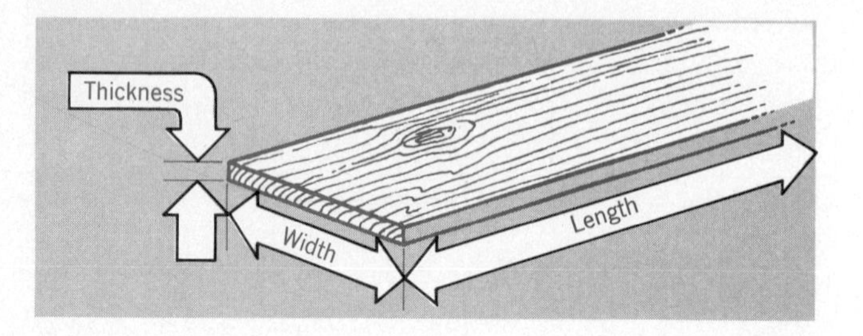

Number of board feet
$= \text{thickness in inches} \times \text{width in feet} \times \text{length in feet}$

The thickness and width are not the actual measured dimensions but are always rounded up to the nearest inch. For example, a 2×4 is not actually 2 inches thick and 4 inches wide as the name implies, but is $1\frac{3}{4}$ in. thick and $3\frac{1}{2}$ in. wide. A 1 by 12 plank is actually $\frac{3}{4}$ inches thick and $11\frac{1}{2}$ inches wide. The wood is cut roughly to 2 by 4 or 1 by 12 and then smoothed leaving it slightly smaller. In calculating board feet we use the given dimensions, 2 inches by 4 inches, and not the actual measured dimensions, $1\frac{3}{4}$ inches by $3\frac{1}{2}$ inches.

How many board feet of lumber are in an 8 foot long 2 by 4?

$$\text{Number of board feet} = 2 \text{ in} \times \frac{4}{12} \text{ ft} \times 8 \text{ ft} = \frac{64}{12} \text{ ft} = 5.3 \text{ bd ft}$$

(2 in: Thickness in inches; $\frac{4}{12}$ ft: Width in feet; 8 ft: Length in feet)

63 Volume = height × (area of base)

$$= 1 \text{ ft} \times \left(3.14 \times \frac{7}{12} \text{ ft} \times \frac{7}{12} \text{ ft}\right)$$

$$= \left(3 \times \frac{7}{12} \times \frac{7}{12}\right) \text{ft}^3 = 1.02 \text{ ft}^3$$

$$\cong 1 \text{ ft}^3 \text{ rounded}$$

Now turn to **64** for a set of practice problems on calculating area, volume and perimeter.

Measurement

PROBLEM SET 6

64 Answers are on page 510.

1. What is the area, in square meters, of a piece of cloth 50 cm wide by 350 cm long?

   1.75 cm²
   ______

2. At \$9.50 per square yard, what is the cost to carpet a room 12 ft 8 in. by 16 ft 6 in.?

   \$220.61
   ______

3. Which has the greater area, a square 2 m on each side or a triangle 2 m high with a base 3.5 m long?

   The square
   ______

4. What is the area, in hectares, of a lot 600 m long and 80 m wide?

   4.8 hectares
   ______

5. What is the area, in acres, of a lot 250 ft long and 85 ft. wide?

   ______

6. What is the volume of a room 10 ft wide, 12 feet long, and 8 feet high?

   ______

7. What is the volume of a block of ice 1 ft by 1 ft by 2 ft? What would it weigh if the density of ice is 0.034 lb/in^3?

   ______

8. How many cubic yards of dirt are needed to fill a hole 5 ft deep, 3 ft wide, and 12 ft long?

   ______

9. What is the volume of a cylindrical hot water tank 15 cm in radius and 160 cm high?

   ______

May 18, 1978
Date

Stephanie Catanzaro
Name

050/003
Course/Section

10. Find the circumference of a circle 10 cm in diameter.

____

11. Find the perimeter of each of the plots of land shown below.

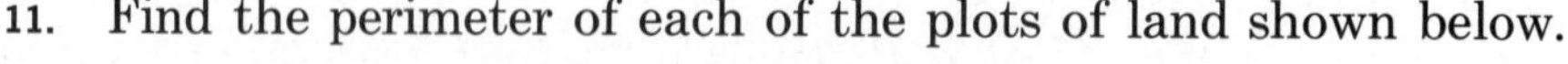

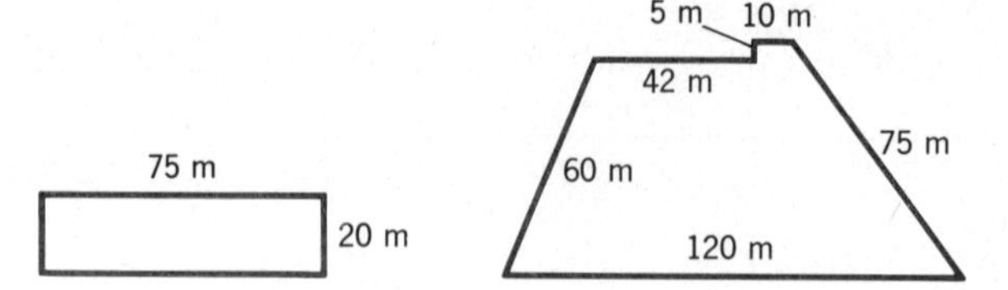

12. (a) How many cubic yards of concrete are needed to build the sidewalk shown below if it is 6 in. thick?______

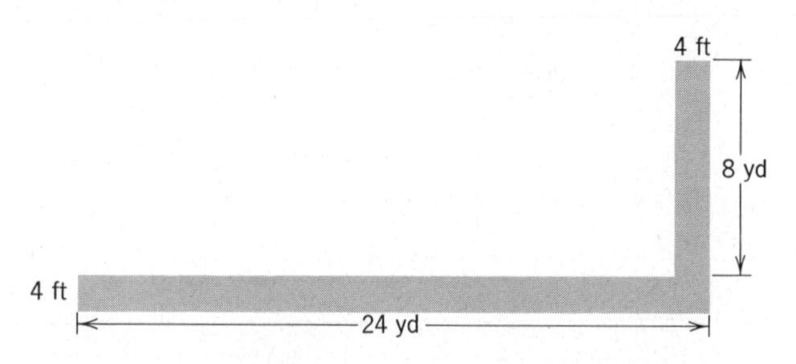

(b) What is the area of this sidewalk?______

(c) What length of wood is needed around the border to hold the concrete?______

13. What is the area in acres of a football field 60 yd wide and 120 yd long?

____

14. If grass seed costs 89¢ per lb, and if 1 lb will cover 120 sq ft, what will it cost to seed a lawn 25 ft wide and 60 ft long?

____

15. Find a 5¢ piece, measure its diameter, and calculate its area in square cm.

____

16. What is the difference in area between a square 8 in. on a side and a circle 8 in. in diameter.

____

17. What is the volume of water in a swimming pool 30 yd long, 12 yd wide, and an average of 5 ft deep.

____

18. (a) What is the volume of a cube 10 m on each edge? ______

    (b) What is the total surface area of the cube? ______

    (c) What is the total length of all the edges of the cube? ______

19. How many cubic ft of food will a freezer hold if its inside dimensions are 36 inches by 22 inches by 18 inches?

____

20. What is the area of triangle whose height is 6 ft and whose base is 10 ft long?

____

When you have had the practice you need turn to **65** for a self-test on measurement.

May 18, 1978
Date

Stephanie Catanzaro
Name

050/003
Course/Section

Unit 5
Measurement

Self-Test

65

1. 5 ft 8 in. = ________ yd
2. 10 km = ________ mi
3. 6 pints = ________ liters
4. 3.5 kg weighs = ________ lb
5. 55 mph = ________ km/hr
6. 1450 m = ________ ft
7. $3\frac{1}{2}$ lb + 5 lb 6 oz + 20 oz = ________ lb
8. $4\frac{1}{2}$ gal − 6 qt = ________ qt
9. 50° C = ________ ° F
10. 6 ft = ________ m
11. 12 gal = ________ liters
12. 60 miles = ________ km
13. 125 lb = ________ kg
14. 4 cm × 8 cm = ________
15. 36 mi ÷ 4 hr = ________
16. What is the cost of 25 lb of flour selling for 13¢ per pound?
17. If one gallon of milk costs $1.03, what is the cost of an 8-oz glass of milk?
18. A dress pattern requires $3\frac{3}{4}$ yd of fabric. How many meters of this fabric should you buy? ________ m
19. Find the area of a room 12 ft 8 in. wide and 16 ft long. ________ yd^2
20. Calculate the volume of a rectangular box 3 ft high, 2 ft 4 in. wide, and $4\frac{1}{2}$ ft long. ________ yd^3
21. What is the volume of a cylinder 120 cm high and having a base 16 cm in radius? ________ cm^3

May 23, 1978
Date

Stephanie Catanzaro
Name

050/003
Course/Section

22. Calculate the area of a circle of radius 8 in. ________ in^2

23. What is the perimeter of a rectangle 2 ft 8 in. long and 1 ft 4 in. wide? ________ ft

24. What is the total cost of carpeting a room 15 ft 6 in. by 12 ft, if the carpet costs $8 per square yard? ________ $

25. A famous footrace is run each year from London to the seacoast town of Brighton 84.5 km away. If the winner runs it at a rate of 10 mph, what will be his time? ________

Answers are on page 515.

SUPPLEMENTARY PROBLEM SET 1A

Unit Conversion

Measurement

A. Complete these statements:

1. 5 lb = ______ oz
2. 40 days = ______ weeks
3. 200 in. = ______ ft
4. 4 weeks = ______ hr
5. 18 ft = ______ yd
6. 500 sec = ______ min
7. 35 oz = ______ lb
8. 5 gal = ______ qt
9. 4 yd = ______ ft
10. 1 week = ______ sec
11. 6 ft = ______ in.
12. 2 cups = ______ tbsp
13. 10 pt = ______ qt
14. 30 qt = ______ gal
15. 3 barrels = ______ qt
16. 40 fl oz = ______ cup
17. 20 tsp = ______ cup
18. 100 days = ______ yr
19. 6 oz = ______ grains
20. 20 cups = ______ gal
21. 3 cups = ______ tbsp
22. 150 sec = ______ week
23. $3\frac{1}{2}$ pt = ______ fl oz
24. $8\frac{1}{4}$ gal = ______ qt
25. 6.8 ft = ______ in.
26. $2\frac{3}{4}$ miles = ______ ft
27. $4\frac{1}{2}$ cups = ______ oz
28. $10\frac{1}{2}$ qt = ______ gal
29. $4\frac{1}{2}$ oz = ______ grains
30. $\frac{1}{2}$ gal = ______ fl oz
31. 2 fl oz ______ fl dram
32. 2 lb = ______ grains

Date

Name

Course/Section

SUPPLEMENTARY PROBLEM SET 2A

Arithmetic with Measurement Numbers

Measurement

A. Add or subtract as shown:

1. 6 ft + 16 in. = ____
2. 3 min + 75 sec = ____
3. 3 days + 11 hr = ____
4. 8 lb + 1 lb 4 oz = ____
5. 5 ft − 1 ft 8 in. = ____
6. 6 qt − 2 qt 3 pts = ____
7. 3 hr − 2 hr 10 min = ____
8. 30 mph − 8 mph = ____
9. 5 hr + 35 min = ____
10. $3\frac{1}{2}$ qt + 6 pt = ____
11. 10 ft − 3 ft 8 in. = ____
12. 14 lb − 4 lb 9 oz = ____
13. 6 lb 10 oz + 3 lb 8 oz = ____
14. 6 lb − 12 oz = ____
15. $2\frac{1}{2}$ hr − 40 min = ____
16. $4\frac{1}{2}$ ft + 9 in. = ____
17. $3\frac{1}{4}$ lb − 10 oz = ____
18. 1000 in. − 20 ft 6 in. = ____

B. Multiply and divide as shown:

1. 4 ft × 8 ft = ____
2. 5 in. × 8 in. = ____
3. 9 ft × 12 ft = ____
4. 20 in. × 30 in. = ____
5. 1 ft 3 in. × 2 ft 6 in. = ____
6. 100 mi ÷ 2.5 hr = ____
7. 80 mi ÷ 1 hr 45 min = ____
8. 45.6 mi ÷ 32 min = ____
9. 55 mph × 3 hr = ____
10. 45 mph × $2\frac{1}{2}$ hr = ____
11. 60 mph × 1 hr 15 min = ____
12. 50 mph × 3 hr 25 min = ____
13. 210 mi ÷ 20 gal = ____
14. $175.08 ÷ 42 = ____
15. 68 cents/gal × 5 gal = ____
16. 13 ft 6 in. ÷ 9 = ____
17. 5 ft 6 in. × 3 = ____
18. 10 lb ÷ $1.65 = ____

Date ____

Name ____

Course/Section ____

SUPPLEMENTARY PROBLEM SET 3A

Metric System: Length and Speed Units

Measurement

A. Think metric: Circle the measurement closest to the first one given:

		(a)	(b)	(c)
1.	400 m	4000 yd	$\frac{1}{4}$ mi	4 mi
2.	5 km	500 m	5 mi	3 mi
3.	2 cm	0.8 in.	1 in.	2 in.
4.	50 km	30 mi	100 mi	10 mi
5.	3 in.	1 cm	7.6 cm	1.1 cm
6.	1000 km	0.6 mi	6 mi	600 mi
7.	5 mm	$\frac{1}{2}$ cm	2 in.	$\frac{1}{2}$ in.
8.	40 ft	120 cm	1.2 m	12 m
9.	100 km/hr	100 mph	60 mph	600 mph
10.	20 km/hr	12 mph	20 mph	30 mph
11.	55 mph	55 km/hr	90 km/hr	15 km/hr
12.	1500 m	1 mile	1000 yd	10 mi

B. Convert to the units shown using either unity fractions or the metric minder:

1. 5 ft = _______ m
2. 6 in. = _______ cm
3. 24 ft = _______ m
4. 5 miles = _______ km
5. 3 in. = _______ mm
6. 60 mi = _______ km
7. 4000 ft = _______ m
8. 100 km = _______ miles
9. 10 cm = _______ in.
10. 9 m = _______ ft
11. 250 mi _______ km
12. 75 km/hr = _______ mph

Date

Name

Course/Section

SUPPLEMENTARY PROBLEM SET 4A

Metric System: Weight–Mass Units

Measurement

A. Think metric: Circle the measurement closest to the first one given:

		(a)	(b)	(c)
1.	100 kg	22 lb	220 lb	2.2 lb
2.	10 g	0.3 oz	3 oz	300 oz
3.	5 kg	10 lb	2 lb	4 lb
4.	100 mg	1000 g	10 g	$\frac{1}{10}$ g
5.	10 lb	2 kg	5 kg	2 g
6.	4 oz	7 g	110 g	10 g
7.	2 kg	4 lb	2 lb	4.5 lb
8.	50 mg	5 g	0.5 g	0.05 g
9.	400 g	1 lb	2 lb	2.2 lb
10.	150 lb	70 kg	15 kg	1.5 kg
11.	1000 kg	2.2 tons	220 lb	2200 lb
12.	16 lb	7 kg	35 kg	3.5 kg

B. Convert to the units shown using either unity fractions or the METRIC MINDER:

1. 50 kg = ______ lb
2. 120 g = ______ lb
3. 300 g = ______ oz
4. 20 lb = ______ kg
5. 150 lb = ______ kg
6. 4200 lb = ______ kg
7. 10 oz = ______ g
8. 1 lb 9 oz = ______ g
9. 500 mg = ______ oz
10. 8 lb 4 oz = ______ kg
11. $\frac{1}{4}$ lb = ______ g
12. 1.2 kg = ______ g

Date

Name

Course/Section

SUPPLEMENTARY PROBLEM SET 5A

Metric System: Volume and Temperature Units

Measurement

A. Think metric: circle the measurement closest to the first one given:

		(a)	(b)	(c)
1.	10° C	10° F	−10° F	50° F
2.	−20° C	−10° F	4° F	−4° F
3.	35° F	2° C	85° C	10° C
4.	−10° F	14° C	−14° C	4° C
5.	4 gal	1.7 liters	17 liters	170 liters
6.	3 pt	160 liters	16 liters	1.6 liters
7.	40 liters	9.5 gal	8 gal	10 gal
8.	4 qt	0.4 liters	40 liters	4 liters
9.	2 tsp	2 ml	0.1 liter	10 ml
10.	$\frac{1}{2}$ liter	500 ml	1000 ml	1 qt
11.	10 fl oz	30 ml	300 ml	10 ml
12.	200 ml	0.2 liter	2 liter	.02 liter

B. Convert to the unit shown:

1. 40 ml = ______ cup
2. 5 qt = ______ liter
3. 15 liters = ______ qt
4. 300 ml = ______ liter
5. 2 liter = ______ fl oz
6. 1250 cc = ______ qt
7. 8 cups = ______ liter
8. $8\frac{1}{2}$ liters = ______ fl oz
9. $1\frac{1}{2}$ liters ______ cc
10. 30° C = ______ ° F
11. 60° F = ______ ° C
12. −15° F = ______ ° C

Date ______

Name ______

Course/Section ______

SUPPLEMENTARY PROBLEM SET 6A

Calculating Area, Volume, and Perimeter

Measurement

A. Find the area of each of the following geometric figures:

1. A square 6 in. on each side ______
2. A 25 ft by 12 ft rectangle ______
3. A triangle with base 8 in. and height 12 in. ______
4. A rectangular floor 9 ft 6 in. by 14 ft 4 in. ______
5. A square lot 106 ft on each side ______
6. A rectangle 6 m wide and 120 m long ______
7. A triangle with base 1.25 m and height 200 cm ______

B. Find the volume of each of the following:

1. A cube 3 ft on each edge. ______
2. A rectangular room 8 ft high by 14 ft long by 9 ft wide. ______
3. A rectangular box 3 cm by 4 cm by 9 cm. ______
4. A rectangular block 4.5 cm by 3.1 m by 9.0 cm. ______
5. A cylinder 10 cm in radius and 100 cm high. ______
6. A triangular prism whose base has an area of 15 cm^2 and whose length is 1.2 m. ______

C. Find the perimeter of each of the following:

1. A square 10 ft on each side. ______
2. A rectangle 6 ft by 4 ft. ______
3. A rectangular lot 281 ft long and 106 ft wide. ______
4. A triangle with sides 8 in., 9 in., and 3 in. ______
5. A circle with diameter 4 ft. ______
6. A polygon with sides 1.2 m, 80 cm, 2.1 m, 160 cm, and 40 cm. ______

Date ______

Name ______

Course/Section ______

Introduction to Algebra

PREVIEW 6

Objective	Sample Problems		Where To Go for Help	
Upon successful completion of this program you will be able to:	Identify each of the following as a term, expression, variable, constant or equation.		Page	Frame
1. Understand and use basic algebra words such as *term, expression, factor, variable, constant,* and *equation.*	(a) $2x$	______	415	1
	(b) The 3 in $3x^2$	______		
	(c) The x in $3x^2$	______		
	(d) $3x + 5$	______		
	(e) $x + 4 = 9$	______		
2. Perform the basic algebra operations.	(a) $2a + 3a =$	______	422	9
	(b) $2b \times 3b =$	______		
	(c) $2x - (x + 2) =$	______		
	(d) $3y(x - 2) =$	______		
	Write each phrase as an algebraic expression.			17
3. Translate simple English sentences into algebraic expressions.	(a) four times a number	______	431	
	(b) three less than a number	______		
	(c) the product of a number and twice the number.	______		
	(d) the sum of what two consecutive numbers is a perfect square?	______.		
	If $x = 2$, $a = 5$, $b = 6$ find the value of			23
4. Calculate the numerical value of literal expressions.	(a) $\frac{2b - x}{a} =$	______	441	
	(b) $x^2 + 3x =$	______		
	(c) $3abx^2 =$	______		

5. Recognize and solve linear equations in one variable.	(a) $3x - 4 = 11$ $\quad x =$	________	451	**30**
	(b) $2x = 18$ $\quad x =$	________		
	(c) $2x + 7 = 43 - x$ $\quad x =$	________		
	(d) 5 less than twice a number is equal to 9. Find the number.	________		
6. Solve problems using ratio and proportion.	(a) In the first 12 games of the basketball season Joe Dokes scored 252 points. How many can he expect to score in the entire 22 game season?	________	471	**45**
	(b) If the ratio of my height to his height is 8/7 and I am 6 ft. tall, how tall is he?	________		

(Answers to these problems are at the bottom of this page.)

If you are certain you can work all of these problems correctly, turn to page 483 for a self-test. If you want help with any of these objectives or if you cannot work one of the sample problems, turn to the page indicated. Superstudents who want to be certain they learn all of this will turn to frame 1 and begin work there.

Date ____________

Name ____________

Course/Section ____________

PREVIEW 6

Answer to Sample Problems

1. (a) *term*
   (b) a *constant* or *coefficient*
   (c) *variable*
   (d) *expression*
   (e) *equation*
2. (a) $5a$
   (b) $6b^2$
   (c) $x - 2$
   (d) $3xy - 6y$
3. (a) $4x$
   (b) $N - 3$
   (c) $n(2n)$ or $2n^2$
   (d) $a + (a + 1) = b^2$
4. (a) 2
   (b) 10
   (c) 360
5. (a) 5
   (b) 9
   (c) 12
   (d) 7
6. (a) 462 points
   (b) $5\frac{1}{4}$ ft.

Introduction to Algebra

1 This unit could have been titled "What You Always Wanted to Know About Algebra But Were Afraid to Ask." It explains the beginning ideas, words, symbols, and operations you need to solve simple algebra problems. These are the skills most college math and science teachers and textbooks assume you know, the ideas they never bother to explain or review. We're going to talk to you about algebra but we do care, and we promise not to mess up your mind. Even Sally might enjoy it.

A mathematical statement in which letters are used to represent numbers is called a *literal* expression. Letters can represent single numbers or entire sets of numbers. Algebra is the arithmetic of literal expressions. It is a kind of symbolic arithmetic that enables us to find answers to problems by simple operations with letters rather than by repeated and difficult arithmetic with numbers.

Any letters will do. In mathematics, English, Greek, and even Hebrew alphabets are used, including lower case, capital, and even script letters. The letters

a, A, **A**, $\mathfrak{A}$ or $\mathcal{A}$

are all different and each represents a *different* quantity in algebra even though all are the same letter of the alphabet. If you want to represent the distance a car travels by the letter d, you should be consistent and use the symbol d and not confuse it with D, **D**, $\mathfrak{D}$ or $\mathcal{D}$.

People who use algebra often choose the symbols they use on the basis of their memory-jogging value. Time is represented by t, distance by d, cost by c, area by A, and so on. The symbol is chosen so as to remind you of its meaning.

In the sentence

"The boy jumped over the fence."

(continued on the next page)

the word *boy* stands for a person or a number of people. Many more specific words may be substituted to give true statements:

John jumped over the fence.
Bill jumped over the fence.
Dave jumped over the fence.

or in general,

☐ jumped over the fence.

The word *boy* or the box ☐ are *variables*. They stand for a particular person or a set of persons.

The equation $3 + 2 = 5$ means that the number named on the left $(3 + 2)$ is the same as the number named on the right (5).

The algebra equation $\square + 3 = 7$ is a statement that some number ☐ added to 3 is equal to 7. The symbol ☐ is a variable representing a particular number.

What is the value of ☐ in the equation above that makes it a true statement?

Try it. Guess if you must and then check your guess. Look in **2** for the answer.

---

2 $\square + 3 = 7$

$\square = 4$

—of course. Congratulations, you have just solved your first algebra equation.

Normally this equation would be written

$x + 3 = 7$ We use letters instead of ☐ because letters are easier to write and because we may need to use several different symbols.

Often an algebra statement contains fixed numbers or constants. These may be written as actual numbers or represented by letters.

In $5x^2 + 3x + 4$ the 5, 3, and 4 are constants.

A constant that multiplies a variable is called a *coefficient*. The 5 in $5x^2$ and the 3 in $3x$ are both coefficients. The 2 in $5x^2$ is a constant called an *exponent*.

In the algebra statement $6t^2 - 3t + 4$

6 is a ______________

t is a ______________

2 is a ______________

3 is a ______________

4 is a ______________

Check your answers in **3**.

---

3 6 is a *coefficient* or *constant*

t is a *variable*

2 is an *exponent*

3 is a *coefficient,* and it is also a *constant*

4 is a *constant*

Most of the usual arithmetic symbols have the same meaning in algebra that they have in arithmetic. For example, the addition (+) and subtraction (−) signs are used in exactly the same way. However, the *multiplication* sign (×) of arithmetic looks like the letter x and to avoid confusion we have other ways to show multiplication in algebra. The product of two algebra quantities a and b, "a times b," may be written using

a raised dot $a \cdot b$

parentheses $a(b)$ or $(a)b$ or $(a)(b)$

or with nothing at all ab

Obviously this last way of showing multiplication won't do in arithmetic; we cannot write "two times four" as "24"—it looks like twenty four. But it is a quick and easy way to write a multiplication in algebra.

Placing two quantities side by side to show multiplication is not new and it is not only an algebra gimmick; we use it every time we write 20¢ or 4 feet.

20¢ = 20 × 1¢
4 feet = 4 × 1 foot

Write the following multiplications using no multiplication symbols.

(a) 8 times a = ______

(b) 2 times s times t = ______

(c) 3 times x times x = ______

Check your answer in **4**.

---

4 (a) 8 times a = $8a$
(b) 2 times s times t = $2st$
(c) 3 times x times x = $3x^2$

Notice that powers are written just as in arithmetic

$x \cdot x = x^2$, $x \cdot x \cdot x = x^3$, and so on.

We use parentheses, ( ), in arithmetic to show that some complicated quantity is to be treated as a unit. For example,

$2 \cdot (3 + 14 - 6)$

means that the number 2 multiplies *all* of the quantity in the parentheses. In exactly the same way in algebra, parentheses ( ), brackets [ ], or braces { }, indicate that whatever is enclosed in them should be treated as a single

(continued on the next page)

quantity. An expression such as $(3x^2 - 4ax + 2by^2)^2$ should be thought of as $(\text{something})^2$. The expression

$$(2x + 3a - 4) - (x^2 - 2a)$$

should be thought of as (first quantity) − (second quantity).

Parentheses are the punctuation marks of algebra. Like the period, comma, or semicolon in regular sentences, they tell you how to read an equation and get its correct meaning.

Write using algebra notation:

(a) 8 times $(2a + b) =$ ______

(b) $(a + b)$ times $(a - b) =$ ______

(c) add x to $(y - 2)$ and multiply the sum by $(x^2 + 1)$. = ______

The correct answers are in 5.

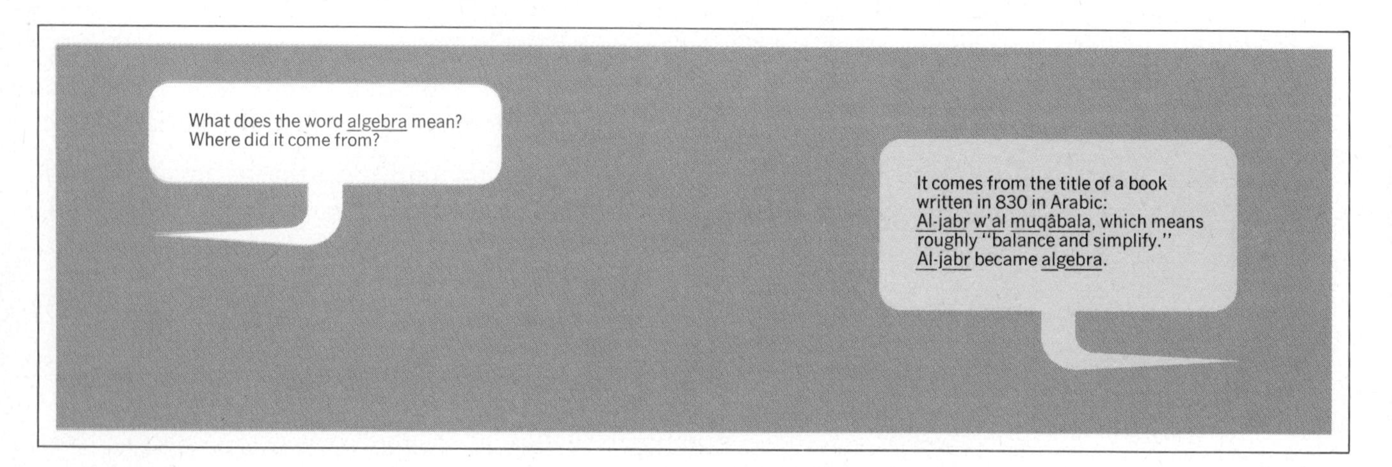

---

5 (a) $8(2a + b)$
(b) $(a + b)(a - b)$
(c) $[x + (y - 2)](x^2 + 1)$ or $(x + y - 2)(x^2 + 1)$

In arithmetic we would write "48 divided by 2" as

$2\overline{)48}$ or $48 \div 2$ or $48/2$

These notations are used very seldom in algebra. Most often division is written as a fraction.

"x divided by y" is written $\dfrac{x}{y}$

"$(2n + 1)$ divided by $(n - 1)$" is written $\dfrac{(2n + 1)}{(n - 1)}$ or $\dfrac{2n + 1}{n - 1}$.

Write using algebra notation:

(a) a divided by b = ______

(b) $(x + 2)$ divided by $(y + 3)$ = ______

(c) Multiply e by h and divide the product by $2d$ = ______

Turn to **6** to check your answer.

---

6 (a) $\dfrac{a}{b}$ (b) $\dfrac{x + 2}{y + 3}$ (c) $\dfrac{eh}{2d}$

Several words are used very often in algebra, and you should use them correctly.

An *expression* is a general name for any collection of numerals and letters connected by arithmetic operation signs. For example,

$2x^2 + 4$

is an algebra expression.

$(b - 1)$, $(x + y)$, and $2(x^2 - 3ab)$ are all algebra expressions.

An algebra expression may be a product of several factors or a sum of two or more terms. When two variables, constants or expressions are multiplied together, each multiplier is called a *factor* of the product. For example,

a and b are factors of the product ab.

$2x$ and $(x + 1)$ are factors of the product $2x(x + 1)$.

The expression $(k - 1)(2k + 1)$ is the product of factors $(k - 1)$ and $(2k + 1)$.

A *term* is a general name for a constant, variable, or product of constants and variables.

2, x, and $2x$ are terms.

x^2, $3x^2$, $5ab$, $2\pi r$, and $\dfrac{3}{c}$ are all algebra terms.

Notice that terms do not contain addition or subtraction signs.

(*continued on the next page*)

The sum of two or more terms is called an expression. For example,

$x + 2$	is an expression.
$3x^2 + 2x + 1$	is an expression.
$5ab + c^2 + 3a$	is an expression.

If an equation is an algebra sentence and parentheses are punctuation marks, then an expression is a phrase or group of words, and a term is a single word.

In the product $3(2x + y)$

$2x$ is a ______.

$2x + y$ is a ______ of the product.

$3(2x + y)$ is a ______.

2 and x are ______ of $2x$.

Complete these, then turn to **7**.

---

7 $2x$ is a *term.*
$2x + y$ is a *factor* of the product.
$3(2x + y)$ is an *expression.*
2 and x are *factors* of $2x$.

For each of the following expressions, name the parts shown.

(a) $3xy^2 - 1$ $3xy^2$ is a ______. 1 is a ______.

(b) $4a(c^2 - 1)$ $4a$ is a ______. $(c^2 - 1)$ is a ______.

(c) $a + b$ a and b are ______.

The correct answers are in **8**.

SUBSCRIPTS

Mathematicians, scientists, and others who use algebra in their work often use subscripts to help name variables and constants. You may find these strange looking and tricky to pronounce. Here is a bit of help.

Suppose you measured the air temperature every morning at a certain time. You might label the temperature variable as T and call the temperature on the first day T_1 (say "tee-sub-one"), the temperature on the second day T_2, the temperature on the third day T_3, and so on. The little numbers slightly below the letters are called subscripts. Any sort of subscripts can be used and, usually we choose subscripts

(*continued on page 421*)

that have some memory jogging value. For example, if we measured all the weights of the boys in a sixth grade class, we could call Al's weight W_A (say "W-sub-A") or W_{Al}, Bill's weight might be W_B or W_{Bill}, Sam's weight would be W_S or W_{Sam}, and so on.

Subscripts can be handy reminders of what the variable letter represents.

8 (a) *Term* *Term*
(b) *Factor* *Factor*
(c) *Terms*

The equals sign (=) is used in algebra and has the same meaning there that it has in arithmetic. The arithmetic equation

$4 + 1 = 3 + 2$

means that the quantity on the left of the equals sign (4 + 1) has the same value as the quantity on the right (3 + 2).

In algebra the equality sign is used to compare algebraic quantities.

The *algebra equation*

$x + 3 = 7$

means that the expression on the left ($x + 3$) has the same value as the quantity on the right (7). This algebraic equation can be either true or false. For most values of x it is false. Only if x is equal to 4 is it true. We shall study algebra equations later in this unit.

There are five simple arithmetic operations that are used in beginning algebra. Knowing how to handle these five will enable you to solve any simple algebra problem. The operations are:

Operation 1. *Adding and subtracting like terms:*

$$2a + 3a = 5a$$

Operation 2. *Adding and subtracting expressions:*

$$(2a - b) + (a + 2b) = 3a + b$$

Operation 3. *Multiplying simple factors:*

$$(2ax)(3x) = 6ax^2$$

Operation 4. *Multiplying expressions:*

$$(a + b)(2a - b) = 2a^2 + ab - b^2$$

Operation 5. *Dividing similar expressions:*

$$\frac{6a^2b^3}{2ab} = 3ab^2$$

Let's look at them in order. Turn to **9** to begin.

9 Operation 1. *Adding and subtracting like terms.*

Example: $2a + 3a = 5a$

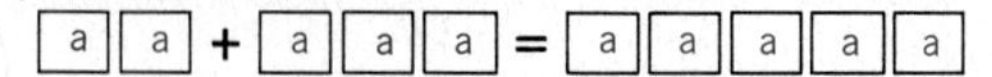

It will help if you think of this as

2 *things* + 3 *things* = 5 *things* ... Apples

or

2¢ + 3¢ = 5¢

We are adding like quantities, "a"s or *things* or ¢.

$2a$ — $\boxed{a}\ \boxed{a} = 2\,\boxed{a} = 2a$

Coefficient — Variable part

$3a^2b$ — $\boxed{a^2b}\ \boxed{a^2b}\ \boxed{a^2b} = 3\,\boxed{a^2b} = 3a^2b$

Coefficient — Variable part

When the variable part is the same for each term, we add coefficients and attach the variable part to the sum.

$3xy^2 + 4xy^2 = 7xy^2$

The variable part of each term is xy^2

Subtraction is precisely the same.

$8aqx - 3aqx = 5aqx$

$8 - 3$ — Variable part of each term

Try these for practice:

(a) $12d^2 + 7d^2 =$ ______

(b) $2ap + ap + 5ap =$ ______

(c) $3(a + 1) + 9(a + 1) =$ ______

(d) $3p + 2m + 4p =$ ______

Turn to **10** to check your answers.

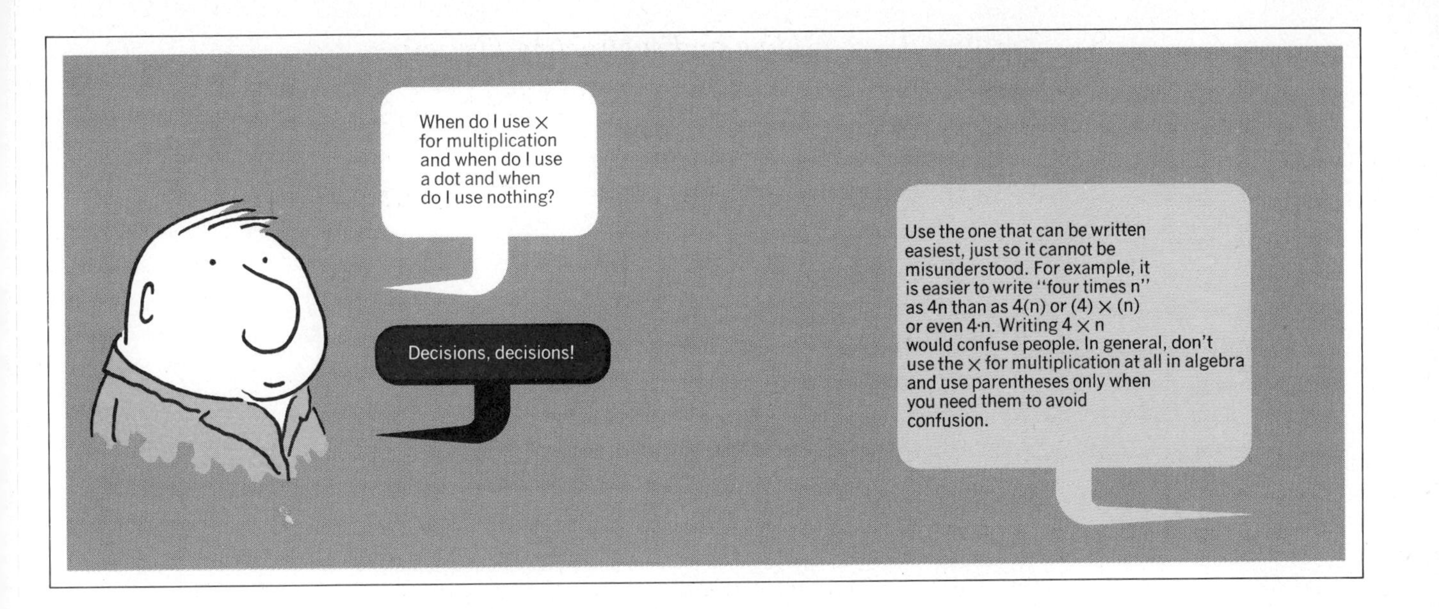

---

10 (a) $12d^2 + 7d^2 = 19d^2$

(b) $2ap + ap + 5ap = 8ap$

The coefficient of ap is 1. Add these terms as

$2ap + 1ap + 5ap = 8ap$ $\boxed{ap}\ \boxed{ap} + \boxed{ap} + \boxed{ap}\ \boxed{ap}\ \boxed{ap}\ \boxed{ap}\ \boxed{ap}$

(c) $3(a + 1) + 9(a + 1) = 12(a + 1)$

The common variable part is $(a + 1)$. Again think of it as

3 *things* + 9 *things*

where a *thing* is $(a + 1)$

(d) $3p + 2m + 4p = 7p + 2m$

We cannot combine the $2m$ term with the others because the variable parts are not the same.

Operation 2. *Adding and subtracting expressions.*

Example: $(a + b) + (a + d) = 2a + b + d$

Parentheses are used in algebra to group together numbers and variables that are to be treated as a unit. Adding or subtracting expressions involves two steps:

Example:

First, remove all parentheses, $(a + b) + (a + d)$
$= a + b + a + d$
Second, add like terms $= a + a + b + d$
$= 2a + b + d$

Removing parentheses can be a very tricky business, and it will help greatly if you remember what you learned about signed numbers in Unit 3, starting in frame **35**, page 225.

(*continued on the next page*)

$+(a + b) = \quad +a + b$	Think of this as $(+1)(a + b)$ or $(+1)a + (+1)b = +a + b$
$+2(a + b) = +2a + 2b$	Think of this as $(+2)(a + b)$ or $(+2)a + (+2)b = +2a + 2b$
$-(a + b) = \quad -a - b$	Think of this as $(-1)(a + b)$ or $(-1)a + (-1)b = -a - b$
$-2(a + b) = -2a - 2b$	
$-(a - b) = \quad -a + b$	Think of this as $(-1)(a - b)$ or $(-1)a - (-1)b = -a + b$

Careful! Tricky Stuff ⇨

Once the parentheses have been removed, add like terms as explained in Operation 1.

Add these expressions, following the explanations given above.

$(2a + b) - (a + 3b) =$ ______

Check your work in **11**.

---

11 First, carefully remove parentheses.

$$(2a + b) = 2a + b$$
$$-(a + 3b) = -a - 3b$$

Second, add like terms.

$$(2a + b) - (a + 3b) = 2a + b - a - 3b = 2a - a + b - 3b$$
$$= a - 2b$$

Notice that once the parentheses are removed the terms should be rearranged so that similar terms are together.

Try these problems for practice.

(a) $(3y - w) + (y + 1) =$ ______

(b) $(2x + 1) + (x - 4) =$ ______

(c) $(a + b) - (a - b) =$ ______

(d) $(3x - 5) - (x - 2) =$ ______

(e) $(p + 1) - (p - 2) =$ ______

Turn to **12** to check your answers.

---

12 (a) $4y - w + 1$
(b) $3x - 3$
(c) $2b$
(d) $2x - 3$
(e) 3

Operation 3. *Multiplying simple factors.*

Example: $(2x)(3xy) = 6x^2y$

First, remember that $2x$ means $(2)(x)$

Second, recall that

$x \cdot x = x^2$
$x \cdot x \cdot x = x^3$

and so on.

Third, remember that the order in which you do multiplications does not make a difference.

$2 \cdot 3 \cdot 4 = (2 \cdot 4) \cdot 3 = (3 \cdot 4) \cdot 2$

and

$a \cdot b \cdot c = (a \cdot c) \cdot b = (c \cdot b) \cdot a$

For example,

$(2a)(a) = 2a^2$

Think of this as

$$\begin{aligned}(2a)(1a) &= 2 \cdot a \cdot 1 \cdot a \\ &= 2 \cdot 1 \cdot a \cdot a \\ &= 2a^2 \quad \text{Multiply each like factor separately.}\end{aligned}$$

Similarly,

$$\begin{aligned}(3x^2yz)(2xy) &= 3 \cdot x^2 \cdot y \cdot z \cdot 2 \cdot x \cdot y \\ &= (3 \cdot 2)(x^2 \cdot x)(y \cdot y)(z) \\ &= 6x^3y^2z\end{aligned}$$

If you need to review the multiplication of exponent numbers, hop back to Unit 1, p. 81–86, starting in frame **59**.

Try these problems for practice in multiplying simple factors.

(a) $(x)(y)$ = ______ (b) $(2x)(3x)$ = ______

(c) $(2x)(3y)$ = ______ (d) $(2ab)(a)$ = ______

(e) $(3a^2b)(4ab)$ = ______ (f) $(4xyz)(2ax^2)$ = ______

Check your answers in **13**.

13 (a) xy (b) $6x^2$
(c) $6xy$ (d) $2a^2b$
(e) $12a^3b^2$ (f) $8ax^3yz$

Operation 4. *Multiplying expressions.*

Example: $(a + b)(a + d) = a^2 + ab + ad + bd$

In order to multiply two expressions, treat the second expression as a unit and multiply by each of the terms in the first expression.

$$\begin{aligned}(a + b)(a + d) &= (a)(a + d) + (b)(a + d)\\ &= a \cdot a + a \cdot d + b \cdot a + b \cdot d\\ &= a^2 + ad + ba + bd\end{aligned}$$

Another example:

$$\begin{aligned}(2a + b)(a - 2b) &= (2a)(a - 2b) + (b)(a - 2b)\\ &= (2a)(a) - (2a)(2b) + (b)(a) + (b)(-2b)\\ &= 2a^2 - 4ab + ab - 2b^2\\ &= 2a^2 - 3ab - 2b^2\end{aligned}$$

Notice that in line 3 we wrote ab for ba. The factors can be multiplied in any order.

In the last line we have combined like terms:

$-4ab + ab = -3ab$

Here is a bit of practice in multiplying expressions.

(a) $(x + y)(x + y) =$ ______ (b) $(a + b)(a - b) =$ ______

(c) $(y - 1)(y - 2) =$ ______ (d) $(2x + 3)(x - 4) =$ ______

(e) $(2ab - c)(c + ab) =$ ______ (f) $3(2a + b)(a - b) =$ ______

The answers are in **14**.

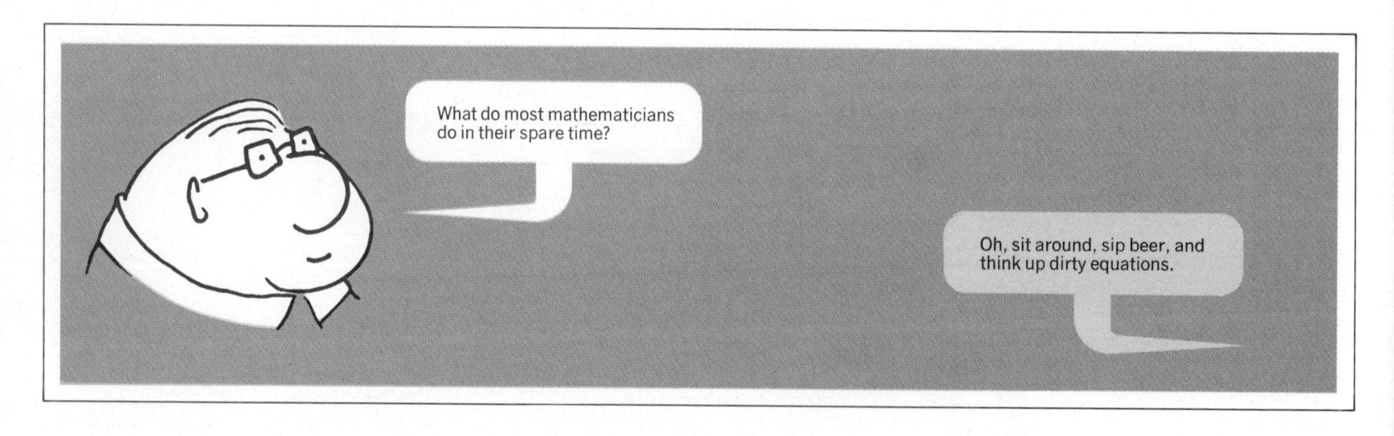

14 (a) $x^2 + 2xy + y^2$ (b) $a^2 - b^2$
(c) $y^2 - 3y + 2$ (d) $2x^2 + 5x - 12$
(e) $abc - c^2 + 2a^2b^2$ (f) $6a^2 - 3ab - 3b^2$

Operation 5. *Dividing similar expressions.*

Example: $\dfrac{4a^2b}{2a} = 2ab$

In algebra the division of one expression by another is most often shown by writing a fraction. For example,

$$(4a^2b) \div (2a) = \frac{4a^2b}{2a}$$

It is often possible to simplify algebra fractions by eliminating common factors from numerator and denominator.

$$\frac{4a^2b}{2a} = \frac{2 \cdot 2 \cdot a \cdot a \cdot b}{2 \cdot a} = \frac{2 \cdot \cancel{2} \cdot \cancel{a} \cdot a \cdot b}{\cancel{2} \cdot \cancel{a}} = 2ab$$

The first step is to write both numerator and denominator as products of constants and variables. Once this is done we may cancel common factors.

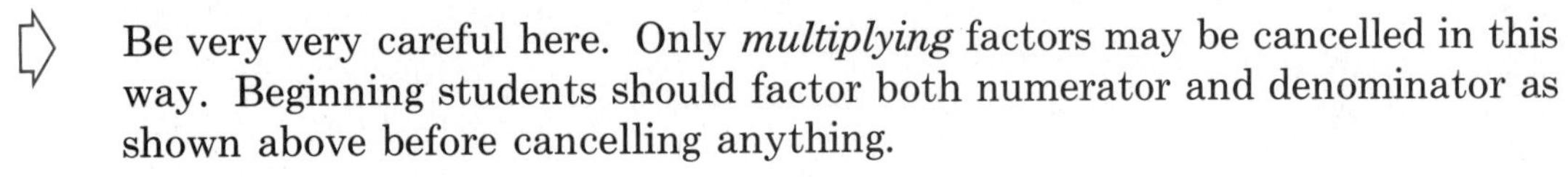

Be very very careful here. Only *multiplying* factors may be cancelled in this way. Beginning students should factor both numerator and denominator as shown above before cancelling anything.

Another example,

$$\frac{6x^2y^2}{2x^3y} = \frac{3 \cdot 2 \cdot x \cdot x \cdot y \cdot y}{2 \cdot x \cdot x \cdot x \cdot y} = \frac{3 \cdot \cancel{2} \cdot \cancel{x} \cdot \cancel{x} \cdot y \cdot \cancel{y}}{\cancel{2} \cdot \cancel{x} \cdot \cancel{x} \cdot x \cdot \cancel{y}}$$

$$= \frac{3y}{x}$$

Try these practice problems.

(a) $\dfrac{xy^2}{2x}$ = ______ (b) $\dfrac{ab}{bc}$ = ______

(c) $\dfrac{9a^2x}{3ax^2}$ = ______ (d) $\dfrac{3abc}{6a^2b^2}$ = ______

(e) $\dfrac{a + b}{a + c}$ = ______ (f) $\dfrac{3(a + b)}{(a + b)}$ = ______

Check your answers in **15**.

15 (a) $\dfrac{y^2}{2}$ (b) $\dfrac{a}{c}$

(c) $\dfrac{3a}{x}$ (d) $\dfrac{c}{2ab}$

(e) $\dfrac{a + b}{a + c}$ There is no way to simplify this further. You *cannot* cancel the as because they are not multiplying factors.

(f) 3 The multiplying factor $(a + b)$ appears in both numerator and denominator.

Turn to **16** for a set of practice problems on what you have learned so far in this unit.

Introduction to Algebra

PROBLEM SET 1

16 Algebra Words and Operations

The answers are on page 510.

A+B only

A. Complete each of the following statements using the words *constant, variable, coefficient, exponent, expression, factor,* or *equation.*

1. In the ______ $2x^3$, 2 is a ______, x is a ______, and 3 is a ______.
2. In the ______ $3a^4b + 2ab$, b is a ______, and $2ab$ is a ______. 3 and 2 are ______.
3. x and y are the ______ of xy.
4. $x + 2 = y$ is an ______, 2 is a ______, and x is a ______.
5. $(y - 1)$ is a ______ of the ______ $(x + 2)(y - 1)$.
6. In the ______ $2a + 3b + 4c$, a, b, and c are ______, 2, 3, and 4 are ______, and $2a$ is a ______.

B. Write out the following in algebra notation:

1. x divided by y = ______
2. A minus 3 = ______
3. 6 times x times y = ______
4. a plus b^2 = ______
5. 3 plus y = ______
6. x^2 times y^2 = ______
7. 2 times $(2a + b)$ = ______
8. $(a + 1)$ divided by b = ______
9. x minus y = ______
10. 3 times x times x times y = ______
11. $(a + b)$ times $(a - b)$ = ______
12. A plus B minus C = ______

C. Calculate

1. $2M + 3M =$ ______
2. $(2R)(4R) =$ ______
3. $2x + (x - 2) =$ ______
4. $3a - a + 2b =$ ______
5. $8p^2 - 4p^2 + p^2 =$ ______
6. $2(a + 1) + 7(a + 1) =$ ______
7. $(2a + 6) - (a + 2) + 1 =$ ______
8. $(x + 2y) - (2x - 3y) =$ ______

Date: June 1, 1978

Name: Stephanie Catanzaro

Course/Section: 050/003

(continued on the next page)

9. $3y - (y - 2) =$ ____

10. $(3ab)(2a^2bc) =$ ____

11. $(3x)(4z) =$ ____

12. $(a)(10a + 1) =$ ____

13. $(a + b)(a - b) =$ ____

14. $(a + b)(a + b) =$ ____

15. $(a - b)(a - b) =$ ____

16. $(M + 1)(M - 1) =$ ____

17. $\dfrac{6x^2y}{9xy} =$ ____

18. $\dfrac{abc}{bcd} =$ ____

19. $\dfrac{2a + 3a}{ab} =$ ____

20. $\dfrac{3x + 4x}{3x^2 + 4x^2} =$ ____

21. $4a + 6 - b - a + (2b - 1) =$ ____

22. $x - (1 - x) =$ ____

23. $(1 + y) - (1 - y) =$ ____

24. $(a + 1)^2 =$ ____

25. $(a + 1)^3 =$ ____

26. $(a - 1)^2 =$ ____

27. $\dfrac{2x^2y^3z^2}{xyz} =$ ____

28. $x^2 - 2x + (1 - x)^2 =$ ____

29. $2xy(x - y) =$ ____

30. $3R \div R^2 =$ ____

When you have had the practice you need, either return to the preview test on page 413 or continue in **17** with the translation of English statements into algebra expressions.

TRANSLATING ENGLISH TO ALGEBRA

17 Algebra is most useful as a tool for applying mathematical logic to real situations and solving real problems. However, in order to use algebra for practical problem solving, you must be able to translate simple English sentences and phrases into mathematical equations or expressions. For example, the phrase

"Three times a certain number . . ."

becomes $3n$

and

"Five times a number added to four"

becomes $5n + 4$.

Any letter may be used, of course; there is nothing special about the n except that it reminds us that it represents a *n*umber.

Can you translate these?

(a) "The sum of two and a number"
(b) "The product of six and a number"
(c) "A number divided by seven"
(d) "Five less than a number"

Try it. The correct translations are in **18**.

---

18 (a) "The sum of two and a number" $\longrightarrow$ $2 + n$

(b) "The product of six and a number" $\longrightarrow$ $6n$

(c) "A number divided by seven" $\longrightarrow$ $\dfrac{n}{7}$

(d) "Five less than a number" $\longrightarrow$ $n - 5$

Be careful on problem (d). The phrase "five less than a number" does *not* translate to $5 - n$. If you get confused on this, substitute numbers in the English phrase and in the algebra expression and see if they agree.

"5 less than 8" is 3 and $n - 5 = 8 - 5 = 3$

Of course any letter can replace n in the expressions above. For example, these algebra phrases are also correct:

(a) $2 + a$ or $2 + B$ or $2 + \mathcal{R}$

(b) $6k$ or $6M$ or $6\mathcal{P}$

(c) $\dfrac{d}{7}$ or $\dfrac{Q}{7}$ or $\dfrac{x}{7}$

(d) $b - 5$ or $E - 5$ or $\mathcal{S} - 5$

On the next page is a very handy little dictionary of English words and phrases with their mathematical translations.

(continued on the next page)

English Term	Math Translation	Example
Equals Is, is equal to, was The same as . . . What is left is . . . The result is . . . Gives, makes Leaves, leaving	$=$	$a = b$
Plus Sum of Increased by More than	$+$	$a + b$
Minus (b) Less (b) Decreased by (b) Take away (b) Reduced by (b) Diminished by (b) (b) less than (a) (b) subtracted from (a)	$-$	$a - b$
Times Multiplied by Product of	$\times$	$a \times b$ or ab
Divided by (b) Quotient of	$\div$	$a \div b$ or $\frac{a}{b}$
Twice Double	$\times 2$	$2a$

Use this dictionary to translate these English phrases to mathematical expressions.

1. "Two more than a number" ____________
2. "A number reduced by five" ____________
3. "d more than k" ____________
4. "g less than m" ____________
5. "q subtracted from p" ____________
6. "The product of a and b" ____________
7. "x divided by y" ____________

8. “14 increased by z” ____________

9. “Twice a number plus four” ____________

10. “The product of two numbers minus 5” ____________

11. “Six less than a number” ____________

12. “Six less a number” ____________

13. “Six more than a number” ____________

14. “A number less six” ____________

15. “Half of a number” ____________

Check your translations in **20**.

---

19 Four times Tom's age is 72.

$4 \quad \times \quad T \quad = 72$

$4T = 72$

Follow these steps when you must translate English sentences into algebra equations:

Example

Step 1. Cross out all unnecessary words.

“~~We noticed that~~ Bill's age plus 8 ~~years~~ is ~~exactly~~ equal to three times Ron's age.”

Step 2. Make a word equation using parentheses.

(Bill's age) (plus) (8) (is equal to) (three) (times) (Ron's age)

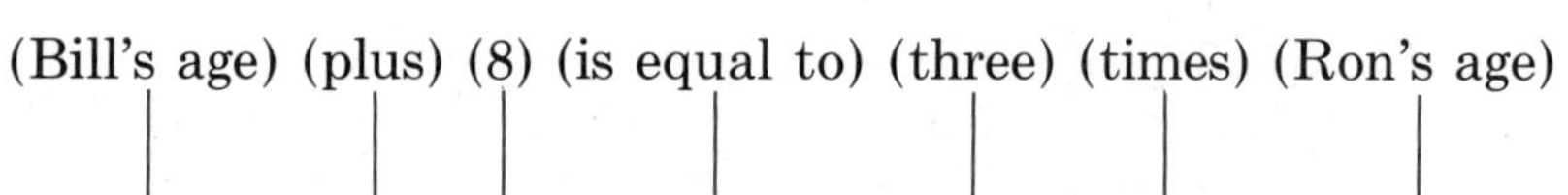

Step 3. Substitute a letter or an arithmetic symbol for each parentheses.

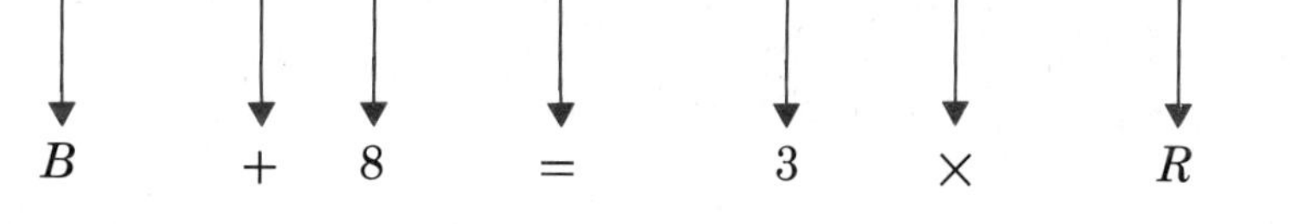

$B \quad + \quad 8 \quad = \quad 3 \quad \times \quad R$

Step 4. Combine and simplify.

$$B + 8 = 3R$$

(continued on the next page)

Here's another example:

> "It is an interesting coincidence that the cost of the desk is just \$8 more than double the cost of the chair."

Step 1. "~~It is an interesting coincidence that the cost of the~~ desk is ~~just~~ \$8 more than double ~~the cost of the~~ chair."

Step 2. (Desk) (is) (8) (more than) (double) (chair)

Step 3. $D \quad = \quad 8 \quad + \quad 2\times \quad C$

Step 4. $D = 8 + 2C$

Translate the following sentence to an algebra equation using the four-step procedure shown above.

> "The surprising thing about it is that this year's bill is only \$3 more than twice last year's bill."

Check your work in **21**.

20

1. $x + 2$	2. $N - 5$	3. $d + k$
4. $m - g$	5. $p - q$	6. ab
7. $\frac{x}{y}$	8. $z + 14$	9. $2N + 4$
10. $ab - 5$	11. $N - 6$	12. $6 - N$
13. $N + 6$	14. $N - 6$	15. $\frac{1}{2}N$ or $\frac{N}{2}$

So far we have only translated English phrases, but complete English sentences can also be translated into mathematical language. For example, the sentence

"John's age is equal to Kevin's age."

translates to $J \quad = \quad K$

Each word or phrase in the sentence is translated directly into a mathematical term, variable, constant, expression, or arithmetic operation sign.

Translate this sentence into mathematical form as we did above.

"Four times Tom's age is 72."

Do it here . . .

. . . then check your translation in **19**.

21

"~~The surprising thing about it is that~~ this ~~year's bill~~ is ~~only~~ \$3 more than twice last ~~year's bill.~~"

(This)	(is)	(3)	(more than)	(twice)	(last)
↓	↓	↓	↓	↓	↓
T	$=$	3	$+$	$2\times$	L

$$T = 3 + 2L$$

The best way to learn the art of translating English to mathematics is by doing it. Turn to **22** for a set of practice problems designed to exercise your mental muscles.

Introduction to Algebra

PROBLEM SET 2

22

Translating English to Algebra

The answers are on page 510.

A. Write each of these phrases as an algebra term or expression:

1. "A number increased by 4" ____________
2. "A number decreased by 5" ____________
3. "Three more than d" ____________
4. "Three less than h" ____________
5. "The sum of a number and 8" ____________
6. "The difference between x and y" ____________
7. "Twice a number" ____________
8. "M divided by 3" ____________
9. "The product of x and y" ____________
10. "A times the sum of B and C" ____________
11. "R multiplied by 9" ____________
12. "K less 10" ____________
13. "K more than 10" ____________
14. "x reduced by 5" ____________
15. "15 decreased by b" ____________
16. "8 less than H" ____________
17. "Take away a from b" ____________
18. "x subtracted from y" ____________
19. "The sum of R and half of T" ____________
20. "Triple a number" ____________

June 1, 1978
Date

Stephanie Catanzaro
Name

050/003
Course/Section

B. Write each of these phrases as an algebra expression:

1. The sum of a number and half the number ______________
2. The sum of two numbers decreased by 4 ______________
3. One less than twice a number ______________
4. Twice the square of a number ______________
5. Six more than three times a number ______________
6. Twice a number less five ______________
7. Twice a number plus two ______________
8. The sum of three times a number and five ______________
9. The sum of two numbers divided by 6 ______________
10. Four times a number minus seven ______________
11. The square of the radius ______________
12. The base times the height divided by 2 ______________
13. Three times the square root of a number ______________
14. One-third the sum of two numbers ______________
15. Some number decreased by 5 ______________
16. Three less than one-half a number ______________
17. The sum of the squares of two consecutive numbers ______________
18. Twice the sum of a number and its square ______________
19. The product of mass and velocity ______________
20. One-half of the gravitational constant times the square of the time ______________

C. Write each of these sentences as an algebra equation:

1. "Three times a number divided by four is equal to six."

____

2. "The sum of two numbers is 31."

____

3. "Sally's age is five years less than Charlie's age."

____

4. "The product of Al's age and Bill's age is ten times Cathy's age."

____

5. "Five times a number decreased by seven is 88."

____

6. "The difference between the cost of the shirt and the jacket is twice the cost of the dress."

____

7. "Four times a number less three gives twice another number."

____

8. "Bill's age four years ago was the same as twice Tom's age three years from now." (*Hint.* "Four years ago" means -4. "Three years from now" means $+3$.)

____

9. "The sum of six times a number and its square is twenty-seven."

____

10. "The area of a circle is equal to π times the radius squared."

____

11. "The volume of a cylinder is equal to $\frac{1}{4}$ of its height times π times its diameter squared."

____

Date

Name

Course/Section

12. "The book costs $3 more than the sum of the costs of the paper and the notebook."

____

13. "Some number increased by 10 is equal to 16."

____

14. "The perimeter of a rectangle is equal to twice the length plus twice the width."

____

15. "The volume of a cube is equal to the cube of its edge."

____

16. "The sum of two numbers decreased by four is equal to twice their product."

____

17. "The Fahrenheit temperature is equal to 32 plus the product 1.8 times the Celsius temperature."

____

18. "The product of one number and the square of another is six."

____

19. "One-half the sum of two numbers is equal to seven."

____

20. "The sum of the squares of two numbers is four more than their sum."

____

When you have had the practice you need, either return to the preview test on page 413 or continue in **23** for help in calculating the value of literal equations.

EVALUATING LITERAL EQUATIONS

23 A literal or algebraic equation is an equation in which letters are used to represent numbers. In mathematics, science, and in many practical situations that arise in business and work, it is often necessary to find the value of a literal expression when the numerical values of the letter variables are given. This process is known as "evaluating" the equation or expression.

For example, in retail stores the following equation is used:

$M = R - C$ where M is the *markup* on an item,
R is the retail selling price,
and C is the original cost.

Find M if $R = \$25$ and $C = \$21$.

$M =$ ______ $-$ ______ $=$

Check your work in **24**.

---

24 $M = \underline{\$25} - \underline{\$21} = \underline{\$4}$

Easy? Of course. Simply substitute the numbers for the corresponding letters and then do the arithmetic.

Here is a bit of practice for you. Evaluate these expressions by filling in the blanks.

If $x = 2$, $y = 3$, $z = 5$, find the value of

(a) xy $=$ ______ $\cdot$ ______ $=$ ______

(b) x^2 $=$ ______ $\cdot$ ______ $=$ ______

(c) $2y^2$ $=$ ______ $\cdot$ (______ $\cdot$ ______) $=$ ______ $\cdot$ ______ $=$ ______

(d) xz^2 $=$ ______ $\cdot$ (______ $\cdot$ ______) $=$ ______ $\cdot$ ______ $=$ ______

(e) $z - y$ $=$ ______ $-$ ______ $=$ ______

(continued on the next page)

(f) $x^2 + z = (\_\_\_\_\_ \cdot \_\_\_\_\_) + \_\_\_\_\_ = \_\_\_\_\_ + \_\_\_\_\_$

$= \_\_\_\_\_$

(g) $8xy - z^2 = (\_\_\_\_\_ \cdot \_\_\_\_\_ \cdot \_\_\_\_\_) - (\_\_\_\_\_ \cdot \_\_\_\_\_)$

$= \_\_\_\_\_ - \_\_\_\_\_ = \_\_\_\_\_$

(h) $z(2y - x) = \_\_\_\_\_ \cdot [(\_\_\_\_\_ \cdot \_\_\_\_\_) - \_\_\_\_\_]$

$= \_\_\_\_\_ \cdot [\_\_\_\_\_ - \_\_\_\_\_]$

$= \_\_\_\_\_ \cdot \_\_\_\_\_ = \_\_\_\_\_$

Look in **26** for the step by step solutions.

---

25 (a) $3ab = 3 \cdot 3 \cdot 4 = (3 \cdot 3) \cdot 4 = 9 \cdot 4 = 36$

(b) $2a^2c = 2 \cdot 3^2 \cdot 6 = 2 \cdot 9 \cdot 6 = (2 \cdot 9) \cdot 6 = 18 \cdot 6 = 108$

(c) $2a^2 - b^2 = 2 \cdot 3^2 - 4^2 = 2 \cdot 9 - 16 = 18 - 16 = 2$

(d) $2a^2bc = 2 \cdot 3^2 \cdot 4 \cdot 6 = 2 \cdot 9 \cdot 4 \cdot 6 = (2 \cdot 9) \cdot (4 \cdot 6)$
$= 18 \cdot 24 = 432$

(e) $2 \cdot 3 + 4 = 6 + 4 = 10$

The biggest headaches come for most people when they must find the value of literal equations containing parentheses. It will help if you try some number equations before you tackle the algebra. Find the value of each of these:

(a) $(1 + 3) + 2 = \_\_\_\_\_ + \_\_\_\_\_ = \_\_\_\_\_$

(b) $(3 \times 2) + 5 = \_\_\_\_\_ + \_\_\_\_\_ = \_\_\_\_\_$

(c) $4(2 + 3) = \_\_\_\_\_ \cdot \_\_\_\_\_ = \_\_\_\_\_$

(d) $2(1 + 4) + 5 = \_\_\_\_\_ \cdot \_\_\_\_\_ + \_\_\_\_\_ = \_\_\_\_\_ + \_\_\_\_\_$

$= \_\_\_\_\_$

Check your work in **27**.

---

26 (a) $xy = \underline{2} \cdot \underline{3} = \underline{6}$

(b) $x^2 = \underline{2} \cdot \underline{2} = \underline{4}$

(c) $2y^2 = \underline{2} \cdot \underline{3} \cdot \underline{3} = \underline{18}$

(d) $xz^2 = \underline{2} \cdot \underline{5} \cdot \underline{5} = \underline{50}$

(e) $z - y = \underline{5} - \underline{3} = \underline{2}$

(f) $x^2 + z = (\underline{2} \cdot \underline{2}) + \underline{5} = \underline{4} + \underline{5} = \underline{9}$

(g) $8xy - z^2 = (\underline{8} \cdot \underline{2} \cdot \underline{3}) - (\underline{5} \cdot \underline{5})$

$= \underline{48} - \underline{25} = \underline{23}$

(h) $z(2y - x) = \underline{5} \cdot [(\underline{2} \cdot \underline{3}) - \underline{2}]$

$= \underline{5} \cdot [\underline{6} - \underline{2}]$

$= \underline{5} \cdot \underline{4} = \underline{20}$

The important idea here is that you must work carefully, step by step. Here are a few hints to help you in evaluating literal expressions.

Hint 1. If the expression is a simple multiplication such as xy or $2ab$ or $2\pi R$, substitute the number values and multiply.

Example: $2ab = 2 \cdot 5 \cdot 3 = 10 \cdot 3 = 30$ where $a = 5$, $b = 3$

Hint 2. If the expression has a square, cube or other power of some variable, as in xy^2 or $2ab^2c$, find the value of that factor first, then multiply as usual.

Example: $xy^2 = 3 \cdot 4^2 = 3 \cdot 16 = 48$ where $x = 3$, $y = 4$

Hint 3. If the expression is the sum or difference of terms, as in $xy + z^2$ or $3ab^2 - 2ac$, find a numerical value for each term first, then add or subtract terms.

Example: $xy - z^2 = 3 \cdot 4 - 2^2$ where $x = 3, y = 4, z = 2$
$= 12 - 4 = 8$

Use these hints to evaluate the following expressions:

$a = 3$, $b = 4$, $c = 6$

(a) $3ab$ = ______

(b) $2a^2c$ = ______

(c) $2a^2 - b^2$ = ______

(d) $2a^2bc - ab$ = ______

(e) $2a + b$ = ______

Our step-by-step answers are in **25**.

27 (a) $(1 + 3) + 2 = \underline{4} + \underline{2} = \underline{6}$

(b) $(3 \times 2) + 5 = \underline{6} + \underline{5} = \underline{11}$

(c) $4(2 + 3) = \underline{4} \cdot \underline{5} = \underline{20}$

(d) $2(1 + 4) + 5 = \underline{2} \cdot \underline{5} + \underline{5} = \underline{10} + \underline{5} = \underline{15}$

A few additional hints are needed to help you evaluate expressions containing parentheses.

Hint 4. If there is only one set of parentheses, do the operations inside it first, then continue as before.

Example: $(a + b) + c = (2 + 3) + 4 = 5 + 4$ for $a = 2,\ b = 3,\ c = 4$

or $c(a + b) = 4(2 + 3) = 4 \cdot 5 = 20$

or $c(2a + b) - 4 = 4(2 \cdot 2 + 3) - 4 = 4 \cdot 7 - 4 = 28 - 4 = 24$

Hint 5. If the expression contains two sets of parentheses next to each other and separated by an arithmetic operation sign, first do the calculation inside each parentheses, and then combine them.

Example: $(a + b^2) - (2a + b)$

$$= (2 + 3^2) - (2 \cdot 2 + 3) = (2 + 9) - (4 + 3) = 11 - 7 = 4$$

Hint 6. If the expression contains two sets of parentheses next to each other with no arithmetic operation sign between, multiply them.

Example: $(a + b)(b^2 - a) = (2 + 3)(3^2 - 2) = (5) \cdot (9 - 2) = 5 \cdot 7 = 35$

Hint 7. If the expression contains one set of parentheses enclosed inside another set, find the value of the inner set first. Always work from the inside out.

Example: $(a(a^2 + b)) + 4$

$$= (2(2^2 + 3)) + 4$$
$$= (2(4 + 3)) + 4$$
$$= (2 \cdot 7) + 4 = 14 + 4 = 18$$

Parentheses tell you the order in which to do the arithmetic.

Evaluate the following literal equations.

(a) $A = 2(x + y) - 1$ for $x = 2,\ y = 4$
(b) $V = (L + W)(2L + W)$ for $L = 3,\ W = 5$
(c) $I = PRT$ for $P = 100,\ R = .05,\ T = 2$
(d) $B = (a + b) + (2a - c) + b^2$ for $a = 1.5,\ b = 2,\ c = 1$
(e) $C = 2(2a + 1) - (2b - 1)$ for $a = 3,\ b = 4$
(f) $H = 2(a^2 + b^2)$ for $a = 2,\ b = 1$
(g) $D = (2a + (a^2 - b)) + 1$ for $a = 2,\ b = 1$
(h) $A = a(a + a(a + 1)^2) + 1$ for $a = 2$

Follow the hints given above and work carefully, step by step. Check your work in **28**.

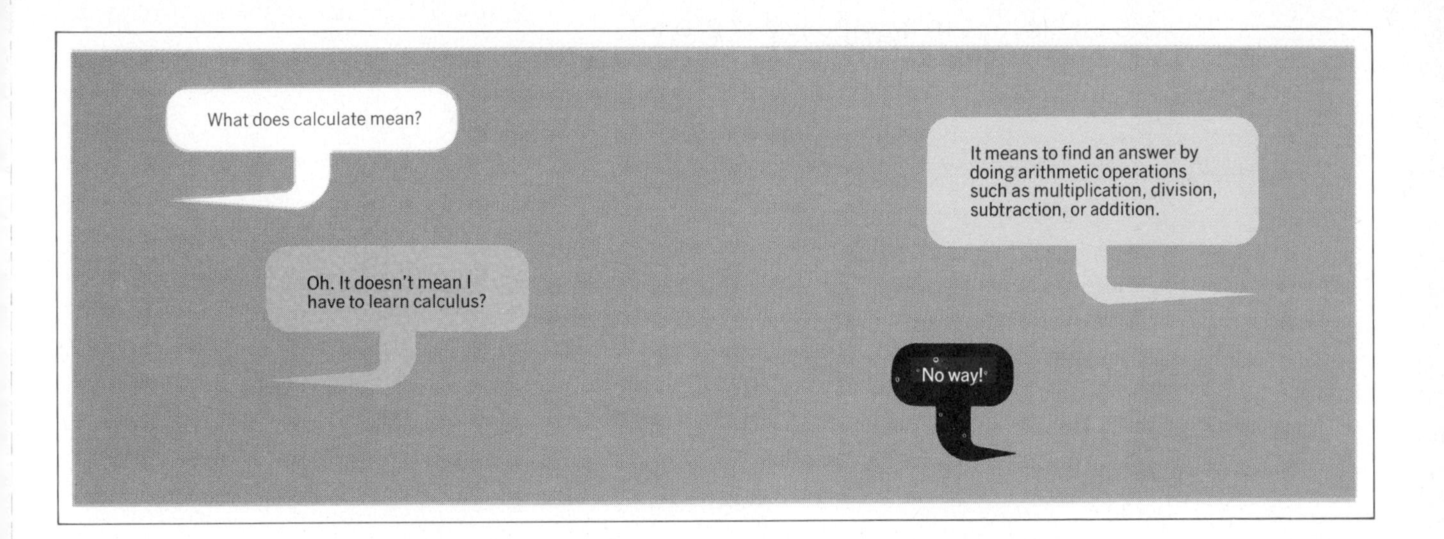

28 (a) $A = 2(2 + 4) - 1 = 2 \cdot 6 - 1 = 12 - 1 = 11$ See Hint 4.

(b) $V = (3 + 5)(2 \cdot 3 + 5) = (8)(6 + 5) = 8 \cdot 11 = 88$ See Hint 6.

(c) $I = 100 \cdot .05 \cdot 2 = (100 \cdot .05) \cdot 2 = 5 \cdot 2 = 10$ See Hint 1.

(d) $B = (1.5 + 2) + (3 - 1) + 2^2$
$= 3.5 + 2 + 4 = 9.5$ See Hint 3.

(e) $C = 2(2 \cdot 3 + 1) - (2 \cdot 4 - 1) = 2(6 + 1) - (8 - 1)$
$= 2 \cdot 7 - 7 = 14 - 7 = 7$ See Hint 5.

(f) $H = 2(2^2 + 1^2) = 2(4 + 1) = 2(5) = 10$ See Hint 4.

(g) $D = (2 \cdot 2 + (2^2 - 1)) + 1 = (4 + (4 - 1)) + 1 = (4 + 5) + 1$
$= 9 + 1 = 10$ See Hint 7.

(h) $A = 2(2 + 2(2 + 1)^2) + 1 = 2(2 + 2(3)^2) + 1$
$= 2(2 + 2 \cdot 9) + 1$
$= 2(2 + 18) + 1$
$= 2(20) + 1 = 40 + 1 = 41$ See Hint 7.

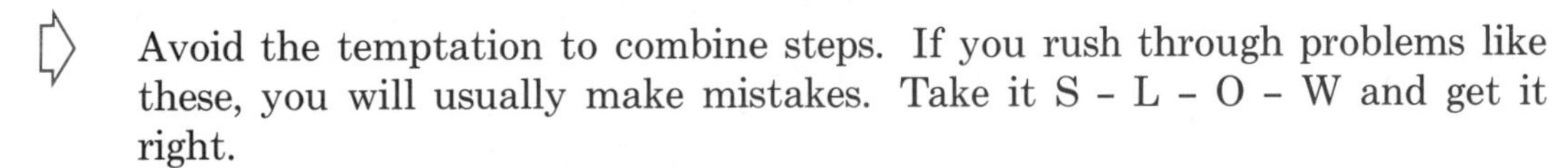

Avoid the temptation to combine steps. If you rush through problems like these, you will usually make mistakes. Take it S - L - O - W and get it right.

Now turn to **29** for a set of practice problems on evaluating literal expressions.

Introduction to Algebra

PROBLEM SET 3

29

Evaluating Literal Expressions

The answers are on page 511.

A. Evaluate each of these expressions by completing the step-by-step procedures shown: (Use $x = 2$, $y = 3$, $z = 5$, $w = 6$):

1. $3x$ = ____ · ____ = ____

2. y^2 = ____ · ____ = ____

3. wx = ____ · ____ = ____

4. $2yz$ = ____ · ____ · ____ = ____

5. $z - y$ = ____ − ____ = ____

6. $x + w$ = ____ + ____ = ____

7. $2x + y$ = (____ · ____) + ____ = ____ + ____ = ____

8. $y^2 + x^2$ = (____ · ____) − (____ · ____) = ____ − ____
 = ____

9. $3w - 7$ = (____ · ____) − ____ = ____ − ____ = ____

10. $2w - 3y$ = (____ · ____) − (____ · ____) = ____ − ____
 = ____

11. $3xy^2 - 2w + 4$ = (____ · ____ · (____ · ____)) − (____ · ____)
 + ____
 = (____ · ____ · ____) − (____ · ____) + ____
 = ____ − ____ + ____ = ____

12. $\frac{1}{x} + \frac{y}{2} + 1$ = ____ + ____ + ____ = ____

13. $y^2x - x^2y$ = (____ · ____) · ____ − (____ · ____) · ____
 = ____ · ____ − ____ · ____ = ____ − ____
 = ____

Date ____________

Name ____________

Course/Section ____________

14. $2y - 1 = \_\_\_\_ \cdot \_\_\_\_ - \_\_\_\_ = \_\_\_\_ - \_\_\_\_ = \_\_\_\_$

15. $2(y - 1) = \_\_\_\_ \cdot (\_\_\_\_ - \_\_\_\_) = \_\_\_\_ \cdot \_\_\_\_ = \_\_\_\_$

16. $2x + y - z + 1 = \_\_\_\_$

17. $\dfrac{yz}{x} + w = \_\_\_\_$

18. $2x + 3y - 1 = \_\_\_\_$

19. $2(x + y) + 3(w - z) = \_\_\_\_$

20. $3x^2y - 1 = \_\_\_\_$

B. Find the value of each of these literal equations for $x = 4$, $y = 10$, $z = 7$, $w = 3$:

1. $A = 2x^2 - y = \_\_\_\_$

2. $B = 3(x + y) - 2z = \_\_\_\_$

3. $C = 2x + 3y - z - 5w = \_\_\_\_$

4. $D = x^2 + y^2 = \_\_\_\_$

5. $E = (x + y)^2 - (z - w)^2 = \_\_\_\_$

6. $F = (x + 2(y + 2)) + 2 = \_\_\_\_$

7. $G = ((2x + y) + 3w) - 5 = \_\_\_\_$

8. $H = (x + y)(2w + 1) = \_\_\_\_$

9. $J = ((x + 1)^2(y - 6)^2) + 5 = \_\_\_\_$

10. $K = \dfrac{2(y - x) + 3w}{z} = \_\_\_\_$

11. $L = 2xyz - 10 = \_\_\_\_$

12. $M = ((x + 1) + 2(y + 1)(w - 1) + 1) = \_\_\_\_$

13. $N = (w + w(w + w(w + 1))) = \_\_\_\_$

14. $P = 3(2x^2 + 1)(y + 2) = \_\_\_\_$

15. $Q = (2z - w) + (2y - 5)^2 = \_\_\_\_$

16. $x^2 + w^2 + (x + w)^2 = \_\_\_\_$

17. $3x^2y - w$ $= \_\_\_\_\_\_$

18. $(x + 1)(x - 1)$ $= \_\_\_\_\_\_$

19. xyz $= \_\_\_\_\_\_$

20. $\dfrac{y + x}{y - w}$ $= \_\_\_\_\_\_$

C. Brain Boosters:

1. Evaluate each of these expressions for $A = 5$, $B = 2$:

   (a) $3A - B + 1 = \_\_\_\_\_\_$ (b) $3(A - B) + 1 = \_\_\_\_\_\_$

   (c) $3(A - B + 1) = \_\_\_\_\_\_$ (d) $3A - (B + 1) = \_\_\_\_\_\_$

   (e) $3(A - (B + 1)) = \_\_\_\_\_\_$

2. The volume of a cylinder is given by the equation $V = \dfrac{\pi D^2 H}{4}$.

   Calculate the volume of a cylindrical tank with height, $H = 12$ ft., diameter, $D = 6$ ft. Use $\pi = 3.14$ $V = \_\_\_\_\_\_$

3. Evaluate $(W + H(A + A(A + A)T)T)$, where $W = 4$, $H = 2$, $A = 3$, $T = 1$.

4. The distance a free-falling object drops in T seconds is $D = \frac{1}{2}gT^2$ meters, where g is roughly 10. How far would a stone fall in 10 seconds? Use $g = 9.8$. $D = \_\_\_\_\_\_$

5. The Celsius and Fahrenheit temperature scales are related by the equation

   $$C = \frac{5(F - 32)}{9}$$

   What is the Celsius temperature corresponding to $F = 140°$?
   $C = \_\_\_\_\_\_$

6. The area of a trapezoid is $T = \dfrac{(A + B)H}{2}$, where A and B are the lengths of its parallel sides and H is its height. Find the area of the trapezoid shown. $T = \_\_\_\_\_\_$

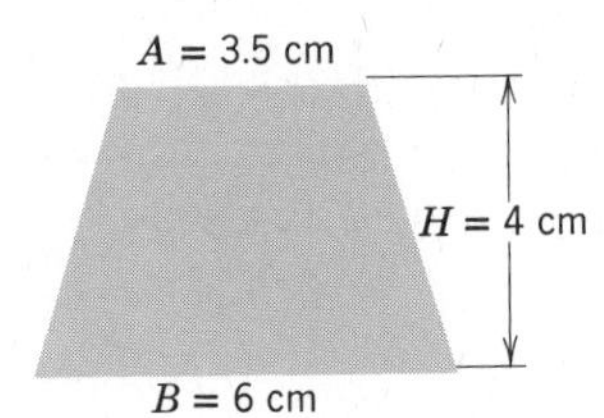

Date ______

Name ______

Course/Section ______

7. The following equation appeared on a chemistry exam:

$$P = \frac{n\,R\,(T + 273)}{V}$$

Find the value of P when $n = 5$, $R = .08$, $T = 27$, $V = 3$. $P =$ ______

8. The interest I paid on a loan can be calculated from the equation

$$I = P \cdot R \cdot T$$

where P is the *principal* or amount of the loan, R is the *rate* of interest per year, and T is the *time* in years. What interest would you pay on a \$2000 loan (P) at 8% (R) interest for 3 years (T)? $I =$ ______

9. The volume of a sphere can be calculated from the equation $V = (\frac{4}{3})\pi R^3$. Calculate the volume of a sphere from which the radius $R = 6$ in. (Use $\pi \cong 3$.) $V =$ ______

10. The longest side or hypotenuse H of a right triangle can be calculated from the legs A and B by using the equation $H = \sqrt{A^2 + B^2}$. Find H if $A = 12$ and $B = 5$. $H =$ ______

11. The nth triangular number is given by the equation $T_n = \frac{1}{2}n(n + 1)$. Find the first four triangular numbers ($n = 1, 2, 3, 4$). Find the 10th triangular number.

$T_1 =$ ______ $T_2 =$ ______ $T_3 =$ ______ $T_4 =$ ______ $T_{10} =$ ______

12. The weight of a metal cylinder can be found from the equation $W = D(ab - \pi r^2)h$, where $a = 8$ cm, $b = 6$ cm, $r = 2$ cm, $h = 10$ cm, $D = 8$ g/cm^3, and $\pi \cong 3$. $W =$ ______

When you have had the practice you need, either return to the preview test on page 413 or continue in **30** and learn how to solve algebra equations.

---

30 An arithmetic equation such as $3 + 2 = 5$ means that the number named on the left $(3 + 2)$ is the same as the number named on the right (5). An algebra equation $x + 3 = 7$ is a statement that the sum of some number (x) and 3 is equal to 7. The number named on the left $(x + 3)$ will be equal to the number named on the right (7) if we choose the correct value for x. x is a variable and you should think of it as a symbol that holds a place for a number, a blank to be filled. Many numbers might be put in the space, but only one makes the equation a true statement.

Find the missing numbers in these arithmetic equations:

(a) $37 + \_\_\_\_ = 58$

(b) $\_\_\_\_ - 15 = 29$

(c) $4 \times \_\_\_\_ = 52$

(d) $28 \div \_\_\_\_ = 4$

Puzzle them out, then turn to **31**.

---

31 (a) $\underline{37} + 21 = 58$

(b) $\underline{44} - 15 = 29$

(c) $4 \times \underline{13} = 52$

(d) $28 \div \underline{7} = 4$

Is it obvious that we could have written these equations as

$37 + A = 58$

$B - 15 = 29$

$4C = 52$

$\frac{28}{D} = 4$?

(Of course, any letters would do in place of A, B, C and D.)

How did you solve these equations? You probably "eye-balled" them—mentally juggled the other information in the equation until you found a number that made the equation true. Solving algebra equations is very similar except that we can't "eye-ball" it entirely. We need certain and systematic ways of solving the equation that will produce the correct answer quickly every time.

In this section you will learn first what a solution to an algebra equation is—how to recognize it if you stumble over it in the dark—then what a linear equation looks like, and finally how to solve linear equations.

Each value of the variable that makes an equation true is called a *solution* of the equation. For example, the solution of $x + 3 = 7$ is $x = 4$.

(continued on the next page)

Check: $4 + 3 = 7$

The solution of $2x - 9 = 18 - 7x$ is $x = 3$.

Check:

$$\begin{aligned} 2\cdot 3 - 9 &= 18 - 7\cdot 3 \\ 6 - 9 &= 18 - 21 \\ -3 &= -3 \end{aligned}$$

It may happen that more than one value of the variable makes the equation true. For example, the equation

$x^2 + 6 = 5x$

is true for $x = 2$

(**Check:** $2^2 + 6 = 5\cdot 2$
$4 + 6 = 10$)

and also for $x = 3$

(**Check:** $3^2 + 6 = 5\cdot 3$
$9 + 6 = 15$)

The collection of numbers that makes a given equation true is called the *solution set* of that equation. The solution set of the equation $x^2 + 6 = 5x$ is the pair of numbers 2 and 3.

What is the solution of $17 - x = 5$?

Work it out, then turn to **32**.

---

32 The solution of $17 - x = 5$ is $x = 12$

(**Check:** $17 - \underline{12} = 5$)

In this introduction to algebra we shall only work with *linear equations in one variable.* This kind of equation contains only one variable and that variable appears only to the first power. There will be no x^2, x^3, x^4, $\sqrt{x}$, or other power of x. Only constants and x will appear in the equation. For example,

$$\begin{aligned} 2x + 3 &= 4 \\ 3a &= 17 \\ 2 - p &= 7 \end{aligned}$$

and $\quad y + 2 = 3 - y$

are all linear equations in one variable.

$x^2 - 1 = 3$ is not a linear equation because it has an x^2 term.

$p + q = 2$ is a linear equation in two variables.

Which of the following is *not* a linear equation?

(a) $2x + 3y = 4$ Go to **33**.
(b) $x^2 = 10 + x$ Go to **34**.
(c) $2x - 1 = 7$ Go to **35**.

---

33 Not quite.

The equation $2x + 3y = 4$ *is* a linear equation in two variables x and y. We know it is a linear equation because the variables x and y both appear only as first powers. There are no other powers of x or y, no x^2, x^3, y^2, $y^3 \sqrt{x}$, $\sqrt{y}$, and so on.

In this book we shall work with linear equations in one unknown only.

Return to **32** and choose a different answer.

---

34 Right!

$x^2 = 10 + x$ is *not* a linear equation because the variable x appears as a second power: x^2. In a linear equation the variable appears only as x, no other powers.

Find the solutions of these equations:

(a) $4 + x = 11$ (b) $x - 1 = 6$
(c) $x + 2 = 9$ (d) $8 - x = 1$

(Of course you can solve them by simple mental arithmetic. Try it.)

Check your work in **36**.

---

35 You are mistaken.

The equation $2x - 1 = 7$ is a linear equation in one variable. It is linear because the variable appears only as x, never as x^2, x^3, or any other power.

Does that explanation help?

Now scoot back to **32** and choose a better answer.

$x = 7$. Every equation has the same solution.

If you replace the variable x in each equation by the number 7, all four equations will be true.

Equations with the exact same solution are called *equivalent* equations.

The equations $2x + 7 = 13$ and $3x = 9$ are equivalent because substituting the value 3 for x makes them both true.

We say that a linear equation in one variable or one unknown x is solved if it can be put in the form

$x = \square$ where $\square$ is some real number.

For example, the solution to the equation

$2x - 1 = 7$

is

$x = 4$

because

$$\begin{aligned} 2 \cdot 4 - 1 &= 7 \\ 8 - 1 &= 7 \\ 7 &= 7 \end{aligned}$$

is a true statement.

Equations as simple as the one above are easy to solve by guessing, but guessing is not a very happy way to do mathematics. We need some sort of rule that will enable us to rewrite the equation to be solved ($2x - 1 = 7$, for example) as an equivalent solution equation ($x = 4$).

The general rule is to treat every equation as a balance of the two sides.

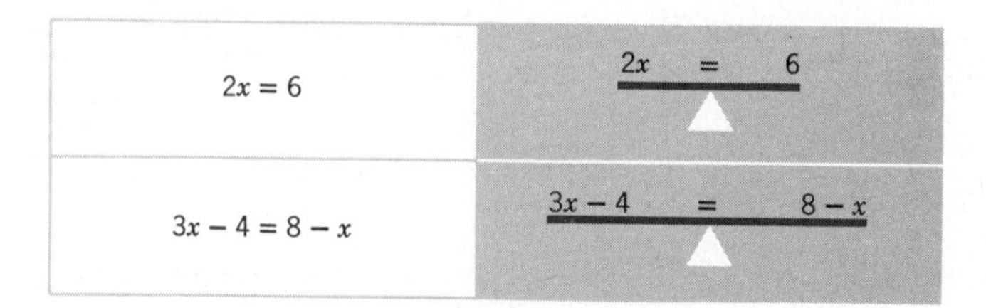

Any changes made in the equation must not disturb this balance. Any operation performed on one side of the equation must also be performed on the other side. Two kinds of balancing operations may be used.

1. Adding or subtracting a number on both sides of the equation does not change the balance.

and

2. Multiplying or dividing both sides of the equation by a number (but not zero) does not change the balance.

Example

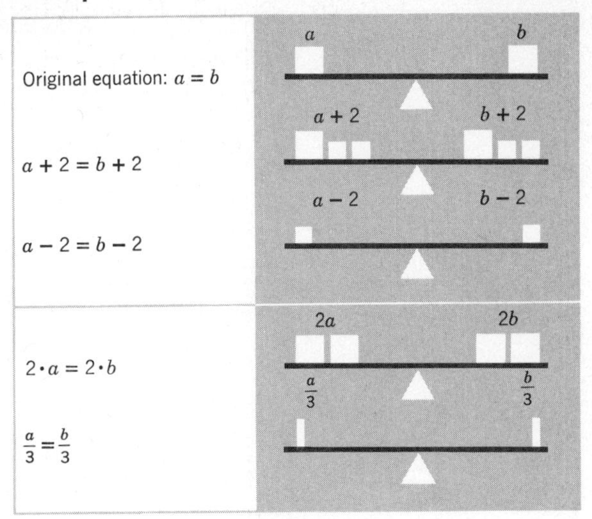

Let's work through an example.

Solve: $x - 4 = 2.$

Step 1: We want to change this equation to an equivalent equation with only x on the left, so we add 4 to each side of the equation.

$$(x - 4) + 4 = (2) + 4$$

Step 2: Combine terms.

$$x \underbrace{- 4 + 4}_{0} = 2 + 4$$

$$x = 6 \quad \text{Solution}$$

Check:

$$6 - 4 = 2$$
$$2 = 2$$

Use these balancing operations to solve the equation

$8 + x = 14$

Check your work in **37**.

ALGEBRA EQUATIONS

There are three kinds of algebra equations:

1. *Identities* are equations that are always true, no matter what the value of the variable in the equation. For example,

$$a + a = 2a$$
$$a + b = b + a$$

or

$$x + 0 = x$$

are always true for ordinary values of the variables a, b, and x.

2. *False equations* are never true, no matter what value the variable is given. For example,

$$3 + 2 = 4$$

and

$$x + 1 = x$$

are never true.

Some equations are not just false, they are *meaningless*. For example,

$$\frac{x}{0} = y$$

has no meaning because division by zero is not defined for ordinary real numbers.

3. *Conditional equations* are true for some certain value or values of the variable.

$x + 3 = 7$	is true for $x = 4$
$6x - 5 = 13$	is true for $x = 3$
$x^2 - 3x + 2 = 0$	is true for $x = 1$ and $x = 2$

Conditional equations are by far the most interesting and useful. The equations you are asked to solve in algebra courses or in any sort of applications of algebra are conditional equations.

37 **Solve:** $8 + x = 14$

Step 1: We want to change this equation to an equivalent equation with only x on the left, so we subtract 8 from each side of the equation.

$$(8 + x) - 8 = (14) - 8$$

Step 2: Combine terms.

$$x + \underbrace{8 - 8}_{0} = 14 - 8 \qquad (8 + x = x + 8 \quad \text{of course})$$

$x = 6$ Solution

Check:
$$8 + 6 = 14$$
$$14 = 14$$

Solve these in the same way.

(a) $x - 4 = 10$
(b) $12 + x = 27$
(c) $11 - x = 2$
(d) $x + 6 = 2$
(e) $8 = 3 + x$

The step-by-step solutions are in **39**.

---

38 **Solve:** $3x = 39$

Step 1: $\frac{3x}{3} = \frac{39}{3}$ We want to change this equation to an equivalent equation with x alone on the left, so we divide both sides by 3.

$$\frac{3x}{3} = \left(\frac{3}{3}\right)x = x$$

Step 2: $x = \frac{39}{3}$

$x = 13$ Solution

Check:
$$3 \cdot 13 = 39$$
$$39 = 39$$

Solve these in the same way.

(a) $7x = 35$
(b) $\frac{1}{2}x = 14$
(c) $\frac{2x}{3} = 6$

The step-by-step solutions are in **40**.

---

39 (a) **Solve:**

$$x - 4 = 10$$
$$(x - 4) + 4 = (10) + 4 \quad \text{Add 4 to each side.}$$
$$x \underbrace{- 4 + 4}_{0} = 10 + 4 \quad \text{Combine terms.}$$
$$\boxed{x = 14} \quad \text{Solution}$$

Check:

$$14 - 4 = 10$$
$$10 = 10$$

(b) **Solve:**

$$12 + x = 27$$
$$(12 + x) - 12 = (27) - 12 \quad \text{Subtract 12 from each side.}$$
$$x + \underbrace{12 - 12}_{0} = 27 - 12 \quad \text{Combine terms.}$$
(Note that $12 + x = x + 12$)
$$\boxed{x = 15} \quad \text{Solution}$$

Check:

$$12 + 15 = 27$$
$$27 = 27$$

(c) **Solve:**

$$11 - x = 2$$
$$(11 - x) - 11 = (2) - 11 \quad \text{Subtract 11 from each side.}$$
$$-x + \underbrace{11 - 11}_{0} = 2 - 11 \quad \text{Combine terms.}$$
$$-x = -9$$
$$\boxed{x = 9} \quad \text{Solution} \quad \text{Multiply each side by } -1.$$

Check:

$$11 - 9 = 2$$
$$2 = 2$$

(d) **Solve:**

$$x + 6 = 2$$
$$(x + 6) - 6 = (2) - 6 \quad \text{Subtract 6 from each side.}$$
$$x + \underbrace{6 - 6}_{0} = 2 - 6 \quad \text{Combine terms.}$$
$$\boxed{x = -4} \quad \text{Solution} \quad \text{The solution is a negative number.}$$

Check:

$$-4 + 6 = 2$$
$$2 = 2$$

(e) **Solve:**

$$8 = 3 + x$$
$$(8) - 3 = (3 + x) - 3 \quad \text{Subtract 3 from each side.}$$
$$8 - 3 = x + \underbrace{3 - 3}_{0} \quad \text{Combine terms.}$$
$$5 = x$$
or $\boxed{x = 5}$ Solution — $5 = x$ is the same as $x = 5$.

Check:

$$8 = 3 + 5$$
$$8 = 8$$

If you found any difficulty in working with negative numbers in problems (c) or (d) you should review the section on signed numbers in Unit 3, page 225, frame **35**

Here is a slightly different problem.

Solve: $2x = 14$

Step 1: We want to change this equation to an equivalent equation with x alone on the left, so we divide both sides by 2.

$$\frac{2x}{2} = \frac{14}{2}$$

Step 2: $x = \frac{14}{2}$ $\quad \frac{2x}{2} = \left(\frac{2}{2}\right)x = x$

$x = 7$ Solution

Check: $2 \cdot 7 = 14$

$14 = 14$

Try this problem.

Solve: $3x = 39$

Look in **38** for the solution.

---

40 (a) **Solve:** $7x = 35$

Step 1: $\frac{7x}{7} = \frac{35}{7}$ Divide both sides by 7.

Step 2: $x = \frac{35}{7}$ $\quad \frac{7x}{7} = \left(\frac{7}{7}\right)x = x$

$x = 5$ Solution

Check: $7 \cdot 5 = 35$
$35 = 35$

(b) **Solve:** $\frac{1}{2}x = 14$

Step 1: $(\frac{1}{2}x)2 = (14)2$ Multiply both sides by 2

Step 2: $(x \cdot \frac{1}{2})2 = 14 \cdot 2$ $\quad \frac{1}{2} \cdot x = x \cdot \frac{1}{2}$

$x \cdot \underbrace{\frac{1}{2} \cdot 2}_{1} = 28$ $\quad \frac{1}{2} \cdot 2 = \frac{2}{2} = 1$

$x = 28$

Check: $\frac{1}{2} \cdot 28 = 14$
$\frac{28}{2} = 14$
$14 = 14$

(continued on the next page)

(c) **Solve:** $\frac{2x}{3} = 6$

Step 1: $\left(\frac{2x}{3}\right)3 = (6)3$ Multiply both sides by 3.

$2x = 6 \cdot 3$ $\left(\frac{2x}{3}\right)3 = \frac{2 \cdot x \cdot \cancel{3}}{\cancel{3}} = 2 \cdot x$

$2x = 18$

Step 2: $\frac{\cancel{2}x}{\cancel{2}} = \frac{18}{2}$ Divide both sides by 2. $\frac{2x}{2} = x$

$x = 9$ Solution

Check: $\frac{2 \cdot 9}{3} = 6$

$\frac{18}{3} = 6$

$6 = 6$

Solving some linear equations involves both kinds of operations: addition/subtraction and multiplication/division. For example,

Solve: $2x + 6 = 14$

Step 1: We want to change this equation to an equivalent equation with only x or terms with x on the left, so subtract 6 from both sides.

$(2x + 6) - 6 = (14) - 6$

$2x + \underbrace{6 - 6}_{0} = 14 - 6$ Combine terms.

$2x = 8$ Now change this to an equivalent equation with only x on the left.

Step 2: $\frac{2x}{2} = \frac{8}{2}$ Divide both sides of the equation by 2.

$x = \frac{8}{2}$ $\frac{\cancel{2}x}{\cancel{2}} = x$

$x = 4$ Solution

Check: $2 \cdot 4 + 6 = 14$

$8 + 6 = 14$

$14 = 14$

Try this one to test your understanding of the process.

Solve $3x - 7 = 11$

Check your work in **41**.

To solve $\frac{1}{2}x + 4 = 8$ can I multiply by 2 and get $x + 4 = 16$?

No. When you multiply by 2 you must multiply **all** of the expression on the left by 2:
$2(\frac{1}{2}x + 4) = 2(8)$ or $x + 8 = 16$
Careful.

---

41 **Solve:** $3x - 7 = 11$

Step 1: $(3x - 7) + 7 = (11) + 7$ Add 7 to each side.

$$3x \underbrace{- 7 + 7}_{0} = 11 + 7$$

$$3x = 18$$

Step 2: $\frac{\not{3}x}{\not{3}} = \frac{18}{3}$ Divide both sides of the equation by 3.

$x = 6$ Solution

Check:
$$3 \cdot 6 - 7 = 11$$
$$18 - 7 = 11$$
$$11 = 11$$

Here are a few practice problems. Solve them as we have done above.

(a) $7x + 2 = 51$
(b) $18 - 5x = 3$
(c) $15 = 4x - 1$
(d) $3(x + 4) = 27$

Our step-by-step solutions are in 42.

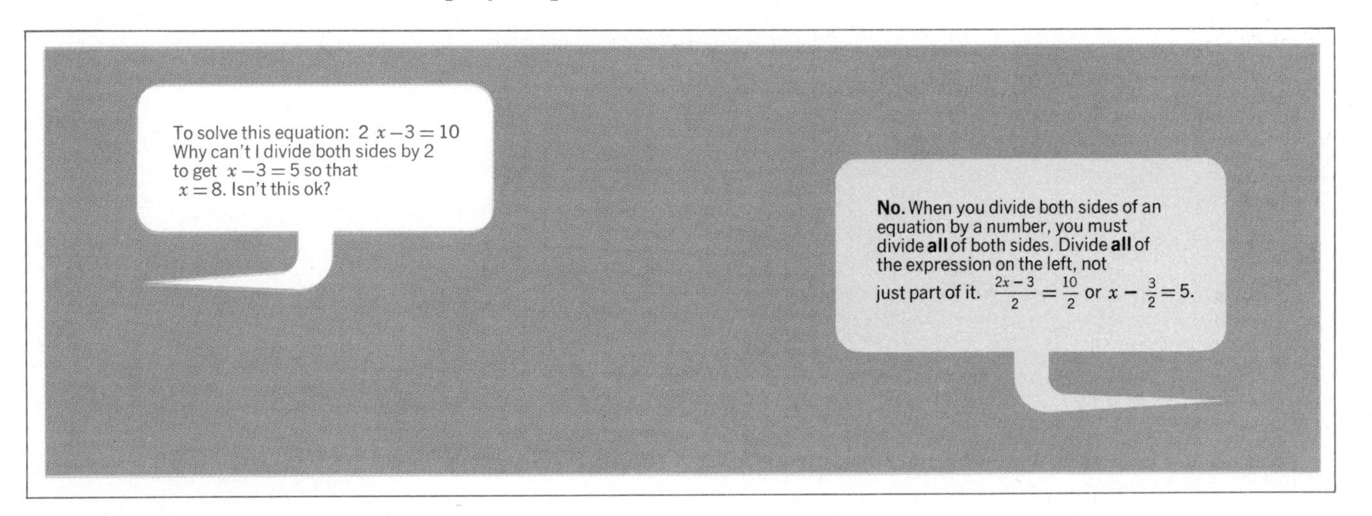

To solve this equation: 2 x − 3 = 10 Why can't I divide both sides by 2 to get x − 3 = 5 so that x = 8. Isn't this ok?
No. When you divide both sides of an equation by a number, you must divide all of both sides. Divide all of the expression on the left, not just part of it. (2x − 3)/2 = 10/2 or x − 3/2 = 5.

42 (a) **Solve:** $7x + 2 = 51$

Change this equation to an equivalent equation with only an x term on the left.

Step 1: $(7x + 2) - 2 = (51) - 2$ Subtract 2 from each side.

$7x + \underbrace{2 - 2}_{0} = 51 - 2$ Combine terms.

$7x = 49$

Step 2: $\frac{7x}{7} = \frac{49}{7}$ Divide both sides by 7.

$x = \frac{49}{7}$

$x = 7$ Solution

Check: $7 \cdot 7 + 2 = 51$

$49 + 2 = 51$

$51 = 51$

(b) **Solve:** $18 - 5x = 3$

Step 1: $(18 - 5x) - 18 = (3) - 18$ Subtract 18 from each side.

$-5x + \underbrace{18 - 18}_{0} = 3 - 18$ Rearrange terms.

$-5x = -15$ Multiply both sides by -1.

$5x = 15$

Step 2: $\frac{\not{5}x}{\not{5}} = \frac{15}{5}$ Divide both sides by 5.

$x = 3$ Solution

Check: $18 - 5 \cdot 3 = 3$

$18 - 15 = 3$

$3 = 3$

(c) **Solve:** $15 = 4x - 1$

Step 1: $(15) + 1 = (4x - 1) + 1$ Add 1 to each side.

$15 + 1 = 4x \underbrace{- 1 + 1}_{0}$ Combine terms.

$16 = 4x$

or

$4x = 16$

Step 2: $\frac{\not{4}x}{\not{4}} = \frac{16}{4}$ Divide both sides by 4.

$x = 4$ Solution

Check: $15 = 4 \cdot 4 - 1$

$15 = 16 - 1$

$15 = 15$

(d) **Solve:** $3(x + 4) = 27$

Step 1: $\frac{\not{3}(x + 4)}{\not{3}} = \frac{27}{3}$ Divide both sides by 3

$$x + 4 = 9$$

$$(x + 4) - 4 = (9) - 4$$ Subtract -4 from each side

$$x \underbrace{+ 4 - 4}_{0} = 9 - 4$$ Combine terms

$x = 5$ Solution

Check:

$$3(5 + 4) = 27$$
$$3 \cdot 9 = 27$$
$$27 = 27$$

If several terms in the equation have x as a factor, we must combine these terms. For example,

Solve: $2x = 20 - 3x$ Change to an equivalent equation with all x terms on the left.

Step 1: $(2x) + 3x = (20 - 3x) + 3x$ Add $3x$ to both sides.

$$(2x) + 3x = 20 \underbrace{- 3x + 3x}_{0}$$ Combine terms.

$$5x = 20$$

Step 2: $\frac{\not{5}x}{\not{5}} = \frac{20}{5}$ Divide both sides by 5

$x = 4$ Solution

Check:

$$2 \cdot 4 = 20 - 3 \cdot 4$$
$$8 = 20 - 12$$
$$8 = 8$$

Solving a few problems will help groove the procedure into your tired little mind. Try these:

(a) $2x + 3 = 7x + 8$
(b) $4x - 3 = 21 - 2x$
(c) $2(x + 5) = 3(x - 2)$

Check your answers in **43**.

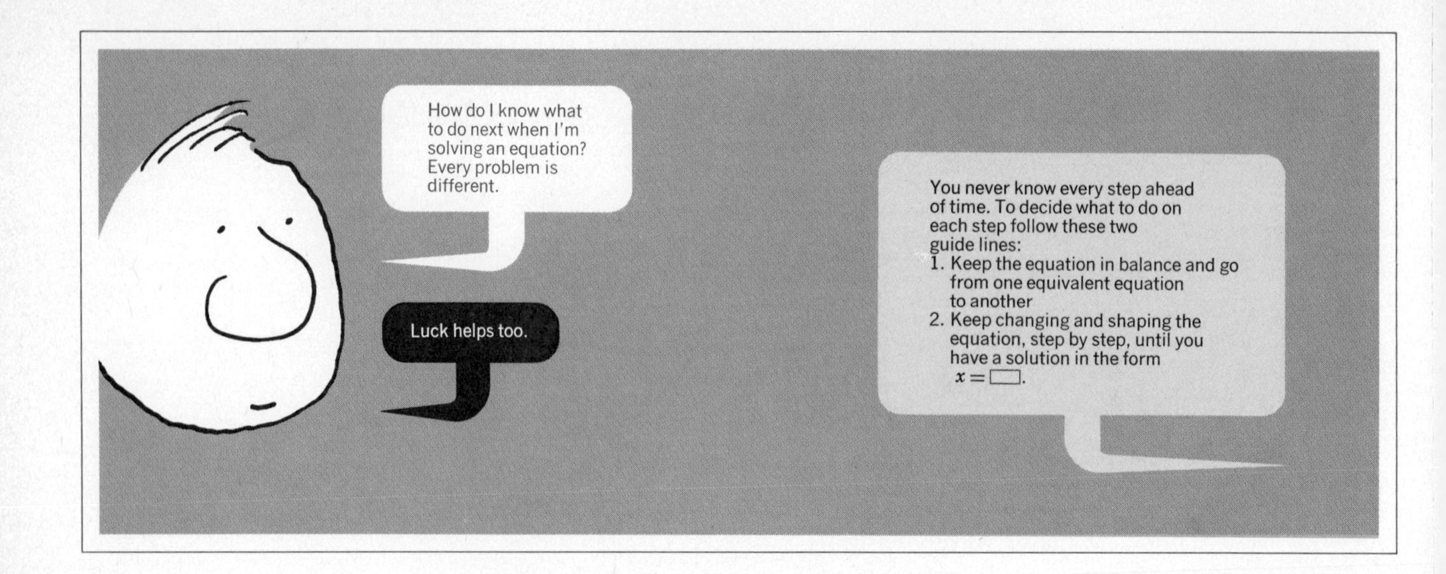

---

43 (a) **Solve:** $2x + 3 = 7x + 8$

Step 1: $(2x + 3) - 3 = (7x + 8) - 3$ — Subtract -3 from each side of the equation.

$$2x \underbrace{+ 3 - 3}_{0} = 7x + \underbrace{8 - 3}_{5}$$ Combine terms.

$$2x = 7x + 5$$

Step 2: $(2x) - 7x = (7x + 5) - 7x$ — Subtract $7x$ from each side of the equation.

$$2x - 7x = 5 + \underbrace{7x - 7x}_{0}$$ Combine terms.

$$-5x = 5$$

Step 3: $\dfrac{-\cancel{5}x}{-\cancel{5}} = \dfrac{5}{-5}$ — Divide each side by -5

$x = -1$ Solution

Check: $2(-1) + 3 = 7(-1) + 8$

$$-2 + 3 = -7 + 8$$

$$1 = 1$$

(b) **Solve:** $4x - 3 = 21 - 2x$

Step 1: $(4x - 3) + 3 = (21 - 2x) + 3$ — Add 3 to each side of the equation.

$$4x \underbrace{- 3 + 3}_{0} = 21 - 2x + 3$$ Combine terms.

$$4x = -2x + \underbrace{21 + 3}_{24}$$ Combine terms.

$$4x = -2x + 24$$

Step 2: $(4x) + 2x = (-2x + 24) + 2x$ — Add $2x$ to both sides.

$4x + 2x = -2x + 24 + 2x$

$4x + 2x = 24 \underbrace{- 2x + 2x}_{0}$ — Combine terms.

$6x = 24$

Step 3: $\frac{\not{6}x}{\not{6}} = \frac{24}{6}$ — Divide both sides by 6.

$x = 4$ Solution

Check: $4 \cdot 4 - 3 = 21 - 2 \cdot 4$

$16 - 3 = 21 - 8$

$13 = 13$

(c) **Solve:** $2(x + 5) = 3(x - 2)$

Step 1: $2x + 10 = 3x - 6$ — Remove the parentheses.

Step 2: $(2x + 10) + 6 = (3x - 6) + 6$ — Add 6 to each side of the equation.

$2x \underbrace{+ 10 + 6}_{16} = 3x \underbrace{- 6 + 6}_{0}$ — Combine terms.

$2x + 16 = 3x$

Step 3: $(2x + 16) - 2x = (3x) - 2x$ — Subtract $2x$ from each side of the equation.

$2x + 16 - 2x = 3x - 2x$

$16 \underbrace{+ 2x - 2x}_{0} = x$ — Combine terms.

$16 = x$

or $x = 16$ Solution

Check: $2(16 + 5) = 3(16 - 2)$

$2 \cdot 21 = 3 \cdot 14$

$42 = 42$

Remember:

1. Do only legal operations. Add or subtract the same quantity from both sides of the equation. Multiply or divide both sides of the equation by the same non-zero quantity.
2. Always check your answer.

For a set of practice problems on solving equations turn to 44.

Introduction to Algebra

PROBLEM SET 4

44 Solving Simple Equations

The answers are on page 512.

A. Solve:

1. $x + 3 = 14$ $x =$ ______
2. $x - 21 = 13$ $x =$ ______
3. $10 = 3 - a$ $a =$ ______
4. $b + 34 = 41$ $b =$ ______
5. $p - 17 = 8$ $p =$ ______
6. $q + 41 = 53$ $q =$ ______
7. $40 = s + 15$ $s =$ ______
8. $4b = 22$ $b =$ ______
9. $\frac{1}{4}x = 17$ $x =$ ______
10. $26 = 5y$ $y =$ ______
11. $2c + 11 = 31$ $c =$ ______
12. $42 - 3d = 27$ $d =$ ______
13. $67 = 5h - 3$ $h =$ ______
14. $31 = 7k + 3$ $k =$ ______
15. $2x = 39 - 11x$ $x =$ ______
16. $5a = a - 56$ $a =$ ______
17. $2(b + 13) = 36$ $b =$ ______
18. $7s + 5 = 2s - 25$ $s =$ ______
19. $2Q - 11 = Q + 17$ $Q =$ ______
20. $14 - P = 2P - 4$ $P =$ ______
21. $Q + 18 = 30$ $Q =$ ______
22. $50 = A + 32$ $A =$ ______
23. $T + 38 = 61$ $T =$ ______
24. $\frac{1}{2} = F - 1\frac{1}{3}$ $F =$ ______
25. $1\frac{1}{2}N = 6$ $N =$ ______
26. $\frac{5}{8} = x + \frac{1}{4}$ $x =$ ______
27. $2C + \frac{1}{2}C = 15$ $C =$ ______
28. $8 - A = 10 - 2A$ $A =$ ______
29. $2B - 3 = 5B + 18$ $B =$ ______
30. $3(x - 1) = 2(1 - x) + 5$ $x =$ ______

B. Translate these problems into algebra equations and solve them.

1. A number is multiplied by 6, and 10 is subtracted from the product. The result is 62. Find the original number.

______

2. The sum of two numbers is 18. One of them is twice as large as the other. What are the numbers?

______

Date ______

Name ______

Course/Section ______

3. If $\frac{1}{8}$th of a certain number is added to the number itself, the sum is 18. What is the number?

____

4. $\frac{2}{3}$ of a certain number added to $\frac{1}{5}$ of the number is equal to 13. What is the number?

____

5. Solve: $N + 1 = 1 - N$. $N =$ ______

6. If you wanted to solve the equation $2x + 9 = 31$, what would be the first step? (Pick one)
   (a) Add 9 to both sides.
   (b) Subtract 9 from both sides.
   (c) Divide both sides by 2.
   (d) Add $-2x$ to both sides.
   (e) Subtract 31 from both sides.

7. Solve: $\frac{3}{4}x - \frac{2}{3} = \frac{5}{6}$. $x =$ ______

8. John is 10 years older than Bill and the sum of their ages is 45. How old are they?

John ______ Bill ______

9. Lyn is 7 years more than twice as old as Mary. The difference in their ages is 20. How old is Lyn?

____

10. The sum of two consecutive numbers is 15. What is the smaller number?

____

11. Five less than $\frac{1}{2}$ of a number is equal to 6. What is the number?

____

12. Thirteen minus twice a certain number is 9. Find the number.

____

13. $ax - b = c + dx$. Solve for x in terms of the other letters.

$x =$ ______

14. Solve: $\frac{14}{x} = 2$ $x =$ ______

15. Solve: $x + 2(x + 2(x - 1)) = 31$ $x =$ ______

16. Solve: $6x - 1 = 2(x + 4) - 9$. $x =$ ______

17. Eric has 13 more marbles than Dana. Eric also has one-fifth more marbles than Dana. How many marbles does each boy have?

______, ______

18. Solve $(x + 2)^2 = (1 + x)^2 + 7$. $x =$ ______

19. Find two numbers whose sum is 13 and whose difference is 7.

______, ______

20. For what temperature does the Celsius scale read exactly five times the Fahrenheit scale reading?
Hint: $5(F + 40) = g(C + 40)$ C = ______

F = ______

When you have had the practice you need, either return to the preview test on page 413 or continue in **45** with the study of ratio and proportion.

Date

Name

Course/Section

45 It is often necessary in business, industry, or everyday activities to compare two quantities—two costs, two distances, two areas, or two of anything. We could compare them by subtracting to find their difference, but a better comparison, especially if the quantities are very different, is found by dividing to find the ratio of the size of one quantity to the other. A *ratio* is a comparison of the sizes of two quantities of the same kind, found by dividing them. It is a single number usually written as a fraction.

For example, the steepness of a hill can be written as the ratio of its height to its horizontal extent.

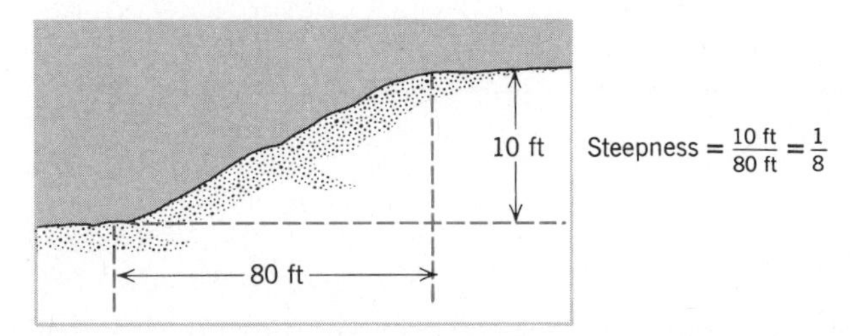

The ratio of the circumference (distance around) of any circle to its diameter is approximately 3.14. This ratio is used so often in mathematics that it has been given a special symbol, the Greek letter π, pronounced "pie."

$\pi = \frac{\text{circumference}}{\text{diameter}}$

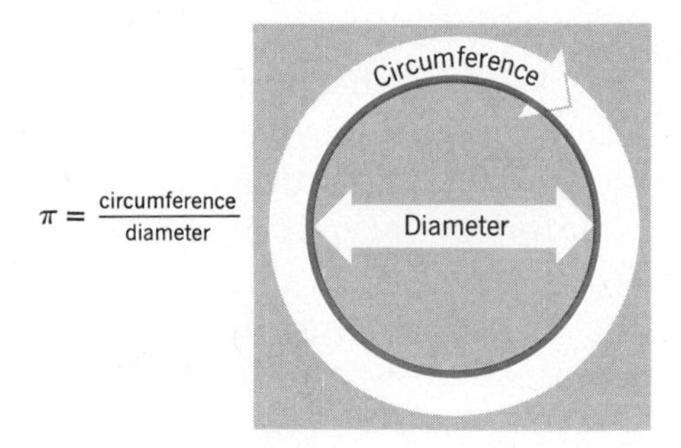

Find the ratio of the areas shown below.

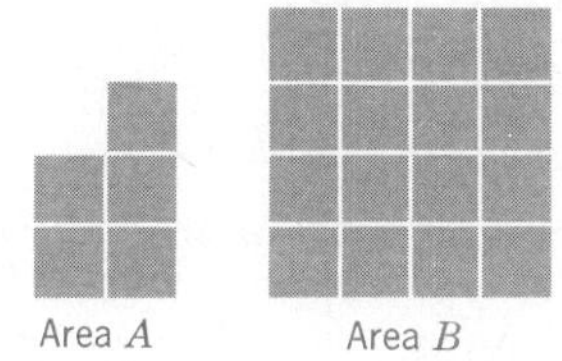

$$\text{Area ratio} = \frac{\text{area } A}{\text{area } B} = \underline{\qquad}$$

Check your answer in **46**.

46 Area ratio $= \dfrac{\text{area } A}{\text{area } B} = \dfrac{5 \text{ unit areas}}{16 \text{ unit areas}} = \dfrac{5}{16}$

The smaller area is $\dfrac{5}{16}$ of the larger area.

A ratio is usually a division of like quantities and therefore the units cancel and do not appear in the ratio number. If necessary, we must convert to get the same units before dividing. For example, what is the ratio of these two lengths?

A — 14 inches

B — 2 feet

Set it up, write out the ratio, and then turn to **48**.

---

47 Ratio $= \dfrac{1 \text{ dime}}{1 \text{ quarter}} = \dfrac{10¢}{25¢} = \dfrac{10}{25} = \dfrac{2}{5}$

Write the numbers in the same units, divide, and reduce the fraction to lowest terms. Just as with any fraction, this ratio could be written as a decimal 0.4, as a percent, 40%, or as some fraction equivalent to $\frac{2}{5}$, such as $\frac{4}{10}$, $\frac{6}{15}$ or $\frac{40}{100}$.

Write the following ratios as fractions reduced to lowest terms.

(a) Sam takes 25 minutes to do a certain job and Bill takes one hour to do the same job. What is the ratio of their job times?

(b) What is the gear ratio of a bicycle using 24 teeth on the rear sprocket and 50 teeth on the front chainwheel?

(c) What is the won-to-lost ratio of a basketball team that wins 8 games and loses 18 in a given season?

(d) What is the pay-off ratio of a game that pays a winner $7 for every $2 he bets?

Check your work in **49**.

---

48 Ratio $= \dfrac{\text{length } A}{\text{length } B} = \dfrac{14 \text{ in.}}{2 \text{ ft.}} = \dfrac{14 \text{ in.}}{24 \text{ in.}} = \dfrac{14}{24} = \dfrac{7}{12}$

Notice that the units cancel and that we have reduced the fraction to lowest terms.

What is the ratio of a dime to a quarter?

Check your answer in **47**.

---

49 (a) Job time ratio $= \frac{25 \text{ min.}}{1 \text{ hr.}} = \frac{25 \text{ min.}}{60 \text{ min.}} = \frac{25}{60} = \frac{5}{12}$

(b) Gear ratio $= \frac{\text{front}}{\text{rear}} = \frac{50 \text{ teeth}}{24 \text{ teeth}} = \frac{50}{24} = \frac{25}{12}$

(c) Won-to-lost ratio = won/lost ratio $= \frac{\text{wins}}{\text{loses}} = \frac{8}{18} = \frac{4}{9}$

The win/lose ratio is 4 to 9.

(d) Pay-off ratio $= \frac{\text{pay-off}}{\text{bet}} = \frac{\$7}{\$2} = \frac{7}{2}$ or $3\frac{1}{2}$

The pay-off is 7 to 2 or $3\frac{1}{2}$ to 1.

A *proportion* is a statement that two ratios are equal. It can be an equation written in numbers and letters or it can be a sentence in words.

The arithmetic equation $\frac{3}{5} = \frac{21}{35}$ is a proportion.

The algebra equation $\frac{11}{4} = \frac{x}{5}$ is a proportion.

The most interesting kind of proportion is that in which one of the ratios is not completely known. For example, consider this problem:

"If the ratio of my height to his height is 7 to 8, and he is 6 feet tall, how tall am I?"

Ratio of our heights $= \frac{\text{my height}}{\text{his height}} = \frac{7}{8}$

and this ratio is equal to the actual height ratio

$\frac{7}{8} = \frac{?}{6 \text{ ft}}$

This last equation is a proportion.

Write this proportion as an algebra equation and solve it.

Check your work in **50**.

I get the ideas of ratio and proportion confused.
Many people do. Remember that a **ratio** is a number, usually a fraction, and a **proportion** is an equation showing the equality of two ratios.

50 $\dfrac{7}{8} = \dfrac{H}{6}$

where H is a height in feet.

Multiply both sides of the equation by 6.

$$\left(\frac{7}{8}\right)6 = \left(\frac{H}{6}\right)6$$

$$\frac{7 \cdot 6}{8} = H$$

$$\frac{42}{8} = H \qquad \text{or} \qquad H = \frac{42}{8} = 5\frac{2}{8}$$

$$H = 5\tfrac{1}{4} \text{ ft}$$

A very easy way to solve proportions is to remember that

If $\dfrac{a}{b} = \dfrac{c}{d}$ then $ad = bc$

The *cross products* of the proportion are equal

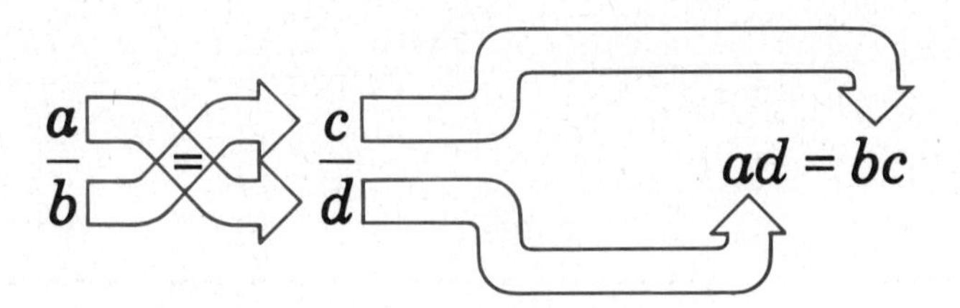

For example,

$\dfrac{2}{3} = \dfrac{18}{27}$ means

$2 \cdot 27 = 3 \cdot 18$

$54 = 54$

The equation above $\dfrac{7}{8} = \dfrac{H}{6}$ becomes $7 \cdot 6 = H \cdot 8$

or

$42 = 8H$

which is easy to solve.

Use the idea of cross products to solve the following proportions.

(a) $\dfrac{1}{2} = \dfrac{x}{9}$ ______ $\cdot$ ______ $= x \cdot$ ______ $x =$ ______

(b) $\dfrac{w}{7} = \dfrac{3}{4}$ $w \cdot$ ______ $=$ ______ $\cdot$ ______ $w =$ ______

(c) $\dfrac{3}{y} = \dfrac{2}{5}$ ______ · ______ = ______ · y $y =$ ______

(d) $\dfrac{7}{16} = \dfrac{21}{z}$ ______ · $z =$ ______ · ______ $z =$ ______

(e) $\dfrac{6}{5} = \dfrac{2}{T}$ $T =$ ______ (f) $\dfrac{21}{12} = \dfrac{k}{6}$ $k =$ ______

(g) $\dfrac{2\frac{1}{2}}{3\frac{1}{2}} = \dfrac{Q}{2}$ $Q =$ ______ (h) $\dfrac{0.4}{1.5} = \dfrac{12}{R}$ $R =$ ______

The answers are in **51**.

---

51 (a) $\dfrac{1}{2} = \dfrac{x}{9}$ $1 \cdot 9 = x \cdot 2$ $9 = 2x$ $x = 4\dfrac{1}{2}$

(b) $\dfrac{w}{7} = \dfrac{3}{4}$ $w \cdot 4 = 3 \cdot 7$ $4w = 21$ $w = 5\dfrac{1}{4}$

(c) $\dfrac{3}{y} = \dfrac{2}{5}$ $3 \cdot 5 = 2 \cdot y$ $2y = 15$ $y = 7\dfrac{1}{2}$

(d) $\dfrac{7}{16} = \dfrac{21}{z}$ $7 \cdot z = 21 \cdot 16$ $7z = 336$ $z = 48$

(e) $T = 1\frac{2}{3}$ (f) $K = 10\frac{1}{2}$ (g) $Q = 1\frac{3}{7}$ (h) $R = 45$

Use the ideas of ratio and proportion to solve this problem:

> A telephone pole casts a shadow 9 ft long. A yard stick beside the pole casts a shadow 16 in. long. What is the height of the telephone pole?

Our solution is in **52**, check it when you are finished.

WHICH IS THE BETTER BUY?

Stroll through any grocery store and you will find yourself swamped with numbers: \$1.56 per pound, 3 for 25¢, 6 cans for 89¢, and so on. Suppose you decide to get the best possible buy for your money in your purchases, how do you wade through all those numbers? How do you decide which product is the better buy?

Let's look at an example. A look at the detergent shelf reveals three sizes of your favorite brand:

(a) Economy size box 1 lb 9 oz for 49¢
(b) Giant size box 3 lb 1 oz at 89¢
(c) Jumbo family size on sale at 5 lb 4 oz box for \$1.76

Which is the better buy? To decide, follow these steps:

First, ask yourself if you need it at all. Something you do not need is a bad buy at any cost.

Second, calculate the ratio of cost/weight, the *per unit cost* for each item. For the three buys above

(a) 49¢ ÷ 25 oz = 2.0¢/oz
(b) 89¢ ÷ 49 oz = 1.8¢/oz
(c) 176¢ ÷ 84 oz = 2.1¢/oz

Keep all ratios in the same units.

Third, if all three are of the same quality, the second one is the better buy.

Sometimes you can save a little time by looking at the numbers carefully before you do the arithmetic. For example, two of the giant size boxes weighs a total of 6 lb 2 oz and sells for \$1.78, only 2¢ more than the jumbo size box, which weighs 14 oz less. Obviously the giant size is a better buy than the jumbo size, and you need compare only with the economy size. A quick inspection has saved you a little arithmetic.

52 Ratio of heights $= \dfrac{\text{height of pole}}{\text{height of stick}} = \dfrac{h \text{ ft}}{3 \text{ ft}}$

Ratio of shadow lengths $= \dfrac{\text{shadow of pole}}{\text{shadow of stick}} = \dfrac{9 \text{ ft}}{16 \text{ in.}} = \dfrac{108 \text{ in.}}{16 \text{ in.}}$

The proportion is

$$\frac{h}{3} = \frac{108}{16}$$

Cross multiply

$$16 \cdot h = 3 \cdot 108$$
$$16h = 324$$
$$h = 20\tfrac{1}{4} \text{ ft}$$

Notice that we set up ratios of like quantities and that top and bottom of each ratio are in the same units. Both fractions have pole information in the numerator (top) and stick information in the denominator (bottom) of the ratios.

Try another ratio and proportion problem:

> If five men can dig three ditches in one day, how many ditches will two men working at the same rate dig in one day?

Set up the ratios carefully, write a proportion equation, and solve it. Check your answer in **53**.

---

53 Diggers ratio $= \dfrac{5 \text{ men}}{2 \text{ men}} = \dfrac{5}{2}$

Work ratio $= \dfrac{3 \text{ ditches}}{D \text{ ditches}} = \dfrac{3}{D}$

The work corresponding to the five men is placed in the numerator of the fraction to match the first fraction.

$$\frac{5}{2} = \frac{3}{D}$$

$$5 \cdot D = 3 \cdot 2$$
$$5D = 6$$
$$D = \tfrac{6}{5} = 1\tfrac{1}{5}$$
$$D = 1.2 \text{ ditches}$$

Now turn to **54** for a set of practice problems on ratio and proportion.

Introduction to Algebra

PROBLEM SET 5

54

Ratio and Proportion

Answers are on page 512.

A. Solve the following proportion equations using cross products.

1. $\frac{2}{15} = \frac{8}{x}$ $x =$ ______
2. $\frac{6}{26} = \frac{w}{13}$ $w =$ ______
3. $\frac{5}{y} = \frac{30}{7}$ $y =$ ______
4. $\frac{A}{115} = \frac{3}{46}$ $A =$ ______
5. $\frac{1}{6} = \frac{17}{B}$ $B =$ ______
6. $\frac{3}{8} = \frac{D}{32}$ $D =$ ______
7. $\frac{M}{54} = \frac{9}{27}$ $M =$ ______
8. $\frac{3}{12} = \frac{5}{K}$ $K =$ ______
9. $\frac{15}{24} = \frac{P}{8}$ $P =$ ______
10. $\frac{5}{9} = \frac{20}{t}$ $t =$ ______
11. $\frac{2}{Q} = \frac{14}{3}$ $Q =$ ______
12. $\frac{F}{221} = \frac{6}{34}$ $F =$ ______
13. $\frac{756}{5} = \frac{252}{G}$ $G =$ ______
14. $\frac{7}{17} = \frac{x}{442}$ $x =$ ______
15. $\frac{3}{Y} = \frac{112}{1008}$ $Y =$ ______
16. $\frac{A}{18} = \frac{4}{6}$ $A =$ ______
17. $\frac{b}{10} = \frac{108}{40}$ $b =$ ______
18. $\frac{Q}{.02} = \frac{2}{.2}$ $Q =$ ______
19. $\frac{3}{d} = \frac{13}{6}$ $d =$ ______
20. $\frac{17}{30} = \frac{K}{21}$ $K =$ ______
21. $\frac{M}{3\frac{1}{2}} = \frac{8}{7}$ $M =$ ______
22. $\frac{6}{7} = \frac{54}{z}$ $z =$ ______
23. $\frac{4}{s} = \frac{10}{25}$ $s =$ ______
24. $\frac{6}{J} = \frac{1}{7}$ $J =$ ______

Date June 8, 1978

Name Stephanie Catanzaro

Course/Section 050/003

B. Solve the following problems using the concepts of ratio and proportion.

1. What number divided by 8 is the same as 5 divided by 3?

____

2. The number π is the ratio of the circumference of a circle to its diameter. If π is about 3.14, what is the circumference of a circle whose diameter is 5 inches?

____

3. If 6 cans of soup cost $1.12, how much would 5 cans cost?

____

4. An automobile travels 140 miles in 3 hours. At this same rate how far would it travel in $2\frac{1}{2}$ hours?

____

5. Two motorists travel the same route. One averages 60 mph and does the trip in $3\frac{1}{2}$ hours. The other averages 45 mph. What is the trip time for the second motorist?

____

6. In a term paper, 16 pages of handwritten material produce 7 pages of typed copy. How many pages of handwritten material is needed to produce a term paper 20 typed pages long?

____

7. Find AB if AC is 30 ft. $AB =$ _______

25 ft
8 ft
A B C

8. In our college 2430 students are male and 2970 are female. How many women students would you expect to find in a class where there are 18 men?

____

9. A car travels 132 miles on 8 gallons of gasoline. How far would you expect it to travel on a full tank of 12 gallons? __________

____

10. In the first 12 games of the season, our star basketball player scored 252 points. How many can he expect to score in the entire 22 game season?

____

11. Grapefruit are 3 for 39¢. How much would four cost?

____

12. A favorite cookie recipe uses 3 cups of flour and produces 2 dozen cookies. How many cups of flour are needed if we want to make 5 dozen cookies?

____

13. If apples sell for 4 pounds for 39¢, what would 10 pounds cost?

____

14. On a certain map a distance of 1 inch represents a distance of 250 miles. What is the actual distance between two towns if they are separated by $3\frac{1}{2}$ inches on the map?

____

15. The fastest worker in the Apex Hickey Co. turns out 347 hickeys per hour. How many hickeys would he make in 45 minutes? In 2 hours 25 minutes?

____

16. Six compared to 20 is the same as what number compared to 30?

____

17. A basketball player makes 7 of 9 free throws attempted in his first game of the season. If he continues shooting at that rate, how many free throws will he make in 198 tries?

____

Date

Name

Course/Section

18. If we build a model of the solar system where one million miles is represented by 1 foot, how big will the sun be on this model? (Its actual diameter is 864,000 miles.) How far from the sun will the planet Jupiter be? (It is actually 484,600,000 miles away.) How far away will the nearest star be? (Actually 13,000,000,000,000 miles.)

______, ______, ______

19. Solve: $\dfrac{3 + x}{x} = \dfrac{3}{4}$. $x =$ ______

20. Solve: $\dfrac{x - 1}{x + 1} = \dfrac{3}{4}$ $x =$ ______

When you have had the practice you need turn to 55 for a self-test covering the work of Unit 6.

Unit 6 Introduction to Algebra

55

Self-Test

Identify each of the following as a term, expression, variable, constant, coefficient, or equation.

1. $3y$ ________
2. The 4 in $4a^2$ ________
3. $8x + y^2$ ________
4. The Q in $2Q + 1$ ________
5. The 2 in $x + 2$ ________
6. $2x - y = 4$ ________

Write as an algebra expression.

7. Six less than a certain number. ________
8. The sum of a number and twice another number. ________
9. Eight times a number. ________
10. Four more than half a number. ________
11. Simplify by adding like terms.
    $3x + 2y - x - 5y + 4x + 1 =$ ________
12. Multiply: $2b(1 - b) =$ ________
13. Multiply: $2x(2x - 3xy) =$ ________
14. Multiply: $(a + b)(a - b) =$ ________
15. Divide $\dfrac{2M^2 - 18M + 4M^2}{2M} =$ ________
16. Multiply and add like terms:
    $2x(1 - 2y) + 3y(x + 2) + 2xy =$ ________
17. Find $5a^2 - b + 1$ for $a = 3$, $b = 2$ ________
18. Find $\dfrac{\pi D^2 H}{4}$ for $\pi = 3.14$, $D = 10$, $H = 8$ ________
19. Find $\dfrac{3ab^2 - 2a^2b}{a + b}$ for $a = 2$, $b = 3$ ________
20. Solve: $3x + 5 = 17$ ________
21. Solve: $5x = 23$ ________
22. Solve: $19 - x = 2x + 4$ ________
23. Solve: 8 more than half of a number is equal to 15. Find the number. ________
24. The ratio of my weight to his weight is $\frac{3}{4}$. I weigh 147 pounds. What does he weigh? ________
25. If $3\frac{1}{2}$ yards of material cost \$4.20, what will 5 yards cost? ________

Answers are on page 516.

Date ________

Name ________

Course/Section ________

SUPPLEMENTARY PROBLEM SET 1A

Algebra Words and Operations

Introduction to Algebra

A. Write out the following in algebra notation:

1. x^2 minus y^2 ____
2. x^2 plus y^2 ____
3. a divided by b ____
4. b divided by a ____
5. 2 times q times r ____
6. a^2 times b^2 ____
7. a^2 divided by b^2 ____
8. $(x - y)$ plus $(a + b)$ ____
9. $(x - y)$ times $(z + y)$ ____
10. $(s + q)$ divided by $(a - b)$ ____
11. a plus $(b - z)$ minus q ____
12. $(k + 1)$ times 4 ____
13. d plus $(e - g)$ ____
14. a divided by $(c - d)$ ____

B. Calculate the following:

1. $2(2R) =$ ____
2. $3a^2 + 2a^2 - a^2 =$ ____
3. $4y^3 + 2y^3 - y^3 + 4 =$ ____
4. $pt^2 + 2pt^2 + 7pt^2 =$ ____
5. $3a + (a + 2) =$ ____
6. $g + 2g - 4 =$ ____
7. $x^2 - 2x^2 + 6 + 7x^2 =$ ____
8. $(x + y) + 7(x - y) =$ ____
9. $(2s)(6t) =$ ____
10. $(x + y)(x + y) =$ ____
11. $(h - 1)(h + 1) =$ ____
12. $(q + 1)(q + 2) =$ ____
13. $2x(10x + 1) =$ ____
14. $(2ab)(a^2b - ab^2) =$ ____
15. $2(z - 1) + 3(2z + 4) =$ ____
16. $3a - (a - 4) =$ ____

Date ____

Name ____

Course/Section ____

SUPPLEMENTARY PROBLEM SET 2A

Translating English to Algebra

Introduction to Algebra

A. Write each of these phrases as an algebraic expression:

1. two more than a ______
2. five less than q ______
3. x multiplied by 8 ______
4. a number increased by 12 ______
5. a number decreased by 5 ______
6. double a number ______
7. the sum of a number and 4 ______
8. the sum of two numbers ______
9. 8 subtracted from k ______
10. Q less 15 ______
11. R divided by 4 ______
12. the product of s and t ______
13. x less 7 ______
14. 4 more than $(x + 1)$ ______
15. d times the sum of m and n ______
16. take away k from m ______
17. the sum of 4 and half of Q ______
18. 2 less than x ______
19. 4 more than triple a number ______
20. the sum of two numbers decreased by 6 ______

______
Date

______
Name

______
Course/Section

B. Write each of these sentences as an algebra equation:

1. The sum of two numbers is equal to 9. ______
2. The difference between x and y is 12. ______
3. The sum of a number and its square is 20. ______
4. Four times a number divided by 3 is equal to 10. ______
5. Laurie's age is four years more than Eric's age. ______
6. The area of a rectangle is the product of its width and its length. ______
7. The sum of three times a number and its square is 70. ______
8. Four times a number decreased by six is 22. ______
9. The sum of two consecutive numbers is equal to 41. ______
10. The product of two consecutive numbers is equal to 90. ______
11. The product of Pat's age and Mary's age is five times Bob's age. ______
12. John's age is three years less than Bill's age. ______

SUPPLEMENTARY PROBLEM SET 3A

Evaluating Literal Expressions

Introduction to Algebra

A. Evaluate each of these expressions: use $x = 1$, $y = 3$, $z = 4$, and $w = -2$

1. $4x =$ ______
2. $x^2 + y =$ ______
3. $xy + z =$ ______
4. $x^2 + y^2 =$ ______
5. $2xy =$ ______
6. $y^2 + x - 4 =$ ______
7. $3y - 2z =$ ______
8. $2y^2 - w =$ ______
9. $z^2 + w^2 =$ ______
10. $xy^2 + yx^2 =$ ______
11. $3y^2 - w =$ ______
12. $2(x + y) =$ ______
13. $3(y - w) + 1 =$ ______
14. $(x + y)^2 =$ ______
15. $(y + z)^2 - x^2 =$ ______
16. $(z - x)^3 - y^2 =$ ______
17. $2(z - 3) =$ ______
18. $3xyz + 2w =$ ______

B. Evaluate each of these expressions: use $a = 3$, $b = 4$, $c = 7$, and $d = 8$.

1. $a + 2(b + c) =$ ______
2. $d - a(c - b) =$ ______
3. $3a - 2b + 1 =$ ______
4. $2(a + b - 5) =$ ______
5. $a^2 + b^2 - 1 =$ ______
6. $2abc - 10d =$ ______
7. $(b + a)(c - b) =$ ______
8. $2(a + b)^2 =$ ______
9. $3(a^2 + b)(b^2 - c) =$ ______
10. $(a + b)^2(c - b)^2 =$ ______
11. $(b - 2(b - a)) =$ ______
12. $((2a - b) + c) - d =$ ______
13. $3a - (b + 1) =$ ______
14. $a^2 + b^2 - c^2 + 3d =$ ______

C. Word Problems

1. The area of a triangle is given by the equation $A = \frac{1}{2}bh$, where b is the length of the base and h is the height. Find the area when $b = 12$ in and $h = 14$ in.

2. The volume of a cylinder can be calculated from the equation $V = \frac{1}{4}\pi D^2 h$, where π is approximately equal to $\frac{22}{7}$; D, the diameter of the

Date ______

Name ______

Course/Section ______

cylinder, is 6 in.; and h, the height of the cylinder, is 21 in. What is the volume of this cylinder?

____

3. The maximum range R of a batted ball is roughly related to the speed v it has when it leaves the bat. $R = \frac{v^2}{g}$ where $g = 32$. If $v = 96$ ft/sec, find R in feet.

SUPPLEMENTARY PROBLEM SET 4A

Solving Linear Equations

Introduction to Algebra

A. Solve:

1. $a - 6 = 10$ $a =$ ______
2. $x + 4 = 3$ $x =$ ______
3. $y - 1 = 6$ $y =$ ______
4. $3y = 12$ $y =$ ______
5. $1 - B = 7$ $B =$ ______
6. $6 - x = 2$ $x =$ ______
7. $5w = 30$ $w =$ ______
8. $7 + Q = 11$ $Q =$ ______
9. $4 + D = 16$ $D =$ ______
10. $42 = 6A$ $A =$ ______

B. Solve:

1. $2x - 1 = 15$ ______
2. $5y + 2 = 6$ ______
3. $1 + 5z = 21$ ______
4. $2 + 3Q = 23$ ______
5. $2x - 1 = x + 2$ ______
6. $3A + 4 = 5A - 8$ ______
7. $21 = 2B + 3$ ______
8. $64 = 3D - 2$ ______
9. $2 - 5T = 17$ ______
10. $7 + 4T = 19$ ______
11. $2x - 6 = 3$ ______
12. $0 = 7x - 56$ ______
13. $1 = 16 - 3x$ ______
14. $4 = 2x + 4(x - 5)$ ______
15. $\frac{1}{2}x - 1 = 4$ ______
16. $2x - 1 = 3x - 6$ ______

C. Word Problems

1. If 3 is added to twice a certain number the sum equals 17. What is the number?

   ____

2. The sum of a certain number and 16 is equal to three times the number. What is the number?

   ____

3. The sum of a number and 8 is 31. What is the number?

   ____

Date ______________

Name ______________

Course/Section ______________

4. Six less than a certain number is 21. Find the number.

____

5. A 45 in. length of string is cut into four pieces of equal length with a one inch piece remaining. What is the length of each of the four pieces?

____

6. The length of a rectangle is five times its width. What are its dimensions if its perimeter is 60 ft?.

____

7. Twice a certain number minus 7 is equal to 15. Find the number.

____

SUPPLEMENTARY PROBLEM SET 5A

Introduction to Algebra

Ratio and Proportion

A. Solve:

1. $\frac{3}{4} = \frac{x}{12}$ ______
2. $\frac{y}{2} = \frac{17}{5}$ ______
3. $\frac{1}{x} = \frac{3}{9}$ ______
4. $\frac{4}{3} = \frac{2}{y}$ ______
5. $\frac{4}{a} = \frac{5}{1}$ ______
6. $\frac{9}{6} = \frac{12}{Q}$ ______
7. $\frac{60}{5} = \frac{2}{R}$ ______
8. $\frac{2}{73} = \frac{x}{365}$ ______
9. $\frac{y}{11} = \frac{1001}{13}$ ______
10. $\frac{123}{41} = \frac{z}{7}$ ______
11. $\frac{144}{\mathrm{E}} = \frac{36}{3}$ ______
12. $\frac{37}{148} = \frac{5}{x}$ ______
13. $\frac{32}{256} = \frac{y}{48}$ ______
14. $\frac{198}{22} = \frac{\mathrm{K}}{3}$ ______
15. $\frac{729}{81} = \frac{9}{x}$ ______
16. $\frac{2}{y} = \frac{27}{324}$ ______
17. $\frac{z}{14} = \frac{31}{217}$ ______
18. $\frac{138}{23} = \frac{18}{\mathrm{R}}$ ______

B. Word Problems:

1. If it takes 3 hours to walk 11 miles, how many miles can be walked in 4 hours at the same rate?

   ____

2. A car uses 11 gallons of gas to travel the first 210 miles of a trip. How many gallons will be needed to travel the remaining 650 miles?

   ____

3. A road map is designed so that $\frac{7}{8}$ inch equals 20 miles. How many miles does 6 in. represent?

   ____

Date ______________

Name ______________

Course/Section ______________

4. The ratio of two numbers is 5/7 and their sum is 60. What are the numbers?

____

5. A 50 pound bag of cement costs $1.49. How much would 175 pounds of cement cost at the same rate?

____

6. In the Fairview Elementary School the ratio of teachers to pupils is kept at $\frac{2}{37}$. If there are 800 pupils enrolled how many teachers should be hired?

____

Answers

HERE I AM AGAIN...STILL LOOKING FOR THE ANSWERS!

UNIT 1, THE ARITHMETIC OF WHOLE NUMBERS

Page 10

A. 8, 16, 6, 13, 11, 11, 13, 9, 13, 12
14, 17, 6, 12, 15, 9, 11, 12, 16, 9
15, 10, 10, 11, 14, 7, 8, 14, 18, 15
14, 14, 12, 12, 16, 12, 13, 13, 13, 15
11, 17, 11, 12, 11, 15, 13, 12, 18, 12

B. 10, 11, 11, 12, 16, 9, 9, 12, 12, 13
10, 13, 15, 8, 14, 9, 13, 18, 15, 11
11, 16, 14, 10, 17, 9, 14, 12, 7, 10
13, 13, 12, 16, 9, 11, 10, 10, 14, 17
11, 14, 11, 10, 11, 14, 16, 12, 13, 15

C. 11, 12, 14, 17, 18, 21, 11, 20, 18, 15
15, 14, 15, 18, 22, 17, 14, 14, 18, 16
12, 19, 8, 10, 22, 12, 19, 15, 14, 16

Box, page 17

91, 81, 46, 92, 74, 130

Problem Set 1, Page 21

A. 1. 70 2. 104 3. 65 4. 126 5. 80 6. 106
7. 112 8. 72 9. 131 10. 103 11. 105 12. 123
13. 103 14. 124 15. 100 16. 132 17. 52 18. 136

B. 1. 415 2. 393 3. 1113 4. 1003 5. 1530
6. 1390 7. 1016 8. 831 9. 1262 10. 1009
11. 824 12. 806 13. 1241 14. 861 15. 1001

C. 1. 5525 2. 9563 3. 9461 4. 2611
5. 9302 6. 3513 7. 3702 8. 12,599
9. 7365 10. 10,122 11. 6505 12. 11,428
13. 5781 14. 15,715 15. 9403 16. 11,850

D. 1. 25,717 2. 11,071 3. 70,251 4. 21,642
5. 14,711 6. 89,211 7. 47,111 8. 175,728
9. 101,011 10. 180,197 11. 110,102 12. 128,786

E. 1. 1042 2. 5211 3. 2442 4. 6441
5. 7083 6. 16,275 7. 6352 8. 7655
9. 6514 10. 9851

F. 1. 882 lb 2. 371 calories 3. the sums are the same
4. 662,289 miles
5. $10,535 6. They are the same: 1,083,676,269
7. $3 = 3, 15 = 15, 42 = 42$
$16 + 17 + 18 + 19 + 20 = 21 + 22 + 23 + 24$
8. 8454 points 9. 5966 10. 692 11. 646 12. 443

Problem Set 2, Page 31

A. 1. 6 2. 5 3. 7 4. 6 5. 2 6. 5
7. 8 8. 0 9. 4 10. 1 11. 9 12. 8
13. 3 14. 7 15. 0 16. 9 17. 3 18. 7
19. 3 20. 4 21. 8 22. 1 23. 8 24. 4
25. 9 26. 7 27. 9 28. 7 29. 9 30. 4
31. 9 32. 6 33. 3 34. 1 35. 6 36. 6
37. 8 38. 5 39. 5 40. 8 41. 7 42. 7
43. 18 44. 7 45. 7 46. 3 47. 8 48. 9
49. 0 50. 2

B. 1. 13 2. 44 3. 29 4. 16 5. 12
6. 57 7. 19 8. 17 9. 15 10. 28
11. 36 12. 18 13. 22 14. 37 15. 25
16. 26 17. 38 18. 85

C. 1. 189 2. 458 3. 85 4. 877 5. 281
6. 176 7. 154 8. 266 9. 273 10. 198
11. 715 12. 51 13. 574 14. 45 15. 29
16. 145

D. 1. 2809 2. 7781 3. 5698 4. 28,842
5. 12,518 6. 7679 7. 56,042 8. 37,328
9. 4741 10. 9897 11. 9614 12. 26,807
13. 47,593 14. 316,640 15. 22,422 16. 55,459
17. 24,939 18. 7238

E. 1. 1819 2. 284 3. 13,819 4. $7289
5. $155 6. 3303 7. 3 8. $595

9. $98 - 76 + 54 + 3 + 21$
$123 + 45 - 67 + 8 - 9$
$12 + 3 + 4 + 5 - 6 - 7 + 89$
$123 + 4 - 5 + 67 - 89$

10. Sure. 1963 pennies are worth $19.63, in fact.

11. 991 12. 12,318 miles 13. 5394 ft
14. $103.97 15. 8692 sq miles

Page 38

A. 12, 32, 63, 36, 12, 18, 0, 24, 14, 8
48, 16, 45, 30, 10, 9, 72, 35, 18, 0
28, 15, 36, 49, 8, 40, 42, 54, 64, 24
20, 0, 25, 27, 81, 6, 1, 48, 16, 63

B. 16, 30, 9, 35, 18, 20, 28, 48, 12, 63
32, 0, 18, 24, 9, 25, 24, 45, 10, 72
15, 49, 40, 54, 36, 8, 42, 64, 0, 4
25, 27, 7, 56, 36, 12, 81, 0, 2, 56

Problem Set 3, Page 47

A. 1. 42 2. 72 3. 56 4. 63 5. 48
6. 54 7. 72 8. 42 9. 54 10. 63
11. 56 12. 48

B. 1. 87 2. 402 3. 576 4. 243 5. 282
6. 792 7. 320 8. 259 9. 156 10. 294
11. 290 12. 261 13. 564 14. 392 15. 153
16. 161 17. 282 18. 424 19. 308 20. 324
21. 720 22. 1505 23. 1728 24. 2736 25. 5040
26. 2952 27. 7138 28. 1170 29. 1938 30. 2548
31. 1650 32. 1349 33. 4484 34. 1458 35. 928
36. 6232 37. 3822 38. 2030 39. 8930 40. 2752

C. 1. 37,515 2. 74,820 3. 375,750
4. 97,643 5. 297,591 6. 384,030
7. 38,023 8. 108,486 9. 378,012
10. 1,279,840 11. 41,064 12. 4,947,973
13. 30,780 14. 225,852 15. 1,368,810
16. 31,152 17. 397,584 18. 43,381
19. 60,241 20. 5,098,335

D. 1. $338 2. 8760 3. 1196 4. $205
5. 378 6. $130,640 7. 987,654,321
8. (a) 3,999,996 (b) 333 (c) 333,333,333 (d) 333,333
9. (a) 1156 111,556 (b) 4356 443,556
(c) 12,321 1,234,321 (d) 5929 603,729
(e) 3025 308,025 (f) 4422 444,222
10. 111 1111 11,111 111,111 1,111,111 11,111,111
11. 123,458,769 123,547,689 967,854,321 Each has all nine nonzero digits.
12. 986 trees 13. $290.50
14. 5,588,460 miles 15. $9.48

Box, page 53

1. $2\frac{5}{6}$ hours 2. 84.5 points 3. $663.45

Problem Set 4, Page 61

A.
1. 9
2. 12
3. 11, remainder 4
4. 7
5. Not defined
6. 13
7. 7, remainder 2
8. 5
9. 10, remainder 1
10. 7
11. 1
12. Not defined
13. 8
14. 6
15. 4
16. 7
17. 6
18. 9

B.
1. 35
2. 41
3. 23, remainder 6
4. 42
5. 57
6. 46, remainder 2
7. 51, remainder 4
8. 45
9. 112, remainder 1
10. 44
11. 21
12. 27
13. 52
14. 37
15. 88
16. 125
17. 50, remainder 1
18. 67

C.
1. 23
2. 21
3. 20, remainder 2
4. 31, remainder 4
5. 39
6. 50, remainder 2
7. 25
8. 19, remainder 17
9. 9, remainder 1
10. 41
11. 53
12. 43
13. 22
14. 11, remainder 34
15. 12
16. 34
17. 9, remainder 6
18. 71, remainder 5

D.
1. 95, remainder 6
2. 104
3. 96
4. 208
5. 142, remainder 6
6. 107
7. 222, remainder 2
8. 170, remainder 10
9. 32
10. 1000
11. 305, remainder 5
12. 311, remainder 8
13. 84, remainder 41
14. 100, remainder 5
15. 119
16. 61
17. 102, remainder 98
18. 81

E.
1. 7 mph, remainder of 5
2. \$507
3. 207 pounds
4. 38, remainder 2
5. \$51
6. 25 minutes
7. 247 years, 309 days
8. 344, remainder 8
10. (a) 900,991 (b) 111,111,111 (c) 8, 68, 668
11. \$265
12. 1538

Problem Set 5, Page 79

A.
1. $2 \times 2 \times 3$
2. $2 \times 2 \times 2 \times 2$
3. 2×7
4. $2 \times 3 \times 3$
5. $2 \times 2 \times 2 \times 3$
6. $2 \times 2 \times 5$
7. 2×13
8. Prime
9. $2 \times 2 \times 2 \times 2 \times 2$
10. $2 \times 2 \times 3 \times 3$
11. 3×13
12. $2 \times 3 \times 7$
13. $2 \times 2 \times 2 \times 7$
14. $3 \times 3 \times 3 \times 3$
15. 3×7

B.
1. $2 \times 2 \times 2 \times 2 \times 2 \times 3$
2. $2 \times 2 \times 3 \times 7$
3. $2 \times 2 \times 2 \times 17$
4. $2 \times 5 \times 17$
5. $2 \times 2 \times 3 \times 3 \times 7$
6. $2 \times 2 \times 2 \times 2 \times 2 \times 2 \times 2 \times 2$
7. $2 \times 2 \times 2 \times 2 \times 2 \times 3 \times 3$
8. $2 \times 3 \times 5 \times 13$
9. $2 \times 2 \times 3 \times 3 \times 13$
10. $2 \times 3 \times 7 \times 13$
11. $2 \times 2 \times 5 \times 7 \times 7$
12. 37×37
13. 29×47
14. $2 \times 2 \times 3 \times 3 \times 43$
15. 47×67

C. Primes: 2, 5, 3, 31, 23, 37, 53, 19, 67, 61, 89, 17

D. Divisible by 2: 12, 4, 144, 1044, 1390, 72, 102, 2808, 2088, 8280, 8802
Divisible by 3: 9, 12, 231, 45, 144, 261, 1044, 72, 81, 102, 2808, 2088, 8280, 8802, 111
Divisible by 5: 45, 1390, 8280

E. 1. $28 = 1 + 2 + 4 + 7 + 14$
$496 = 1 + 2 + 4 + 8 + 16 + 31 + 62 + 124 + 248$
$8128 = 1 + 2 + 4 + 8 + 16 + 32 + 64 + 127 + 254 + 508 + 1016 + 2032 + 4064$

2. (a) The divisors of 220 are 1, 2, 4, 5, 10, 11, 20, 22, 44, 55, and 110. These divisors add up to 284. The divisors of 284 are 1, 2, 4, 71, and 142. These divisors add up to 220.
(b) $2924 = 1 + 2 + 4 + 5 + 10 + 20 + 131 + 262 + 524 + 655 + 1310$
$2620 = 1 + 2 + 4 + 17 + 34 + 43 + 68 + 86 + 172 + 731 + 1462$

3. 37; all numbers are primes.

4. 2, 3, 6

5. 1, not prime; 11, prime; $111 = 3 \times 37$; $1111 = 11 \times 101$; $11{,}111 = 41 \times 271$; $111{,}111 = 3 \times 7 \times 11 \times 13 \times 37$

6. (a) 36 (b) 27

7.

$50 = 3 + 47$	$51 = 3 + 7 + 41$	$52 = 5 + 47$
$53 = 5 + 7 + 41$	$54 = 7 + 47$	$55 = 3 + 5 + 47$
$56 = 3 + 53$	$57 = 3 + 7 + 47$	$58 = 5 + 53$
$59 = 5 + 7 + 47$	$60 = 7 + 53$	$61 = 3 + 5 + 53$
$62 = 3 + 59$	$63 = 5 + 11 + 47$	$64 = 5 + 59$
$65 = 3 + 5 + 57$	$66 = 7 + 59$	$67 = 3 + 7 + 57$
$68 = 7 + 61$	$69 = 5 + 7 + 57$	$70 = 3 + 67$

8.

$$\begin{array}{r} 638 \\ 475 \\ +253 \\ \hline 1366 \end{array} \qquad \begin{array}{r} 4752 \\ 3658 \\ 4975 \\ +2403 \\ \hline 15788 \end{array} \qquad \begin{array}{r} 314 \\ 926 \\ +705 \\ \hline 1945 \end{array}$$

Problem Set 6, Page 89

A.

1. 16	2. 9	3. 64	4. 125	5. 1000
6. 49	7. 256	8. 36	9. 512	10. 81
11. 625	12. 100,000	13. 8	14. 243	15. 729
16. 1	17. 6	18. 1	19. 256	20. 32

21. 1,000,000	22. 343	23. 64	24. 1296
25. 1024	26. 6561	27. 216	28. 25
29. 27	30. 2401	31. 1	32. 10,000
33. 16	34. 5	35. 4096	36. 1
37. 81	38. 1024	39. 100	40. 7776

B.

1. 196	2. 441	3. 225	4. 15,625
5. 256	6. 3025	7. 3721	8. 64,000
9. 1,000,000	10. 108	11. 576	12. 1125
13. 7938	14. 4851	15. 2025	16. 2744
17. 1296	18. 24,300	19. 2000	20. 90,000
21. 9216			

C. 1. 9 2. 12 3. 4 4. 5 5. 6
6. 10 7. 7 8. 18 9. 1 10. 11
11. 8 12. 3 13. 15 14. 2 15. 20
16. 16

D. 3. (a) $21^2 + 22^2 + 23^2 + 24^2 = 25^2 + 26^2 + 27^2$
$36^2 + 37^2 + 38^2 + 39^2 + 40^2 = 41^2 + 42^2 + 43^2 + 44^2$
(b) $7^2 - 3^2 = 49 - 9 = 40$
$6^2 - 4^2 = 36 - 16 = 20$
(c) $1 + 2 + 3 + 4 + 3 + 2 + 1 = 4^2$
$1 + 2 + 3 + 4 + 5 + 4 + 3 + 2 + 1 = 5^2$
(d) $7 \times 12 = 9^2 + 3$
$8 \times 13 = 10^2 + 4$
(e) $1111^2 = 1{,}234{,}321$
$11{,}111^2 = 123{,}454{,}321$
(f) $1{,}010{,}101^2 = 1{,}020{,}304{,}030{,}201$
$101{,}010{,}101^2 = 10{,}203{,}040{,}504{,}030{,}201$
4. (a) $5^1 = 5$ $5^2 = 25$ $5^3 = 125$ $5^4 = 625$
(b) $25^1 = 25$ $25^2 = 625$ $25^3 = 15{,}625$ $25^4 = 390{,}625$
(c) $625^1 = 625$ $625^2 = 390{,}625$ $625^3 = 244{,}140{,}625$
$625^4 = 152{,}587{,}890{,}625$
(d) $376^1 = 376$ $376^2 = 141{,}376$ $376^3 = 53{,}157{,}376$
$376^4 = 19{,}987{,}173{,}376$

UNIT 2, FRACTIONS

Problem Set 1, Page 117

A. 1. $\frac{7}{3}$, 2. $\frac{22}{5}$, 3. $\frac{15}{2}$, 4. $\frac{94}{7}$, 5. $\frac{35}{4}$
6. $\frac{4}{1}$, 7. $\frac{5}{3}$, 8. $\frac{35}{6}$, 9. $\frac{31}{8}$, 10. $\frac{13}{5}$
11. $\frac{161}{10}$, 12. $\frac{635}{9}$, 13. $\frac{481}{40}$, 14. $\frac{170}{11}$, 15. $\frac{113}{3}$

B. 1. $8\frac{1}{2}$, 2. $7\frac{2}{3}$, 3. $1\frac{3}{5}$, 4. $4\frac{3}{4}$, 5. $6\frac{1}{6}$,
6. $9\frac{1}{3}$, 7. $4\frac{5}{8}$, 8. $4\frac{1}{7}$ 9. $1\frac{9}{25}$, 10. $5\frac{2}{9}$,
11. $52\frac{3}{4}$, 12. $7\frac{9}{23}$, 13. $4\frac{3}{10}$, 14. $20\frac{5}{6}$, 15. $9\frac{4}{15}$

C. 1. $\frac{13}{15}$, 2. $\frac{4}{5}$, 3. $\frac{4}{5}$, 4. $\frac{1}{2}$, 5. $\frac{1}{8}$
6. $\frac{2}{5}$, 7. $\frac{1}{6}$, 8. $\frac{8}{9}$, 9. $\frac{1}{3}$, 10. $\frac{3}{8}$
11. $\frac{7}{20}$, 12. $\frac{3}{8}$, 13. $\frac{1}{6}$, 14. $\frac{4}{7}$, 15. $\frac{5}{9}$

D. 1. $\frac{14}{16}$, 2. $\frac{27}{45}$, 3. $\frac{9}{12}$, 4. $\frac{145}{60}$, 5. $\frac{7}{63}$
6. $\frac{45}{35}$, 7. $\frac{20}{32}$, 8. $\frac{140}{25}$, 9. $\frac{39}{78}$, 10. $\frac{34}{51}$
11. $\frac{363}{44}$, 12. $\frac{82}{14}$, 13. $\frac{66}{72}$, 14. $\frac{185}{50}$, 15. $\frac{516}{54}$

E. 1. 5 laps where each lap is $\frac{1}{8}$ of a mile.
2. 3, $\frac{355}{113}$, $3\frac{1}{7}$, $\frac{256}{81}$
4. Sugar glops at $\frac{7}{10}$ of a pound
5. Too little
6. 100
7. $\frac{25}{10}$
8. $7\frac{15}{30}$, $\frac{20}{30}$, $1\frac{24}{30}$
9. (a) $\frac{7}{30}$ (b) $\frac{1}{16}$ (c) 2 (d) $\frac{35}{44}$
10. $\frac{2}{3}$

Problem Set 2, Page 125

A. 1. $\frac{1}{8}$, 2. $\frac{1}{9}$, 3. $\frac{4}{15}$, 4. $\frac{1}{8}$
5. $\frac{2}{15}$, 6. $\frac{5}{6}$, 7. 3, 8. $\frac{1}{2}$
9. $2\frac{2}{3}$, 10. $\frac{5}{6}$, 11. $\frac{11}{45}$, 12. $\frac{9}{56}$
13. $1\frac{1}{9}$, 14. $10\frac{1}{2}$, 15. $\frac{13}{16}$, 16. $\frac{4}{5}$
17. $2\frac{1}{2}$, 18. 6, 19. 14, 20. $1\frac{3}{7}$

B. 1. 3, 2. 4, 3. 8, 4. $\frac{21}{16}$
5. $3\frac{1}{4}$, 6. 62, 7. $1\frac{1}{21}$, 8. $\frac{1}{3}$
9. 69, 10. 6, 11. $35\frac{3}{4}$, 12. $1\frac{3}{11}$
13. 74, 14. $7\frac{9}{10}$, 15. $9\frac{7}{8}$, 16. $46\frac{2}{3}$
17. $10\frac{3}{8}$, 18. $13\frac{13}{30}$, 19. $21\frac{1}{3}$, 20. 6

C. 1. $\frac{4}{9}$, 2. $\frac{1}{16}$, 3. $\frac{27}{125}$, 4. $10\frac{6}{25}$, 5. $91\frac{1}{8}$
6. $\frac{3}{4}$, 7. $\frac{2}{7}$, 8. $\frac{4}{5}$, 9. $\frac{5}{8}$, 10. $\frac{9}{11}$
11. $\frac{1}{15}$, 12. $\frac{1}{6}$, 13. $1\frac{1}{3}$ 14. 20 15. $31\frac{1}{2}$

D. 1. 1530 miles 2. Bert ate $\frac{1}{2}$ of a pie 3. 110 km
4. 54¢ 6. 69 years old 7. $\frac{7}{8}$ sq miles
8. $16\frac{1}{2}$ mg 9. \$1.09 10. \$1.29

Problem Set 3, Page 137

A. 1. $1\frac{2}{3}$, 2. $1\frac{3}{4}$, 3. 9, 4. $\frac{1}{12}$
5. $\frac{5}{16}$, 6. $\frac{4}{9}$, 7. 24, 8. $\frac{7}{16}$
9. $\frac{8}{13}$, 10. $7\frac{1}{2}$, 11. 1, 12. $1\frac{1}{2}$
13. $\frac{5}{28}$, 14. $1\frac{4}{5}$, 15. $2\frac{2}{5}$, 16. $\frac{1}{12}$

B. 1. 9, 2. $7\frac{1}{3}$, 3. 4, 4. $\frac{3}{4}$
5. $1\frac{1}{3}$, 6. 6, 7. $1\frac{3}{4}$, 8. $2\frac{4}{7}$
9. $1\frac{1}{5}$, 10. $8\frac{1}{3}$, 11. $1\frac{1}{4}$, 12. $1\frac{2}{3}$
13. $\frac{6}{7}$, 14. $5\frac{1}{4}$, 15. $\frac{4}{5}$, 16. $5\frac{1}{3}$

C. 1. 16, 2. $\frac{3}{8}$, 3. $\frac{1}{9}$
4. $3\frac{1}{9}$, 5. 18, 6. 25
7. $\frac{6}{7}$, 8. 6, 9. $1\frac{1}{4}$
10. $\frac{6}{29}$, 11. $17\frac{1}{2}$ 12. 17

D. 1. $10\frac{8}{13}$ 2. $4\frac{1}{2}$ 3. 34 mph 4. $1\frac{3}{8}$
5. First customer—16 eggs ($\frac{1}{2}$ of 31 is $15\frac{1}{2}$ and $15\frac{1}{2} + \frac{1}{2}$ is 16)
Second customer—8 eggs
Third customer—4 eggs
The grocer didn't need to crack any eggs in half.
6. 6 o'clock 7. 26 8. 57 9. 26
10. (a) $1\frac{1}{4}$ (b) 5 (c) $\frac{4}{5}$ (d) $\frac{1}{5}$

Problem Set 4, Page 153

A. 1. $1\frac{2}{5}$, 2. $1\frac{1}{3}$, 3. 1, 4. $\frac{2}{3}$
5. $\frac{1}{4}$, 6. $\frac{1}{2}$, 7. $\frac{5}{12}$, 8. $\frac{3}{8}$
9. $1\frac{1}{8}$, 10. $1\frac{1}{8}$, 11. $\frac{3}{4}$, 12. $\frac{1}{4}$
13. $1\frac{3}{8}$, 14. $1\frac{2}{3}$, 15. $1\frac{4}{7}$
16. $\frac{5}{8}$, 17. $1\frac{1}{4}$, 18. $\frac{9}{20}$

B. 1. $1\frac{5}{8}$, 2. $\frac{1}{8}$, 3. $1\frac{11}{36}$, 4. $\frac{19}{36}$
5. $\frac{29}{48}$, 6. $\frac{29}{35}$, 7. $\frac{53}{192}$, 8. $\frac{215}{216}$
9. $\frac{13}{48}$, 10. $\frac{4}{35}$, 11. $\frac{5}{96}$, 12. $\frac{83}{216}$
13. $1\frac{5}{8}$, 14. $4\frac{1}{36}$, 15. $5\frac{17}{48}$, 16. $4\frac{51}{56}$
17. $\frac{3}{4}$, 18. $\frac{3}{8}$, 19. $1\frac{23}{48}$, 20. $\frac{51}{70}$

C. 1. $1\frac{2}{3}$, 2. $1\frac{13}{16}$, 3. $5\frac{1}{4}$, 4. $1\frac{4}{5}$
5. $20\frac{3}{4}$ 6. $1\frac{89}{120}$ 7. $\frac{33}{40}$ 8. $1\frac{17}{60}$
9. $3\frac{1}{12}$ 10. $2\frac{13}{40}$ 11. $5\frac{11}{48}$ 12. $1\frac{9}{10}$

D. 1. $\frac{15}{56}$ 2. $36\frac{2}{3}$ hours; \$91.67
4. $\frac{7}{8} = \frac{1}{2} + \frac{1}{4} + \frac{1}{8}$ 5. $414\frac{7}{8}$ ft
$\frac{5}{9} = \frac{1}{3} + \frac{1}{6} + \frac{1}{18}$
$\frac{5}{12} = \frac{1}{3} + \frac{1}{12}$
6. $28\frac{2}{5}$; 65 miles per gallon 7. $2\frac{3}{16}$
8. $\frac{1}{4}$ minus $\frac{1}{4}$ of $\frac{1}{4}$ is larger by $\frac{7}{64}$
9. (a) $\frac{5}{6} = \frac{5}{6}$; $\frac{47}{60} = \frac{47}{60}$; $\frac{1}{4} + \frac{1}{5} + \frac{1}{6} + \frac{1}{7}$
$= 1 - \frac{1}{2} + \frac{1}{3} - \frac{1}{4} + \frac{1}{5} - \frac{1}{6} + \frac{1}{7} = \frac{319}{420}$

(b) $$\frac{1}{1 \times 3} + \frac{1}{3 \times 5} + \frac{1}{5 \times 7} + \frac{1}{7 \times 9} = \frac{4}{9}$$

(c) $$\frac{1}{1 \times 2} + \frac{1}{2 \times 3} + \frac{1}{3 \times 4} + \frac{1}{4 \times 5} = \frac{4}{5}$$

10. $\frac{5}{12}$ lb
11. $12\frac{3}{8}$ yd
12. $8\frac{5}{8}$ lb

Problem Set 5, Page 167

A. (a) $=$ (b) $\times$ (c) $+$ (d) $=$
(e) $\square - 6$ (f) $\frac{1}{2} \times \square$ (g) $2 \times \square$ (h) $\square \div 2\frac{1}{2}$
(i) $\square + \frac{2}{3}$ (j) $\square + \frac{2}{5}$ (k) $\square \div \frac{3}{7}$ (l) $\frac{7}{8} \times \square = 1\frac{1}{2}$
(m) $\square - 1\frac{3}{4}$ (n) $\square \times 3\frac{1}{4} = 11\frac{1}{2}$ (o) $\frac{7}{16} \times 3\frac{5}{8}$

B. (a) $1\frac{1}{2}$ (b) $\frac{3}{10}$ (c) $1\frac{3}{5}$ (d) $\frac{16}{21}$
(e) $1\frac{3}{8}$ (f) $1\frac{1}{4}$ (g) $14\frac{2}{5}$

C. (a) 52¢ (b) $8\frac{1}{8}$ miles (c) $5\frac{1}{3}$ gallons
(d) $4\frac{1}{8}$ (e) \$186 (f) \$1.68
(g) $\frac{7}{24}$ (h) $23\frac{1}{7}$ lb

UNIT 3, DECIMALS

Problem Set 1, Page 191

A.

1.	0.8	2.	1.6	3.	1.5	4.	0.6	5.	0.9
6.	1.2	7.	1.6	8.	0.7	9.	1.7	10.	0.1
11.	0.7	12.	3.3	13.	0.5	14.	2.3	15.	1.8
16.	1.4	17.	3.3	18.	5.2	19.	1.9	20.	9.9
21.	4.8	22.	1.4	23.	9.0	24.	18.1	25.	5.5
26.	1.4	27.	17.5	28.	1.6	29.	6.7	30.	0.6
31.	3.6	32.	1.3	33.	1.8	34.	2.6	35.	2.6
36.	0.4								

B.

1.	21.01	2.	$15.02	3.	1.617
4.	27.19	5.	$30.60	6.	6.486
7.	78.17	8.	$151.11	9.	5.916
10.	828.60	11.	16.2019	12.	1031.28
13.	63.7313	14.	238.24	15.	128.3685
16.	45.195	17.	$27.59	18.	70.871
19.	108.37	20.	19.37	21.	$15.36
22.	51.34	23.	1.04	24.	3.86
25.	42.33	26.	6.63	27.	6.52
28.	6.42	29.	$36.18	30.	22.016
31.	2.897	32.	$17.65	33.	6.96
34.	0.3759				

C.

1.	151.461	2.	602.654	3.	95.888
4.	91.15	5.	316.765	6.	14.67755
7.	16.0425	8.	35.4933	9.	19.011
10.	3.3499				

D.

1.	$1.84	2.	4687.8	3.	$13,212
4.	2.556 yards	5.	2.267 inches	6.	0.163826 seconds
7.	$415.35	8.	968.749 meters	9.	$32.14
10.	$12.07				

11. (a) 0.045 (b) 0.81 (c) 3.888 (d) 0.0006

12. $123.95

Box, page 208

1.	40	2.	6400	3.	160,000
4.	3500	5.	12.6	6.	423
7.	4	8.	0.4	9.	0.75
10.	75	11.	125.7	12.	20,000
13.	4.5	14.	3.76	15.	8.21
16.	0.821	17.	0.00821	18.	0.004
19.	0.024	20.	0.06	21.	0.0035

Problem Set 2, Page 209

A.	1.	0.00001	2.	0.1	3.	21.5				
	4.	0.06	5.	0.008	6.	0.014				
	7.	0.09	8.	0.72	9.	0.84				
	10.	0.009	11.	0.00006	12.	4.20				
	13.	1.4743	14.	2.18225	15.	0.5022				
	16.	0.03	17.	0.024	18.	2.16				
	19.	0.01476	20.	3.6225	21.	1.44				
	22.	80.35	23.	0.00117	24.	287.5				
	25.	0.03265	26.	0.0009255	27.	1223.6				
B.	1.	1300	2.	12.6	3.	0.045				
	4.	13	5.	126	6.	450				
	7.	60	8.	2000	9.	10,000				
	10.	0.037	11.	11.2	12.	6.6				
	13.	1900	14.	3256.25	15.	0.11				
C.	1.	3.33	2.	1.43	3.	0.83	4.	0.09	5.	10.53
	6.	0.67	7.	37.50	8.	0.89	9.	0.12	10.	73.40
	11.	2.62	12.	0.12	13.	33.86	14.	474.44	15.	5.00
	1.	33.3	2.	1.1	3.	0.2	4.	6.7	5.	0.2
	6.	0.1	7.	11.1	8.	285.7	9.	0.1	10.	16.0
	1.	14.286	2.	0.023	3.	0.225	4.	65.000	5.	3.462
	6.	13.640	7.	3.815	8.	2.999	9.	571.429	10.	1109.001

D. 1. 0.09 2. 0.0009 3. 0.000009 4. 0.027
5. 1.44 6. 1.728 7. 0.01 8. 0.0001
9. $\frac{1}{81} = 0.012$ (rounded) 10. $\frac{1}{7} = 0.143$ (rounded)
11. 0.223 12. 0.019
13. 0.215 14. 27777.778

E. 1. $84.15 2. 8.00000007
3. 0.012345679, 0.00112233445566, 0.000111222333444
4. $5.40 5. $126.00 6. $26.35
7. 2872.148
8. 1st: (b) $12 \div 0.03 = 400$
2nd: (e) $0.2 \times 0.2 = 0.04$
3rd: The third error is that there are only two errors.
9. Dec. 31, her birthday
10. Yes. It is true for any digit.
11. (a) $2.37 (b) $6.24 (c) $3.04 (d) $7.40
12. 39.1 gallons
13. $148.50
14. 295 miles per hour
15. (a) 145.6 (b) 1456 (c) 14,560 (d) 0.1456
(e) 1.456 (f) 1.456 (g) 0.1456 (h) 0.01456
(i) 145.6 (j) 1456

Problem Set 3, Page 221

A. 1. .50   2. .33   3. .67   4. .25   5. .50   6. .75
7. .20   8. .40   9. .60   10. .80   11. .17   12. .83
13. .14   14. .28   15. .43   16. .13   17. .38   18. .63
19. .88   20. .10   21. .20   22. .30   23. .08   24. .17
25. .25   26. .42   27. .58   28. .92   29. .06   30. .19
31. .31   32. .44   33. .56   34. .69   35. .81   36. .94

B. 1. $\frac{3}{10}$ 2. $\frac{3}{4}$ 3. $\frac{11}{25}$ 4. $\frac{4}{5}$ 5. $\frac{3}{5}$ 6. $\frac{1}{40}$
7. $\frac{2}{5}$ 8. $1\frac{3}{10}$ 9. $2\frac{1}{4}$ 10. $2\frac{1}{20}$ 11. $3\frac{4}{25}$ 12. $1\frac{1}{8}$
13. $3\frac{11}{50}$ 14. $2\frac{1}{25}$ 15. $\frac{3}{40}$ 16. $10\frac{7}{8}$ 17. $\frac{7}{10000}$ 18. $\frac{3}{2500}$
19. $\frac{17}{50}$ 20. $11\frac{21}{2000}$ 21. $6\frac{1}{500}$ 22. $4\frac{23}{200}$ 23. $\frac{7}{20}$ 24. $\frac{191}{200}$

C. 1. 1.2   2. 408   3. 4.5   4. 17   5. 0.5

D. 1. (a) 4.385   (b) 1.77   (c) 1.475   (d) 0.7681
(e) 7.875   (f) 2.296   (g) 1.434   (h) 0.3125
2. 73.575, round to 73.6
3. The same 6 digits repeat:
$\frac{2}{7} = 0.285714$
$\frac{3}{7} = 0.428571$
$\frac{4}{7} = 0.571428$
$\frac{5}{7} = 0.714285$
$\frac{6}{7} = 0.857142$
4. \$.43   5. \$7.88   6. 1.375 grams, $5\frac{1}{5}$ tablets
7. 0.279   8. \$3.84   9. \$5.92   10. \$18.07

Problem Set 4, Page 233

A. 1. -2 2. -8 3. -4 4. 4   5. -19
6. 4   7. -20 8. 10   9. -16 10. 9
11. -24 12. -7 13. -39 14. 94   15. -9
16. -2.1 17. 6.4   18. 2.6   19. 1.44   20. 2.1
21. -10.2 22. .19   23. -2.1 24. -1.5 25. -9.3

B. 1. -21 2. -24 3. 45   4. -2 5. -6
6. 1   7. -56 8. -18 9. -1 10. 22
11. -1.55 12. .8   13. -8.75 14. $-.1$ 15. 3.6
16. 4   17. -27 18. 2.25   19. -216 20. 16
21. -6 22. -6 23. 2   24. -4 25. 2
26. 1   27. -1 28. $-.7$ 29. $-.6$ 30. 90

C. 1. 14,698 feet   2. 263° F   3. 4
4. a. $-\frac{1}{3}$ b. -9 c. 30 d. $-3\frac{1}{2}$

Problem Set 5, Page 245

A.

1. 1.41	2. 1.73	3. 2.24	4. 2.45	5. 3.16
6. 5.48	7. 12.25	8. 8.66	9. 4.47	10. 12.65
11. 3.87	12. 10.25			

B.

1. 8.25	2. 2.74	3. 1.32	4. 9.41	5. 8.45
6. 18.68	7. 35.28	8. 44.72	9. 21.21	10. 24.49
11. 1.67	12. 4.53			

C. 1. 2.35

2. a. $4 + 4 + \sqrt{4} = 10$ b. $4 + \sqrt{4} - 4 = 2$
c. $(4 + 4) - \sqrt{4} = 6$ d. $(4 \times 4) \div \sqrt{4} = 8$
e. $(4 \times 4) - 4 = 12$ f. $(\sqrt{4} \times \sqrt{4}) \times 4 = 16$

3. 11.66 4. 35 in., 25 in. 5. 3.46, 1.86, 1.36, 1

UNIT 4, PERCENT

Problem Set 1, Page 269

A.

1. 40%	2. 10%	3. 95%	4. 3%
5. 30%	6. 1.5%	7. 60%	8. 1%
9. 120%	10. 456%	11. 225%	12. 775%
13. 0.3%	14. 300%	15. 80%	16. 550%
17. 400%	18. 604%	19. 1000%	20. 33.5%

B.

1. 20%	2. 75%	3. 70%	4. 35%	5. 150%
6. 25%	7. 10%	8. 50%	9. $37\frac{1}{2}$%	10. 60%
11. 175%	12. 220%	13. 180%	14. 90%	15. $33\frac{1}{3}$%
16. $216\frac{2}{3}$%	17. $66\frac{2}{3}$%	18. $68\frac{3}{4}$%	19. $191\frac{2}{3}$%	20. 330%

C.

1. 0.07	2. 0.03	3. 0.56	4. 0.15	5. 0.01
6. 0.075	7. 0.90	8. 2.00	9. 0.003	10. 0.0007
11. 0.0025	12. 1.50	13. 0.015	14. 0.063	15. 0.005
16. 0.1225	17. 1.255	18. 0.667	19. 0.305	20. 0.085

D.

1. $\frac{1}{10}$	2. $\frac{13}{20}$	3. $\frac{1}{2}$	4. $\frac{1}{5}$	5. $\frac{1}{4}$
6. $\frac{2}{25}$	7. $\frac{9}{10}$	8. $1\frac{7}{20}$	9. $\frac{3}{100}$	10. $\frac{3}{25}$
11. $\frac{1}{200}$	12. $\frac{3}{10000}$	13. $\frac{9}{200}$	14. $2\frac{1}{5}$	15. $\frac{3}{200}$
16. $\frac{1}{3}$	17. $\frac{31}{400}$	18. $\frac{13}{200}$	19. $\frac{1}{6}$	20. $\frac{1}{32}$

Problem Set 2, Page 285

A.

1. 80%	2. 20%
3. $9	4. 64%
5. 15	6. 107
7. 100%	8. 54
9. 21	10. 60
11. $12\frac{1}{2}$%	12. 150
13. 59	14. 80
15. $66\frac{2}{3}$%	16. 500%
17. 100	18. 52%
19. 500	20. $21.25

B. 1. 225   2. 7755
3. 25%   4. 18
5. $.1974 or $.20   6. 180%
7. 160   8. $2.72
9. 150%   10. $12\frac{1}{2}\%$
11. $26\frac{2}{3}\%$ 12. 4.5
13. 2%   14. $6\frac{2}{3}\%$
15. 17.5   16. 25%
17. 400%   18. 427
19. 22.1   20. 1600

C. 1. 88%   2. 16   3. 50.4   4. No difference
5. 32.4% (rounded)   6. 5.5% (rounded)
7. 27   8. $6900   9. $805.60   10. 46¢
11. (a) $100   (b) $10   (c) $1   (d) $1000
12. $2.17   13. $2400   14. $5.32   15. $35.15

Problem Set 3, Page 301

1. 29% (rounded)   2. $50,000   3. $21.57   4. $200
5. (a) $4680   (b) $2.06   (c) $.72   (d) $11.03   (e) $3.23
6. 16%
7. (a) $279.65   (b) $28.76   (c) $22.37
8. $107.25
9. $819 and 768. The better loan is at Last National.
10. $122.50
11. $198   12. $2250   13. $74   14. 14.9%
15. (a) $120   (b) $140   (c) $180   (d) $210
16. $8.62   17. $2880   18. 27.5%

UNIT 5, MEASUREMENT

Problem Set 1, Page 335

A. 1. 9   2. 1.25   3. 32   4. $1\frac{2}{3}$
5. $2\frac{1}{2}$ 6. 8   7. $8\frac{1}{3}$ 8. $16\frac{2}{3}$
9. 12   10. $3\frac{1}{3}$ 11. 10,080   12. $1\frac{13}{16}$
13. 4   14. 1040   15. 168   16. 5
17. 7200   18. 47   19. 5   20. 180

B. 1. 1440   2. 1750   3. 16   4. $\frac{40}{73} \cong 0.55$
5. $\frac{5}{18} \cong 0.28$ 6. 3520   7. $\frac{5}{8} = 0.625$ 8. $1\frac{9}{16} \cong 1.6$
9. 252   10. 0.001   11. $13\frac{8}{9} = 13.9$ 12. 0.0008
13. $1\frac{1}{2}$ 14. $\frac{5}{8} = 0.625$ 15. 1750   16. 15,840
17. 48   18. 720   19. 320   20. 32

C. 1. 9¢   2. $2.50   3. About 570
4. The giant family size is the better buy.
5. 10¢   6. 3.2¢ per mile   7. $1\frac{1}{4}$ mile   8. 3 hr 3 min
9. 20¢   10. 32   11. 40 tbsp   12. 1760 oz
13. 3600 sec, 86,400   14. 48 pt   15. 2.7 mi   16. 352 turns
17. 55¢   18. 3 gallons   19. 48 oz   20. 336 hrs

Problem Set 2, Page 343

A.

1.	$4\frac{1}{6}$ ft or 50 in.	2.	9 pints	3.	$3\frac{2}{3}$ min or 220 sec
4.	6 ft 7 in.	5.	9 pints	6.	4 lb 8 oz
7.	73 oz	8.	$1\frac{1}{4}$ hr	9.	385 min
10.	13 mph				

B.

1.	7 lb 10 oz	2.	$28\frac{1}{4}$ qt	3.	2 ft 7 in.
4.	2 qt 1 pt	5.	$44\frac{3}{4}$ in.	6.	52 tbsp
7.	About 65 min	8.	About 11 gal	9.	400 sec
10.	About 143 lb				

C.

1.	12 ft^2	2.	25.2 in.^{2}	3.	About 98.7 ft^2
4.	6 mph	5.	188¢	6.	59¢
7.	$\frac{1}{12}$ lb per cent	8.	15 mile per gallon	9.	62.4 lb per ft^3
10.	$2.90 per hr	11.	21 ft 3 in.	12.	25 lb 4 oz
13.	7 ft 6 in.	14.	$7.80	15.	12.5 cups
16.	35 lb				

D.

1.	1870 in.^{2}	2.	0.17 mile per minute	3.	$1.33 per lb
4.	The 98¢ box	5.	sq blip or b^2, blip/zip or bpz	6.	4.8¢ per mile
7.	2.1 lb	8.	0.08 gal/mi	9.	13¢
10.	14 lb 4 oz	11.	42 in.	12.	9 oz
13.	14 oz	14.	9	15.	30 trips plus 6 miles
16.	5 hr 20 min	17.	$\frac{5}{6}$	18.	14.8¢ per oz
19.	$1.35	20.	15¢ per qt		

Problem Set 3, Page 355

A.

1.	a	2.	b	3.	a	4.	b	5.	c
6.	b	7.	c	8.	a	9.	c	10.	b
11.	b	12.	b	13.	a	14.	b	15.	c

B.

1. 8 cm, .08 m, 80 mm (rounded, of course)
2. 122 cm $\cong$ 120 cm = 1.2 m 3. 3.23 km $\cong$ 3 km

4.	4.6 m	5.	4 cm, 40 mm	6.	16,000 m, 16 km
7.	42 mph	8.	78 mph	9.	210 mph
10.	42.2 km	11.	188 cm	12.	1900 m, 1.9 km
13.	30 cm	14.	18 m	15.	201 km

C.

1. 5.7 cm by 7.0 cm 2. 610 m or 0.38 mi
3. 91-61-91
4. (a) 800 km/hr (b) 97 km/hr (c) 10 km/hr (d) .0003 km/hr
5. 19.1 km/hr
6. (a) 0.8 cm (b) 3.5 cm (c) 1 mm
7. 8.90 m
8. 32 km/hr or 20 mph 9. 3.2 m
10. (a) 1.61 km (b) 61 cm (c) 22 cm by 28 cm
    (d) 2.54 cm (e) 800 km

11. (a) 800 m is 5.1 yd short of 880 yd.
    (b) 5 km is 188 yd longer than 3 miles.
    (c) 110 m is 0.3 yd longer than 120 yd.
12. 1.4 in. or about $1\frac{3}{8}$ in.
13. 9.36 yd
14. 7.87 in. or about $7\frac{7}{8}$ in.

Problem Set 4, Page 363

A. 1. a 2. b 3. b 4. c 5. c
6. a 7. b 8. a 9. c 10. b

B. 1. 66 lb 2. 91 kg 3. 0.33 lb, 5.3 oz, 0.15 kg
4. 1.48 kg 5. 19 lb 6. 110 g 7. 280 g, 0.28 kg
8. 0.6 kg, 1.3 lb 9. 45.5 kg 10. 7.3 kg 11. 14 g
12. .02 oz 13. 11 lb 14. 560 g 15. 2640 lb

C. 1. 2200 lb 3. c 4. $8.52
5. 1000 kg, 2200 lb, 1.1 tons
6. (a) 0.06 g (b) 28 g, 455 g (c) 0.06 g
7. 29 kg 8. 0.9 kg 9. 3¢ per lb
10. The American is 6 lb heavier.
11. .26 lb, 0.43 oz, .12 kg
12. 1.3 g

Page 378

1. 248° F 2. 93° C 3. 104° F 4. −15° C 5. 14° F

Problem Set 5, Page 381

A. 1. b 2. a 3. c 4. a 5. a
6. c 7. b 8. a 9. a 10. c
11. b 12. b 13. a 14. a 15. b

B. 1. 120 ml 2. 6.7 tbsp 3. 3.3 liters
4. 2.1 qt 5. $\frac{1}{12}$ cup $\cong$ 0.08 cup 6. 38 liters
7. 170 oz 8. 400 tsp 9. 5.7 liters
10. 10.6 qt

C. 1. 68° F 2. 93° C 3. 14° F 4. 204° C
5. 104° F 6. 58° C 7. 59° F 8. 10° C
9. 284° F 10. −23° C

D. 1. Call a doctor. 2. 185° C
3. 10° C = 50° F, too cold. 288° F, too hot.
4. 480° F
5. (a) 480 ml, .48 liters (b) 30 liters (c) 1.9 liters
   (d) 30 ml (e) 0.24 liter (f) 0.42 qt (g) 3 m^3
6. (a) 8.8 liters (b) 38 liters (c) 8800 ml
7. 12 km per liter = 28 mi per gallon
8. 800 g 9. $1.32 10. About 57° C
11. (a) 37° C, 98.6° F
    (b) 100° C, 212° F
    (c) 0° C, 32° F
12. 0.21

Problem Set 6, Page 395

1. 1.75 m^2 2. \$220.61 3. The square 4. 4.8 hectares
5. 0.49 acres 6. 960 ft^3 7. 2 ft^3, 118 lb 8. $6\frac{2}{3}$ yd^3
9. 0.113 m^3 10. 31.4 cm
11. (a) 190 m (b) 312 m (c) 253 m
12. (a) 7.1 yd^3 (b) 384 ft^2 (c) 200 ft
13. 1.5 acres 14. \$11.13 15. 2.1 cm, 3.47 sq cm
16. the square is about 14 sq in. larger
17. 16,200 ft^3
18. (a) 1000 m^3 (b) 600 m^2 (c) 120 m
19. 8.25 ft^3 20. 30 ft^2

UNIT 6, ALGEBRA

Problem Set 1, Page 429

A. 1. Term, constant or coefficient, variable, exponent
2. Expression, variable, term, constants or coefficients
3. Factors
4. Equation, constant, variable
5. Factor, expression
6. Expression, variables, constants, or coefficients, term

B. 1. $\frac{x}{y}$ 2. $A - 3$ 3. $6xy$ 4. $a + b^2$
5. $3 + y$ 6. x^2y^2 7. $2(2a + b)$ 8. $\frac{a+1}{b}$
9. $x - y$ 10. $3x^2y$ 11. $(a + b)(a - b)$ 12. $A + B - C$

C. 1. $5M$ 2. $8R^2$ 3. $3x - 2$ 4. $2a + 2b$
5. $5p^2$ 6. $9a + 9$ 7. $a + 5$ 8. $-x + 5y$
9. $2y + 2$ 10. $6a^3b^2c$ 11. $12xz$ 12. $10a^2 + a$
13. $a^2 - b^2$ 14. $a^2 + 2ab + b^2$ 15. $a^2 - 2ab + b^2$
16. $M^2 - 1$ 17. $\frac{2x}{3}$ 18. $\frac{a}{d}$ 19. $\frac{5}{b}$
20. $\frac{1}{x}$ 21. $3a + 5 + b$ 22. $2x - 1$
23. $2y$ 24. $a^2 + 2a + 1$ 25. $a^3 + 3a^2 + 3a + 1$
26. $a^2 - 2a + 1$ 27. $2xy^2z$ 28. $2x^2 - 4x + 1$
29. $2x^2y - 2xy^2$ 30. $\frac{3}{R}$

Problem Set 2, Page 437

A. 1. $n + 4$ 2. $n - 5$ 3. $d + 3$ 4. $h - 3$
5. $n + 8$ 6. $x - y$ 7. $2n$ 8. $\frac{M}{3}$
9. xy 10. $A(B + C)$ 11. $9R$ 12. $K - 10$
13. $K + 10$ 14. $x - 5$ 15. $15 - b$ 16. $H - 8$
17. $b - a$ 18. $y - x$ 19. $R + \frac{1}{2}T$ 20. $3n$

B. 1. $n + \frac{1}{2}n$ 2. $a + b - 4$ 3. $2n - 1$ 4. $2n^2$
5. $6 + 3n$ 6. $2n - 5$ 7. $2n + 2$ 8. $3n + 5$
9. $\frac{a+b}{6}$ 10. $4n - 7$ 11. R^2 12. $\frac{1}{2}bh$
13. $3\sqrt{n}$ 14. $\frac{a+b}{3}$ 15. $a - 5$ 16. $\frac{1}{2}n - 3$
17. $n^2 + (n+1)^2$ 18. $2(n + n^2)$ 19. mv 20. $\frac{1}{2}gt^2$

C. 1. $\frac{3n}{4} = 6$ 2. $a + b = 31$ 3. $S = C - 5$
4. $AB = 10C$ 5. $5n - 7 = 88$ 6. $S - J = 2D$
7. $4a - 3 = 2b$ 8. $B - 4 = 2(T + 3)$ 9. $6n + n^2 = 27$
10. $A = \pi R^2$ 11. $V = \frac{1}{4}\pi D^2 H$ 12. $B = 3 + P + N$
13. $x + 10 = 16$ 14. $P = 2L + 2W$ 15. $V = E^3$
16. $a + b - 4 = 2ab$ 17. $F = 32 + 1.8C$
18. $AB^2 = 6$ 19. $\frac{1}{2}(x + y) = 7$ 20. $a^2 + b^2 = a + b + 4$

Problem Set 3, Page 447

A. 1. $3x = 3 \cdot 2 = 6$ 2. $y^2 = 3 \cdot 3 = 9$
3. $wx = 6 \cdot 2 = 12$ 4. $2yz = 2 \cdot 3 \cdot 5 = 30$
5. $z - y = 5 - 3 = 2$ 6. $x + w = 2 + 6 = 8$
7. $2x + y = (2 \cdot 2) + 3 = 4 + 3 = 7$
8. $y^2 + x^2 = (3 \cdot 3) + (2 \cdot 2) = 9 + 4 = 13$
9. $3w - 7 = (3 \cdot 6) - 7 = 18 - 7 = 11$
10. $2w - 3y = (2 \cdot 6) - (3 \cdot 3) = 12 - 9 = 3$
11. $3xy^2 - 2w + 4 = (3 \cdot 2 \cdot 3^2) - (2 \cdot 6) + 4 = 54 - 12 + 4 = 46$
12. $\frac{1}{x} + \frac{y}{2} + 1 = \frac{1}{2} + \frac{3}{2} + 1 = 3$
13. $y^2x - x^2y = 3^2 \cdot 2 - 2^2 \cdot 3 = 18 - 12 = 6$
14. $2y - 1 = (2 \cdot 3) - 1 = 6 - 1 = 5$
15. $2(y - 1) = 2(3 - 1) = 2 \cdot 2 = 4$
16. $2x + y - z + 1 = (2 \cdot 2) + 3 - 5 + 1 = 3$
17. $\frac{yz}{x} + w = \frac{1}{2}(3 \cdot 5) + 6 = 7.5 + 6 = 13.5$
18. $2x + 3y - 1 = (2 \cdot 2) + (3 \cdot 3) - 1 = 4 + 9 - 1 = 12$
19. $2(x + y) + 3(w - z) = 2(2 + 3) + 3(6 - 5) = 2 \cdot 5 + 3 \cdot 1 = 10 + 3 = 13$
20. $3x^2y - 1 = 3 \cdot 2^2 \cdot 3 - 1 = 36 - 1 = 35$

B. 1. 22 2. 28 3. 16 4. 116 5. 180
6. 30 7. 22 8. 98 9. 405 10. 3
11. 550 12. 50 13. 48 14. 1188 15. 236
16. 74 17. 477 18. 15 19. 280 20. 2

C. 1. (a) 14 (b) 10 (c) 12 (d) 12 (e) 6
2. 339 ft^3 3. 46 4. 490 meters
5. 60° C 6. 19 cm^2 7. 40
8. $480 9. 864 $in.^3$ 10. 13
11. $T_1 = 1$, $T_2 = 3$, $T_3 = 6$, $T_4 = 10$, $T_{10} = 55$
12. 2880 g

Problem Set 4, Page 467

A. 1. 11 2. 34 3. −7 4. 7 5. 25
6. 12 7. 25 8. 5.5 9. 68 10. 5.2
11. 10 12. 5 13. 14 14. 4 15. 3
16. −14 17. 5 18. −6 19. 28 20. 6
21. 12 22. 18 23. 23 24. $1\frac{5}{6}$ 25. 4
26. $\frac{3}{8}$ 27. 6 28. 2 29. −7 30. 2

B. 1. 12 2. 6 and 12 3. 16 4. 15
5. 0 6. (b) 7. 2 8. 17.5 and 27.5
9. 33 10. 7 11. 22 12. 2
13. $x = \dfrac{c + b}{a - d}$ 14. $x = 7$ 15. 5
16. 0 17. 78 and 65 18. 2
19. 3 and 10 20. −4° F and −20° C

Problem Set 5, Page 479

A. 1. 60 2. 3 3. $1\frac{1}{6}$ 4. 7.5
5. 102 6. 12 7. 18 8. 20
9. 5 10. 36 11. $\frac{3}{7}$ 12. 39
13. $1\frac{2}{3}$ 14. 182 15. 27 16. 12
17. 27 18. .2 19. $1\frac{5}{13}$ 20. $11\frac{9}{10}$
21. 4 22. 63 23. 10 24. 42

B. 1. $13\frac{1}{3}$ 2. 15.7 in. 3. 94¢ 4. 117 mi
5. $4\frac{2}{3}$ hr 6. 46 7. 9.6 ft 8. 22
9. 198 mi 10. 462 points 11. 52¢ 12. 7.5 cups
13. 98¢ 14. 875 mi 15. 260, 838
16. 9 17. 154 18. 10.4 in., 485 ft, 2460 mi
19. −12 20. 7

Answers for Unit Self-Tests

If you answer any question incorrectly turn to the page and frame indicated for review.

UNIT 1, SELF-TEST

	Page	Frame
1. 83	3	1
2. 5221	3	1
3. 164	3	1
4. 26	25	16
5. 286	25	16
6. 2028	25	16
7. 3695	25	16
8. 1548	36	24
9. 48,348	36	24
10. 55,622	36	24
11. 654,512	36	24
12. 23	51	35
13. 803, R 6	51	35
14. 502	51	35
15. 32	51	35
16. $2^3 \times 3 \times 17$	65	44
17. $2 \times 3 \times 11^2 \times 13$	65	44
18. $2^3 \times 5^3$	65	44
19. $2^6 \times 3 \times 17$	65	44
20. 81	81	59
21. 2000	81	59
22. 7938	81	59
23. 15,129	81	59
24. 15	81	59
25. 18	81	59

UNIT 2, SELF-TEST

	Page	Frame
1. $\frac{115}{16}$	107	1
2. $3\frac{4}{11}$	107	1
3. $\frac{15}{40}$	107	1
4. $\frac{13}{17}$	107	1
5. $\frac{31}{35}$	141	34
6. $1\frac{19}{60}$	141	34
7. $6\frac{7}{24}$	141	34
8. $\frac{5}{12}$	141	34
9. $1\frac{3}{20}$	141	34
10. $3\frac{5}{12}$	141	34
11. 1	121	19
12. $5\frac{1}{7}$	121	19

		Page	Frame
13.	$1\frac{1}{9}$	129	26
14.	$\frac{69}{154}$	129	26
15.	$\frac{1}{4}$	121	19
16.	$5\frac{4}{9}$	121	19
17.	$58\frac{1}{2}$	121	19
18.	$\frac{7}{16}$	157	51
19.	$\frac{2}{5}$	157	51
20.	$\frac{3}{4}$	157	51
21.	2	157	51
22.	$\frac{7}{25}$	157	51
23.	$12\frac{1}{4}$	157	51
24.	$\frac{9}{100}$	157	51
25.	72¢	157	51

UNIT 3, SELF-TEST

		Page	Frame
1.	19.245	183	1
2.	51.16	183	1
3.	8.07	183	1
4.	.47	183	1
5.	16.524	195	10
6.	168.0	195	10
7.	1.90	201	16
8.	189.33	201	16
9.	.06	201	16
10.	$\frac{14}{25}$	213	25
11.	$3\frac{31}{125}$	213	25
12.	.4375	213	25
13.	1.2	213	25
14.	1.2	213	25
15.	7.00	213	25
16.	3	225	35
17.	-2	225	35
18.	-6	225	35
19.	21	225	35
20.	6.51	225	35
21.	-35	225	35
22.	-4	225	35
23.	-5	225	35
24.	2.55	235	44
25.	20.62	235	44

UNIT 4, SELF-TEST

		Page	Frame
1.	$316\frac{2}{3}\%$	261	1
2.	$41\frac{2}{3}$	261	1
3.	8%	263	4
4.	643%	263	4

	Page	Frame
5. .02	266	7
6. 1.125	266	7
7. $\frac{17}{25}$	267	9
8. 120	273	13
9. 7.56	273	13
10. 115.5	273	13
11. 8.71	273	13
12. 6.231	273	13
13. 35%	276	18
14. 150%	276	18
15. 225%	276	18
16. 42%	276	18
17. 29.6%	276	18
18. 80	279	20
19. .71 (rounded)	279	20
20. 51.5¢	289	24
21. $62.00	289	24
22. $59.20	289	24
23. $5.37	291	29
24. $72.50	296	34
25. $1.97	294	32

UNIT 5, SELF-TEST

	Page	Frame
1. $1\frac{8}{9}$ yd $\cong$ 1.9 yd	319	4
2. 6.2 mi	349	32
3. 2.8 liters	367	43
4. 7.7 lb	359	39
5. 88 km/hr	354	37
6. 4760 ft	349	32
7. 10 lb 2 oz	339	22
8. 12 qt	339	22
9. 122° F	372	48
10. 1.83 m $\cong$ 2 m	349	32
11. 45 liters	367	43
12. 97 km	349	32
13. 56.8 kg	359	39
14. 32 cm^2	341	26
15. 9 mph	341	26
16. $3.25	341	26
17. 7¢ per 8-oz glass, rounded up	341	26
18. 3.43 m	349	32
19. 22.5 yd^2	385	53
20. 1.2 yd^3	391	60
21. 96,000 cm^3 $\cong$ .096 m^3	391	60
22. 200 $in.^2$	385	53
23. 8 ft	390	59
24. $165.34	385	53
25. 5.25 hr	341	26

UNIT 6, SELF-TEST

		Page	Frame
1.	Term	415	**1**
2.	Coefficient or constant	415	**1**
3.	Expression	415	**1**
4.	Variable	415	**1**
5.	Constant	415	**1**
6.	Equation	415	**1**
7.	$n - 6$	431	**17**
8.	$a + 2b$	431	**17**
9.	$8n$	431	**17**
10.	$4 + \frac{1}{2}n$	431	**17**
11.	$6x - 3y + 1$	422	**9**
12.	$2b - 2b^2$	422	**9**
13.	$4x^2 - 6x^2y$	422	**9**
14.	$a^2 - b^2$	422	**9**
15.	$3M - 9$	422	**9**
16.	$2x + xy + 6y$	422	**9**
17.	44	441	**23**
18.	628	441	**23**
19.	6	441	**23**
20.	4	451	**30**
21.	4.6	451	**30**
22.	5	451	**30**
23.	14	451	**30**
24.	196 lb	471	**45**
25.	$6.00	471	**45**

Table of Square Roots

Number	Square root	Number	Square root	Number	Square root	Number	Square root
1	1.0000	51	7.1414	101	10.0499	151	12.2882
2	1.4142	52	7.2111	102	10.0995	152	12.3288
3	1.7321	53	7.2801	103	10.1489	153	12.3693
4	2.0000	54	7.3485	104	10.1980	154	12.4097
5	2.2361	55	7.4162	105	10.2470	155	12.4499
6	2.4495	56	7.4833	106	10.2956	156	12.4900
7	2.6458	57	7.5498	107	10.3441	157	12.5300
8	2.8284	58	7.6158	108	10.3923	158	12.5698
9	3.0000	59	7.6811	109	10.4403	159	12.6095
10	3.1623	60	7.7460	110	10.4881	160	12.6491
11	3.3166	61	7.8102	111	10.5357	161	12.6886
12	3.4641	62	7.8740	112	10.5830	162	12.7279
13	3.6056	63	7.9373	113	10.6301	163	12.7671
14	3.7417	64	8.0000	114	10.6771	164	12.8062
15	3.8730	65	8.0623	115	10.7238	165	12.8452
16	4.0000	66	8.1240	116	10.7703	166	12.8841
17	4.1231	67	8.1854	117	10.8167	167	12.9228
18	4.2426	68	8.2462	118	10.8628	168	12.9615
19	4.3589	69	8.3066	119	10.9087	169	13.0000
20	4.4721	70	8.3666	120	10.9545	170	13.0384
21	4.5826	71	8.4261	121	11.0000	171	13.0767
22	4.6904	72	8.4853	122	11.0454	172	13.1149
23	4.7958	73	8.5440	123	11.0905	173	13.1529
24	4.8990	74	8.6023	124	11.1355	174	13.1909
25	5.0000	75	8.6603	125	11.1803	175	13.2288
26	5.0990	76	8.7178	126	11.2250	176	13.2665
27	5.1962	77	8.7750	127	11.2694	177	13.3041
28	5.2915	78	8.8318	128	11.3137	178	13.3417
29	5.3852	79	8.8882	129	11.3578	179	13.3791
30	5.4772	80	8.9443	130	11.4018	180	13.4164
31	5.5678	81	9.0000	131	11.4455	181	13.4536
32	5.6569	82	9.0554	132	11.4891	182	13.4907
33	5.7446	83	9.1104	133	11.5326	183	13.5277
34	5.8310	84	9.1652	134	11.5758	184	13.5647
35	5.9161	85	9.2195	135	11.6190	185	13.6015
36	6.0000	86	9.2736	136	11.6619	186	13.6382
37	6.0828	87	9.3274	137	11.7047	187	13.6748
38	6.1644	88	9.3808	138	11.7473	188	13.7113
39	6.2450	89	9.4340	139	11.7898	189	13.7477
40	6.3246	90	9.4868	140	11.8322	190	13.7840
41	6.4031	91	9.5394	141	11.8743	191	13.8203
42	6.4807	92	9.5917	142	11.9164	192	13.8564
43	6.5574	93	9.6437	143	11.9583	193	13.8924
44	6.6332	94	9.6954	144	12.0000	194	13.9284
45	6.7082	95	9.7468	145	12.0416	195	13.9642
46	6.7823	96	9.7980	146	12.0830	196	14.0000
47	6.8557	97	9.8489	147	12.1244	197	14.0357
48	6.9282	98	9.8995	148	12.1655	198	14.0712
49	7.0000	99	9.9499	149	12.2066	199	14.1067
50	7.0711	100	10.0000	150	12.2474	200	14.1421

Study Cards

These study cards may be used as a memory aide. Each card contains a bit of information you will want to remember. Select the information you need to remember, then cut out the card and use it to train your memory. Carry the card in your pocket or purse and quiz yourself on it at odd moments throughout the day until you can recall the information quickly and correctly.

ADDITION TABLE

+	0	1	2	3	4	5	6	7	8	9
0	0	1	2	3	4	5	6	7	8	9
1	1	2	3	4	5	6	7	8	9	10
2	2	3	4	5	6	7	8	9	10	11
3	3	4	5	6	7	8	9	10	11	12
4	4	5	6	7	8	9	10	11	12	13
5	5	6	7	8	9	10	11	12	13	14
6	6	7	8	9	10	11	12	13	14	15
7	7	8	9	10	11	12	13	14	15	16
8	8	9	10	11	12	13	14	15	16	17
9	9	10	11	12	13	14	15	16	17	18

MULTIPLICATION TABLE

×	0	1	2	3	4	5	6	7	8	9	10
0	0	0	0	0	0	0	0	0	0	0	0
1	0	1	2	3	4	5	6	7	8	9	10
2	0	2	4	6	8	10	12	14	16	18	20
3	0	3	6	9	12	15	18	21	24	27	30
4	0	4	8	12	16	20	24	28	32	36	40
5	0	5	10	15	20	25	30	35	40	45	50
6	0	6	12	18	24	30	36	42	48	54	60
7	0	7	14	21	28	35	42	49	56	63	70
8	0	8	16	24	32	40	48	56	64	72	80
9	0	9	18	27	36	45	54	63	72	81	90
10	0	10	20	30	40	50	60	70	80	90	100

PERFECT SQUARES

$1^2 = 1$	$6^2 = 36$	$11^2 = 121$	$16^2 = 256$
$2^2 = 4$	$7^2 = 49$	$12^2 = 144$	$17^2 = 289$
$3^2 = 9$	$8^2 = 64$	$13^2 = 169$	$18^2 = 324$
$4^2 = 16$	$9^2 = 81$	$14^2 = 196$	$19^2 = 361$
$5^2 = 25$	$10^2 = 100$	$15^2 = 225$	$20^2 = 400$

THE PRIMES LESS THAN 100

2	3	5	7	11
13	17	19	23	29
31	37	41	43	47
53	59	61	67	71
73	79	83	89	97

Rule for adding or subtracting signed numbers:

1. Replace any number of the form $-(-a)$ with $+a$.
2. If both numbers have the *same* sign, add them and give the sum the sign of the original numbers.
3. If the two numbers have *opposite* signs, subtract them and give the difference the sign of the number that is larger.

If	$A \times B = C$
then	$B = C \div A$
and	$A = C \div B$

Percent	Decimal	Fraction
5%	.05	$\frac{1}{20}$
$6\frac{1}{4}\%$	.0625	$\frac{1}{16}$
$8\frac{1}{3}\%$	$.08\overline{3}$	$\frac{1}{12}$
10%	.10	$\frac{1}{10}$
$12\frac{1}{2}\%$	.125	$\frac{1}{8}$
$16\frac{2}{3}\%$	$.1\overline{6}$	$\frac{1}{6}$
20%	.20	$\frac{1}{5}$
25%	.25	$\frac{1}{4}$
30%	.30	$\frac{3}{10}$
$33\frac{1}{3}\%$	$.3\overline{3}$	$\frac{1}{3}$
$37\frac{1}{2}\%$	.375	$\frac{3}{8}$
40%	.40	$\frac{2}{5}$
50%	.50	$\frac{1}{2}$
60%	.60	$\frac{3}{5}$
$62\frac{1}{2}\%$	.625	$\frac{5}{8}$
$66\frac{2}{3}\%$	$.6\overline{6}$	$\frac{2}{3}$
70%	.70	$\frac{7}{10}$
75%	.75	$\frac{3}{4}$
80%	.80	$\frac{4}{5}$
$83\frac{1}{3}\%$	$.8\overline{3}$	$\frac{5}{6}$
$87\frac{1}{2}\%$	.875	$\frac{7}{8}$
90%	.90	$\frac{9}{10}$
100%	1.00	$\frac{10}{10}$

SIGNAL WORDS	TRANSLATE AS
Is, is equal to, equals, the same as	$=$
of, the product of, multiply, times, multiplied by	$\times$
Add, in addition, plus, more, more than, sum, and, increased by, added to	$+$
Subtract, subtract from, less, less than, difference, diminished by, decreased by	$-$
Divide, divided by	$\div$
Twice, double, twice as much	$2\times$
half of, half	$\frac{1}{2}\times$

$$\sqrt{2} \cong 1.4142 \cong \tfrac{7}{5} \text{ or even closer } \tfrac{17}{12}$$
$$\sqrt{3} \cong 1.7321 \cong \tfrac{7}{4} \text{ or even closer } \tfrac{19}{11}$$
$$\sqrt{5} \cong 2.2361 \cong \tfrac{9}{4}$$
$$\sqrt{6} \cong 2.4495 \cong \tfrac{22}{9}$$
$$\sqrt{7} \cong 2.6458 \cong \tfrac{8}{3}$$
$$\sqrt{8} \cong 2.8284 \cong \tfrac{14}{5} \text{ or even closer } \tfrac{17}{6}$$
$$\sqrt{10} \cong 3.1623 \cong \tfrac{19}{6}$$

A MEMORY GIMMICK

If $A \times B = C$, then $A = \dfrac{C}{B}$ and $B = \dfrac{C}{A}$ for any numbers A, B, and C that are not zero.

Do you need a memory jogger? Try this one.

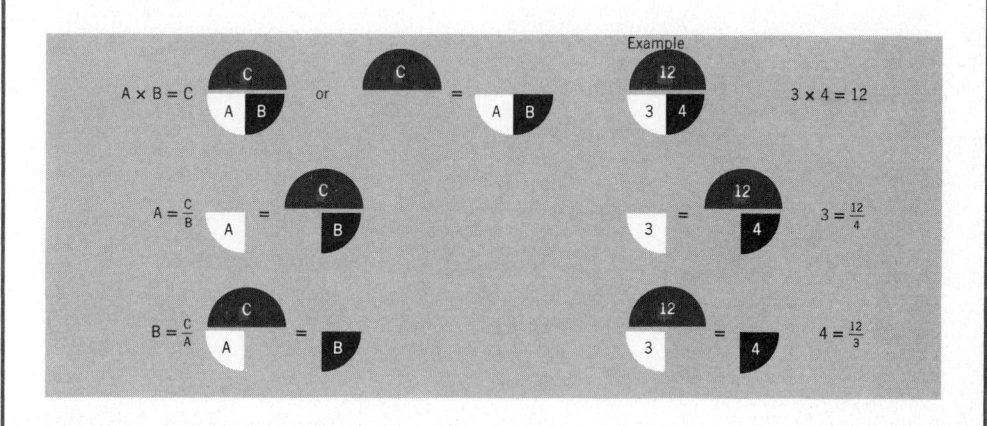

LENGTH

1 foot (ft) = 12 inches (in.)
1 yard (yd) = 3 feet
1 mile (mi) = 5280 feet

TIME

1 minute (min) = 60 seconds (sec)
1 hour (hr) = 60 minutes
1 day (d) = 24 hours
1 week (wk) = 7 days
1 calendar year (yr) = 365 days

WEIGHT

1 ounce (oz) = 437.5 grains (gr)
16 ounces = 1 pound (lb)
2000 pounds = 1 ton (t)

VOLUME

16 fluid ounces (fl oz) = 1 pint (pt)
2 pints = 1 quart (qt)
4 quarts = 1 gallon (gal)
1 barrel (bl) = $31\frac{1}{2}$ gallons

3 teaspoonfuls (tsp) = 1 tablespoonful (tbsp)
2 tablespoonfuls = 1 fluid ounce
8 fluid ounces = 1 cup
2 cups = 1 pint

2 pints = 1 quart
8 quarts = 1 peck
4 pecks = 1 bushel

60 minims (m) = 1 fluid dram (fl dr)
8 fluid drams = 1 fluid ounce (fl oz)

METRIC PREFIXES

Money	Metric Length Unit	Prefix	Multiplier
\$1000	kilometer	kilo-	1000×1 meter
\$ 100	hectometer	hecto-	100×1 meter
\$ 10	decameter	deca-	10×1 meter
\$ 1	meter		1 meter
10¢	decimeter	deci-	0.1×1 meter
1¢	centimeter	centi-	0.01×1 meter
$\frac{1}{10}$¢	millimeter	milli-	0.001×1 meter

METRIC LENGTH

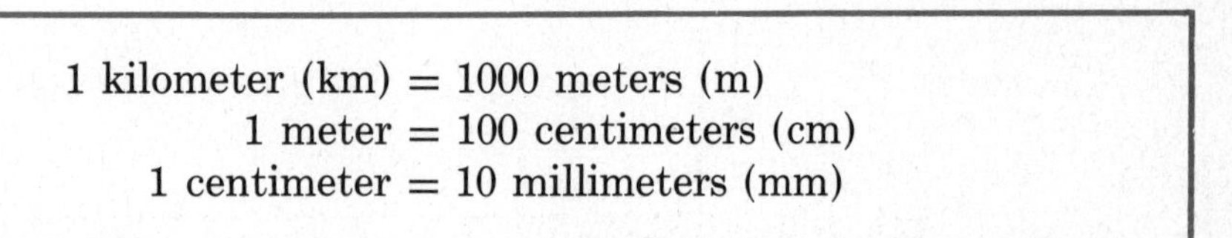
1 kilometer (km) = 1000 meters (m)
1 meter = 100 centimeters (cm)
1 centimeter = 10 millimeters (mm)

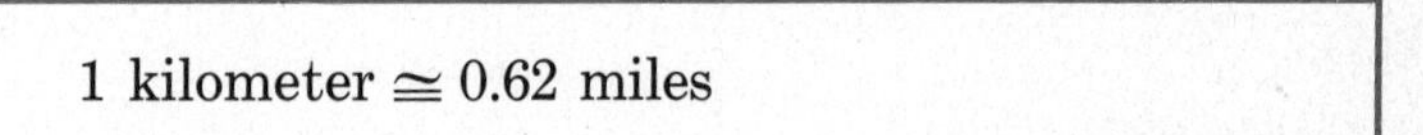
1 kilometer $\cong$ 0.62 miles

1 meter $\cong$ 3.3 feet

1 inch = 2.54 cm

METRIC SPEED

EQUIVALENT SPEEDS (approximate)

km/hr	mph	
100	62	
90	56	Super highway
80	50	
70	43	Main roads
60	38	
50	31	Suburbs
40	25	City driving
30	19	
25	15	School zone
20	12	Run
10	6	Jog

100 km/hr $\cong$ 62 mph

METRIC VOLUME

1 liter = 1000 milliliter (ml)

1 liter $\cong$ 1.06 quart

1 teaspoonful $\cong$ 5 ml
1 fluid ounce $\cong$ 30 ml

METRIC WEIGHT (MASS)

1 kilogram (kg) weighs about 2.20 pounds (lb)

1 ounce (oz) is roughly the weight of 28 grams (g)

$$1 \text{ gram} = \frac{1}{1000} \text{ kilogram}$$

$$1 \text{ milligram} = \frac{1}{1000} \text{ gram}$$

Temperature conversion chart

F = Fahrenheit C = Celsius

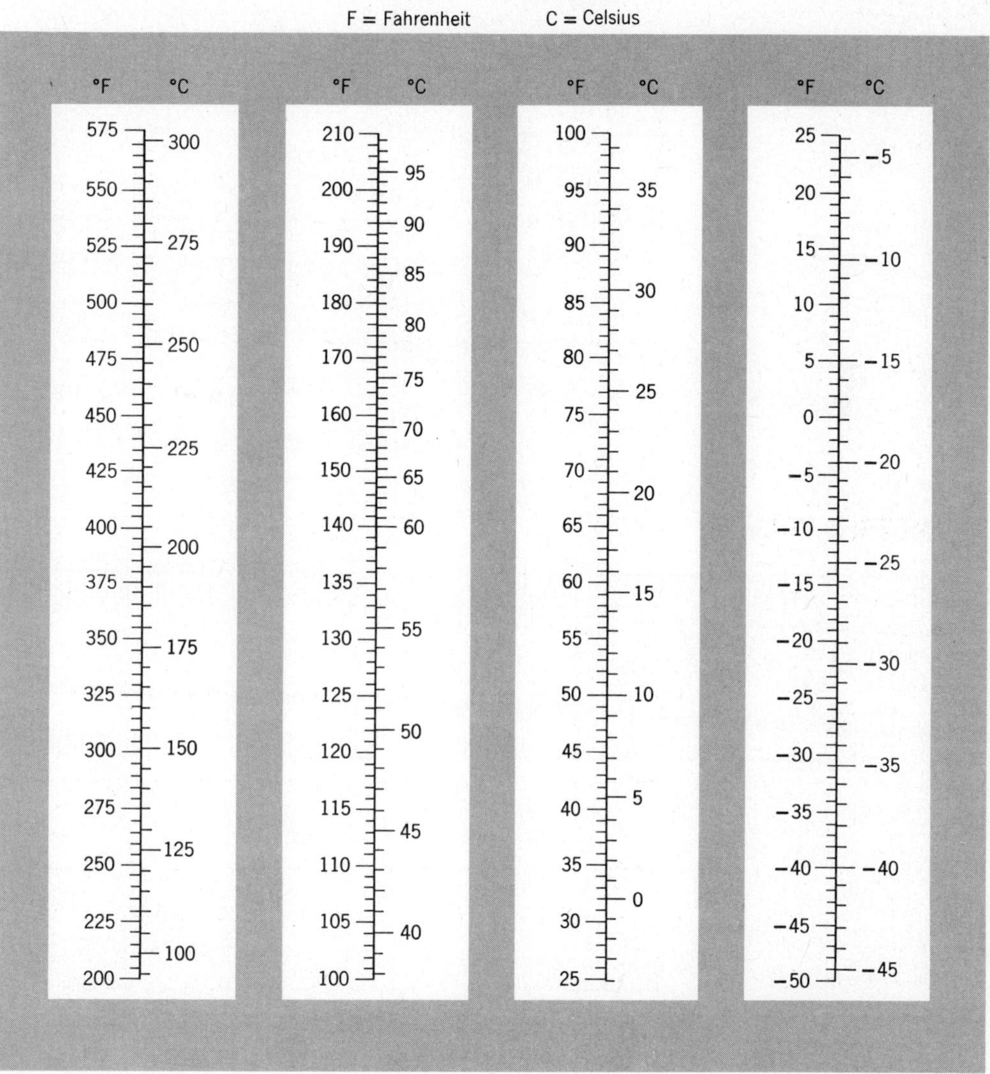

AREA

$$1 \text{ m}^2 = 10{,}000 \text{ cm}^2$$
$$1 \text{ ft}^2 = 144 \text{ in.}^2$$
$$1 \text{ yd}^2 = 9 \text{ ft}^2$$

$$1 \text{ acre} = 4840 \text{ yd}^2$$

$$1 \text{ hectare} = 10{,}000 \text{ m}^2$$

VOLUME

$$1 \text{ m}^3 = 1{,}000{,}000 \text{ cm}^3$$
$$1 \text{ ft}^3 = 1728 \text{ in.}^3$$
$$1 \text{ yd}^3 = 27 \text{ ft}^3$$

Volume of a rectangular solid = height × length × width

CIRCUMFERENCE

Circumference of a circle ≅ 3.14 × diameter

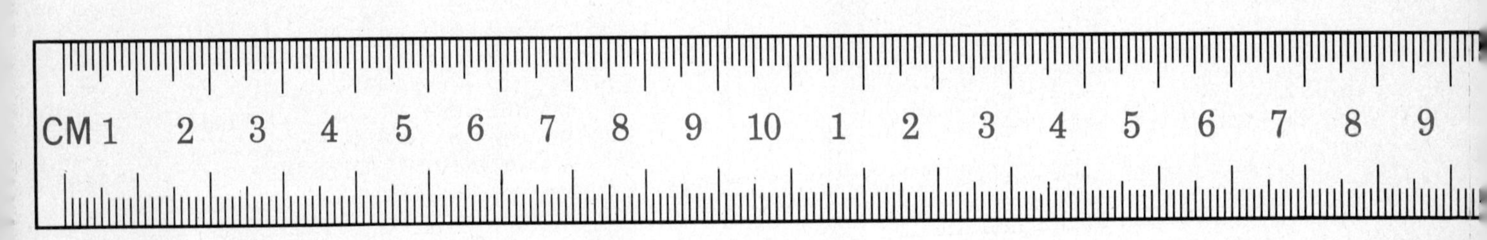

Index

Accuracy, 316
Addend, 8
Addition, of decimal numbers, 186
  of expressions, 423
  of fractions, 141
  of like algebraic terms, 422
  of long lists of numbers, 17
  of measurement numbers, 339
  table, 518
  of whole numbers, 7
Algebra, 415
Algebra equation, 421, 451, 456
Area, of a circle, 368
  definition, 385
  of a triangle, 385
Average, 53

Base, 81, 261, 271
Board feet, 394

Circumference of a circle, 389
Coefficient, 416
Commission, 289
Common multiple, 144
Commutative property, of addition, 8
  of multiplication, 37
Constant, 417
Credit cards, 297
Cross product, 474

Decimal digits, 196
Decimal-fraction equivalents, 219, 268
Decimal numbers, addition of, 186
  division of, 201
  multiplication of, 195
  naming, 197
  subtraction of, 189
Decimal place, 203
Decimal point, 186
Denominator, 108
Difference, 25
Digits, 3
Discount, 291
Discount rate, 291
Dividend, 51
Divisibility rules, 74, 76
Division, of algebraic expressions, 427
  of decimal numbers, 201
  of fractions, 129
  of measurement numbers, 342
  by powers of ten, 207
  by zero, 57
Divisor, 51

Evenly divisible numbers, 65
Expanded form, 5, 184, 187
Exponent, 81, 416
Expression, 419

Factor tree, 73
Factors, 36, 65

Fractional parts of a number, 159
Fractions, addition of, 141
  division of, 129, 130, 134
  equivalent, 112, 165
  improper, 110
  like, 141
  multiplication of, 121
  proper, 110
  reduction to lowest terms, 114
  renaming of, 107
  subtraction of, 151

Installment loans, 296
Interest, 296

Least common denominator, 143
Least common multiple, 142, 145, 149
Length units, 318, 349
Linear equations, 452
List price, 291
Literal expression, 415, 441
Loans, 293

Measurement, 315
Metric rule, 526
Metric prefixes, 350, 523
Metric units, 347
Minuend, 25
Mixed number, 111
Multiple, 143
Multiplicand, 36
Multiplication, of algebraic expressions, 426
  of decimal numbers, 195
  of fractions, 121
  of measurement numbers, 342
  by multiples of ten, 45, 207
  table, 40, 518
  of whole numbers, 36
Multiplier, 36

Naming large numbers, 6
Number line, 225
Numerals, 3
Numerator, 108

Percent, 261, 268, 520
Percentage, 271, 292
Perfect squares, 87, 519
Perimeter, 390
Place value system, 7
Polygon, 390
Powers of ten, 84, 184
Prime factors, 69
Prime numbers, 66, 71, 519
Principal, 297
Product, 36
Proportion, 473
Pythagorean Theorem, 245

Quotient, 51

Rate, 271
Ratio, 471, 473
Repeating decimal, as a decimal, 217
  as a fraction, 220
Roman numerals, 12
Rounding, 204, 324

Sales price, 291
Sales tax, 295
Sieve of Eratosthenes, 68
Signed numbers, addition of, 226, 229, 519
  division of, 231
  multiplication of, 230, 232
  subtraction of, 227, 229, 519
Significant digits, 323, 325
Solution of an equation, 451
Solution set, 452
Square root, of decimal numbers, 235
  of integers, 87
  table, 239, 517, 521
Study cards, 518
Subscripts, 420
Subtraction, of algebraic expressions, 423
  of decimal numbers, 183
  of fractions, 151
Subtraction (*continued*)
  of like algebraic terms, 422
  of measurement numbers, 341
  of whole numbers, 25
Subtrahend, 25
Sum, 8, 186

Temperature, 372, 525
Temperature conversion, 376, 377
Term, 419
Terminating decimal, 214
Time units, 323
Translating English to algebra, 431

Units, compound, 340
  metric, 347, 523
  natural, 318
Unity fractions, 320

Variable, 416
Volume units, 328, 367, 391
Volume, of a cylinder, 393
  defined, 391

Weight units, 327, 359
Word problems, 157, 431